Sustainable Apple Breeding and Cultivation in Germany

Applying a socio-ecological framework, this book explores how the innovative approach of commons-based organic apple breeding can contribute to sustainability in agricultural and food systems more widely.

As fruit breeding and cultivation systems are confronted with a range of sustainability challenges, there are calls for new and innovative breeding approaches beyond mainstream economic solutions that would mitigate these sustainability challenges. Apples, in particular, are facing serious environmental challenges, with the negative environmental impacts of modern conventional breeding and farming, loss of agrobiodiversity, low participation and diminishing diversity of market actors in the wake of privatization and economization trends result in a lack of resilience in current breeding and cultivation systems. Drawing on in-depth case study research on apple production in Germany, this book advances the innovative solution of commons-based apple breeding as a model for developing resilience in fruit breeding and cultivation. It analyzes this approach, comparing it with more conventional practices, and showcases which factors could inhibit the broad implementation of commons-based apple breeding and how they can be overcome to exploit its full potential. Contributing to the great ambition of finding sustainable solutions across all agricultural sectors, this book opens up new and interdisciplinary perspectives on fruit breeding and cultivation, which is a largely neglected issue in contemporary discussions on agriculture and food production.

This book will be of great interest for students and scholars from the fields of sustainable food systems, sustainable agriculture, crop science, and resource management and in particular those that seek inspiration for innovative approaches rooted in sustainability research, social-ecological resilience, and the commons.

Hendrik Wolter is a Research Associate and Lecturer in the Department of Economics and Law at the University of Oldenburg, Germany. He holds a PhD in Social and Political Science and was involved in the transdisciplinary research project EGON (2017–2020), which examined commons-based organic apple breeding.

Earthscan Food and Agriculture

Net Zero, Food and Farming
Climate Change and the UK Agri-Food System
Neil Ward

Conservation Agriculture in India
A Paradigm Shift for Sustainable Production
Edited by A.R. Sharma

Agricultural Commercialization, Gender Equality and the Right to Food
Insights from Ghana and Cambodia
*Edited by Joanna Bourke Martignoni, Christophe Gironde, Christophe Golay,
Elisabeth Prügl and Dzodzi Tsikata*

The Sociology of Farming
Concepts and Methods
Jan Douwe van der Ploeg

Genetically Modified Crops and Food Security
Commercial, Ethical and Health Considerations
Edited by Jasmeet Kour, Vishal Sharma and Imtiyaz Khanday

Climate Neutral and Resilient Farming Systems
Practical Solutions for Climate Mitigation and Adaption
Edited by Sekhar Udaya Nagothu

Sustainable Apple Breeding and Cultivation in Germany
Commons-Based Agriculture and Social-Ecological Resilience
Hendrik Wolter

For more information about this series, please visit: www.routledge.com/books/
series/ECEFA/

Sustainable Apple Breeding and Cultivation in Germany

Commons-Based Agriculture and
Social-Ecological Resilience

Hendrik Wolter

Routledge
Taylor & Francis Group
LONDON AND NEW YORK

earthscan
from Routledge

First published 2023
by Routledge
4 Park Square, Milton Park, Abingdon, Oxon OX14 4RN

and by Routledge
605 Third Avenue, New York, NY 10158

Routledge is an imprint of the Taylor & Francis Group, an informa business

© 2023 Hendrik Wolter

British Library Cataloguing-in-Publication Data
A catalogue record for this book is available from the British Library

Library of Congress Cataloging-in-Publication Data
Names: Wolter, Hendrik, author.
Title: Sustainable apple breeding and cultivation in Germany : commons-based agriculture and social-ecological resilience / Hendrik Wolter.
Other titles: Commons-based agriculture and social-ecological resilience
Description: New York, NY : Routledge, 2023. | Includes bibliographical references and index.
Identifiers: LCCN 2022043839 (print) | LCCN 2022043840 (ebook) | ISBN 9781032409948 (hardback) | ISBN 9781032409962 (paperback) | ISBN 9781003355724 (ebook)
Subjects: LCSH: Apples—Germany. | Apples—Germany—Breeding. | Sustainable agriculture—Germany.
Classification: LCC SB363.2.G3 W65 2023 (print) | LCC SB363.2.G3 (ebook) | DDC 634/.110943—dc23/eng/20220928
LC record available at https://lccn.loc.gov/2022043839
LC ebook record available at https://lccn.loc.gov/2022043840

ISBN: 978-1-032-40994-8 (hbk)
ISBN: 978-1-032-40996-2 (pbk)
ISBN: 978-1-003-35572-4 (ebk)

DOI: 10.4324/9781003355724

Typeset in Goudy
by codeMantra

Contents

Preface   vii
Acknowledgments   ix
List of Acronyms and Abbreviations   xi

Introduction   1

PART I
Research Design and Empirical Context   15

1 Theoretical Framework   17

2 Methodological Framework: Qualitative Research in a
   Transdisciplinary Context   46

3 Empirical Context: Apple Breeding and Cultivation in Germany   53

PART II
Resilience in Fruit Breeding and Cultivation   71

Introduction   73

4 Fruit Breeding and Cultivation as a Social-Ecological System   75

5 Ecosystem Services in Fruit Breeding and Cultivation   92

6 Characteristics of a Resilient Fruit Breeding and Cultivation System   112

   Interim Conclusion   150

PART III
**Evaluation of Apple Breeding Approaches: How Resilient
Are They?**   153

Introduction   155

7 Corporate-Based Apple Breeding   157

8 Public Apple Breeding   167

9 Commons-Based Apple Breeding   177

10 Comparison of Apple Breeding Approaches   191

    Interim Conclusion   199

PART IV
Implementing Commons-Based Organic Apple Breeding   201

Introduction   203

11 Market Structures, Developments and Trends   206

12 Promoting and Inhibiting Factors for Organic Apple
    Cultivation and Breeding   223

13 Business Models for Financing Organic Apple Breeding   238

14 Challenges for Commons-Based Organic Apple Breeding   245

    Interim Conclusion   251

    Conclusion   253

    Appendix A   263
    Appendix B   266
    Appendix C   271
    Index   275

Preface

This book builds on my PhD thesis. I have written this book while being employed and enrolled as a PhD candidate at the Carl von Ossietzky University of Oldenburg. The problem framing and research direction were developed in the context of the transdisciplinary research project "Development of organically bred fruit varieties in commons-based initiatives" (EGON) that was funded by the German Ministry of Science and Culture of Lower Saxony (grant no. 3250). It took place from January 2017 until July 2020. EGON researchers worked closely together with the research project "Right Seeds? Common-based rights on seeds and seed varieties for a social-ecological transformation of crop production" (RightSeeds) that was funded by the German Federal Ministry for Education and Research as part of the program "Research for sustainable development" and took place from October 2016 until July 2022. This book is a research output of the project EGON. All empirical data was collected and analyzed as part of EGON and discussed with the project partners.

Acknowledgments

I had the great pleasure to write my PhD about a very tangible and delicate research topic – apples. Thus, everyone I talked to about my PhD could connect at least in some way to this research. I am sincerely grateful for all the guidance, help, and discussions I experienced throughout this fruitful journey.

In particular, I would like to thank my supervisors for their constant support and guidance during this project. Thank you, Bernd, for giving me the freedom to pursue my own ideas while steering the shape and scope of this project in the right direction. Your advice and trust always prevented me from getting lost. Thank you, Stefanie, for all the constructive and encouraging discussions and feedback rounds that lightened so many sparks in this book. Your support and insight into the subject have been irreplaceable and invaluable.

Special thanks to you, Nina, my writing companion. Without our mutual and steady support throughout the years, all the thinking and writing would not have been as inspiring and funny as it has been. Due to our great writing weeks in remote areas, this book now not only tastes of apples, but also smells of salt and wood. It was my pleasure. Thank you, Sebastian, for many productive and coffee-rich hours writing together, especially in the last writing year. I thank you, Julia, for all the lively discussions and funny office hours together during the last years. Thanks so much, Julie, for the support and patience in the final phase.

I thank the project partners of EGON for so many inspiring discussions throughout the project and beyond: Inde, Matthias, Dirk, and Nick. I learned a lot from you. Thanks to all my interview partners and the participants that took part in the Delphi study. I am grateful for the time you invested. I also thank all the undergraduate assistants that helped EGON and my PhD project to flourish. Thanks to the whole teams from the working groups Ecological Economics and Economy of the Commons for their continuous support during the last years. Thank you, Nicolas and Theresa, for your helpful feedback on parts of the manuscript.

A special thanks goes to you, Anni, for your great support and tolerance during this project. Days that could have been invested in time together and life beyond the academic universe were reserved with solitary hours of thinking, frowning, and writing. Thank you for your patience and constant encouragement.

Acknowledgments

Acronyms and Abbreviations

AGOZV	Verordnung über das Inverkehrbringen von Anbaumaterial von Gemüse-, Obst- und Zierpflanzenarten
AIGN	Associated International Group of Nurseries
BayOZ	Bayerisches Obstzentrum
BLE	Bundesanstalt für Landwirtschaft und Ernährung
BMELV	Bundesministerium für Ernährung, Landwirtschaft und Verbraucherschutz
BS	Bundessortenamt
BZL	Bundesinformationszentrum Landwirtschaft
CAP	Common Agricultural Policy
CAS	Complex Adaptive Systems
CDB	Consortium Deutscher Baumschulen
CICES	Common International Classification of Ecosystem Services
CPR	Common-property Regime
DOSK	Deutsches Obstsorten Konsortium
DUS	Distinctness, Uniformity, and Stability
EEA	European Environmental Agency
ErhaltungsV	Erhaltungssortenverordnung
EU	European Union
EUIPO	EU Intellectual Property Office
FiBL	Forschungsinstitut Biologischer Landbau
FÖKO	Fördergemeinschaft Ökologischer Obstbau
GE	Gene editing
GM	Genetic Modification
IAD	Institutional Analysis and Development Framework
IFOAM	International Federation of Organic Agriculture Movements
INRAE	National Research Institute for Agriculture, Food and Environment
IOBC	International Organization of Biological Control
IPBES	Intergovernmental Science-Policy Platform on Biodiversity and Ecosystem Services
IPCC	Intergovernmental Panel on Climate Change
IPM	Integrated Pest Management

JKI	Julius Kühn-Institut
KOB	Kompetenzzentrum Obstbau Bodensee
LVWO	Staatliche Lehr- und Versuchsanstalt für Wein- und Obstbau
MaBo	Marktgemeinschaft Bodenseeobst
MAL	Marktgemeinschaft Altes Land
MEA	Millennium Ecosystem Assessment
NCP	Nature's Contributions to People
NGO	Non-governmental Organization
ÖON	Öko-Obstbau Norddeutschland Versuchs- und Beratungsring
UN	United Nations
UNEP	United Nations Environmental Programme
UPOV	International Union for the Protection of New Plant Varieties of Plants
PatG	Patentschutzgesetz
PBVO	Pflanzenbeschauverordnung
PELUM	Participatory Ecological Land Use Management
PPB	Participatory Plant Breeding
PflSchG	Pflanzenschutzgesetz
RA	Resilience Alliance
RAYS	Resilience Alliance Young Scholars
RFID	Radio-frequency Identification
SaatG	Saatgutverkehrsgesetz
SDG	Sustainable Development Goal
SESF	Social-ecological System Framework
SortG	Sortenschutzgesetz
TEEB	The Economics of Ecosystems and Biodiversity
TVUL	Turners Vulnerability Framework
WOG	Württembergische Obstgenossenschaft Raiffeisen
ZIN	Züchtungsinitiative Niederelbe
ZO	Institut für Züchtungsforschung an Obst

Introduction

Topic and Problem Context

The earth is a complex *social-ecological system* and contains a diverse range of social-ecological systems, in which humankind and nature are constantly shaping each other (Berkes and Folke 1998b). In the Anthropocene (Steffen et al. 2007), human actions negatively impact the global social-ecological system, for example, in the case of climate change (IPCC 2014) or the dramatic loss of biodiversity across the planet (Pörtner et al. 2021; MEA 2005; Mace et al. 2005; Folke et al. 2004). The thresholds of several "planetary boundaries," which define a safe operating space for humanity, have already been crossed into zones of uncertainty (Steffen et al. 2015). These negative impacts ultimately deteriorate *ecosystem services*, defined as the benefits people obtain from ecosystems (MEA 2005). A growing research community argues that the solution to prevent or stop this deterioration is the development and design of resilient social-ecological systems. *Social-ecological resilience* describes

> the capacity of a social-ecological system to sustain human well-being and a desired set of ecosystem services in the face of disturbance and change, both by buffering shocks and by adapting or transforming in response to change.
> (Biggs et al. 2015, p. 9)

In this sense, social-ecological resilience builds capacities for long-term *sustainability*, understood as human well-being across current and future generations that depends on resilient social-ecological systems and a resilient biosphere as a whole (Folke et al. 2016).

A special type of social-ecological systems is food- and agroecosystems, defined as coupled systems (Berkes and Folke 1998b) or more specifically "human-designed systems" (Hodbod and Eakin 2015, p. 4), in which actors exercise a disproportionate control and influence over ecosystems. Investigations of agroecosystems in particular are increasingly linked with resilience concepts to better understand and evaluate pathways for building more sustainable systems (Peterson et al. 2018; Bennett et al. 2014).

DOI: 10.4324/9781003355724-1

Sustainability Challenges in Plant Breeding

The influence and power of humans within agroecosystems began with the invention of agriculture for the professional production of food,[1] it is "[...] humankind's largest engineered ecosystem" (Zhang et al. 2007, p. 1). This anthropogenic influence is particularly visible in the case of plant breeding. Here, humans directly shape plant organisms according to their specific needs. Originally, crop seeds and cultivars were a common good, belonging to no one in particular and therefore to everyone (Mgbeoji 2006; Shiva 1997). Seeds have been further developed in community seed banks (Francis 2015) or by farmers in on-farm sharing systems (Pautasso et al. 2013), which has resulted in a wide range of diverse varieties (Cleveland and Soleri 2002). Plant breeding was thus an integral part of farming for many centuries.

With the beginning of modern scientific plant breeding, which refers to breeding by scientists off-farm using evolutionary theory and genetic research (Cleveland and Soleri 2002), plant breeders have concentrated on the development of a small number of varieties and aim at high yields under optimal growing conditions. As a result, breeding became more consistent with scientific resource management in line with a "utilitarian and exploitative worldview which assumes that humans have dominion over nature" (Berkes and Folke 1998a, p. 1). However, the preservation and development of a high level of diversity in plant genetic resources is necessary to design resilient agroecosystems (Lammerts van Bueren et al. 2018; Altieri and Nicholls 2017), which contrasts with the utilitarian and exploitative worldview.

Out of carefully developed varieties and using intellectual property rights (e.g., patents or variety protection), the economization of plant breeding culminated in the privatization of plant genetic resources. As a result, breeding and farming are largely subject to economic efficiency and efficacy, ultimately leading to a "genetic erosion" (van de Wouw et al. 2010) and the transformation of crop seeds and varieties from common goods into private goods (Bonny 2017; Kloppenburg 2014; Halewood et al. 2013). A prominent example is the highly criticized market power of global corporations like Cargill or the former company Monsanto,[2] who mostly sell hybrid seeds for the farming of monocultures, creating economic dependencies of farmers on the company's products (Kneen 1999). This business model is consistent with the worldview described above and based on the sale of a limited range of high-yielding, genetically uniform varieties that rely on the intensive use of chemical-synthetic pesticides and fertilizers, leading to environmental damages (Rasmussen et al. 2018).

While the major debate concerning the implications of these developments is focused on grains and vegetables, the fruit sector – sharing similar ethical, economic, and legal problems – remains largely undiscussed. This book addresses this research gap and sheds light on the specific sustainability challenges of fruit breeding and cultivation.

Shifting the Lens to Fruits: Sustainability Challenges in Apple Breeding

Apples are the most widely cultivated tree crops in temperate zones and of high economic and cultural importance for the fruit sector (Brown 2012; Spengler 2019). Although there are certain differences between pomes, stone fruits, and berry fruits as well as between different geographic areas, fruit crops share mainly the same biological characteristics (Janick 2005). Thus, the characteristics and dynamics of the apple sector serve as a fitting representative case for perennial fruit systems in general.[3]

Conventional modern apple breeding is defined as breeding by scientists off-farm, primarily relying on evolutionary theory and genetic research. Within the context of the economic, ecological, and societal developments described in the last subsection, four aspects mainly characterize modern conventional apple breeding:

1 Conventional apple breeding takes place under intensive plant protection conditions and is characterized by a *lack of robustness* within the breeding goals. Robustness is defined as less susceptibility against a wide range of environmental conditions and diseases through high genetic diversity (Wolter et al. 2018).

2 Several modern apple varieties bred in the last decades are primarily derived from five progenitors: *Golden Delicious, Cox Orange, Jonathan, McIntosh, Red Delicious*, and *James Grieve* (Bannier 2011). This narrow genetic basis, which apple breeders all across the world have been working with (Kumar et al. 2010), negatively influences the vitality of current and prospective apple varieties to-be-bred and leads to *inbreeding* (Bannier 2011; Brown 2012; Noiton and Alspach 1996). An exemplary result of these developments is the breakdown of the monogenic dominant resistance against apple scab (Parisi et al. 1993; Bus et al. 2011).

3 Apple breeding tends to rely more and more on laboratorial approaches such as in-vitro breeding (micropropagation of apples in Petri dishes or test tubes) or *marker-assisted breeding*, where genetic markers guide the identification of fitting cultivars for breeding (Hanke and Flachowsky 2017). Especially marker-assisted breeding is perceived as a suitable method for more efficient breeding programs (ibid.). However, because of low cost-efficiency and a high amount of necessary financial resources, markers are not accessible for small breeding initiatives (Brown 2012), leading to economic and technical injustices.

4 Similar to general developments in the breeding sector, newly developed apple varieties are subject to increasing *economization* and *privatization* through club concepts. *Club concepts*[4] describe "cultivars with variety and brand protection [which] are only cultivated by selected farmers (the club) following specific criteria" (Wolter and Sievers-Glotzbach 2019, p. 307). These concepts typically include value chain actors from farming to marketing – breeders and consumers are excluded. Moreover, the whole apple sector is characterized by short-term thinking. An increasing number of varieties are regularly introduced into the market and fail to gain widespread market acceptance (Hanke and Flachowsky 2017).

Overall, these aspects lead to major sustainability challenges in the apple sector, particularly in breeding. Together, these developments limit farmer's access to some existing varieties and inhibit the long-term sustainable development of robust apple varieties. Because of this lack of robustness, fungicides and insecticides are necessary for most apple cultivation models, particularly in intensive apple farming (Granatstein and Peck 2017). These treatments negatively affect ecosystems through greenhouse gas emissions, the agrochemical contamination of water and soil, or the poisoning of organisms (Demestihas et al. 2017; Granatstein and Peck 2017; Power 2010; Dale and Polasky 2007; Zhang et al. 2007). In general, this degrades agrobiodiversity both on the cultivar and ecosystem level and is detrimental to micro-organisms in the soil, birds, and insects (Granatstein and Peck 2017; Wuppertal Institut 2015).

Toward Sustainable Apple Breeding

A solution to these problems could be the implementation of *organic apple farming* on a larger scale, which is generally more sustainable than conventional farming (Sandhu et al. 2010; Niggli and Fließbach 2009; Reganold et al. 2001). However, organic farmers who refuse to use large-scale plant protection measures are severely affected by the consequences of the formerly described issues; they need robust varieties that are developed and tested under organic production conditions (Lammerts van Bueren et al. 2011). The lack of robust varieties with beneficial marketable traits, for example, good taste or visual appeal, presents a challenge for organic apple farmers.

In the general discussion concerning plant breeding, one proposal to stop the loss of agrobiodiversity is a re-orientation to varieties as a common good rather than a private good (Sievers-Glotzbach et al. 2020; Wirz et al. 2017; Halewood et al. 2013; Dedeurwaerdere 2012). Additionally, the model of *participatory plant breeding* (PPB), where breeders and farmers work together on-farm (Ceccarelli 2012), offers a further opportunity to promote commons characteristics. *Commons* describe collective action situations, in which a distinct community of actors collectively own, share, manage, and/or develop a (potentially) diverse range of goods and resources in a particular institutional setting (Ostrom 2005). Contemporary economical, societal, and political developments, however, tend to focus on competition and private property rights (Winterfeld et al. 2012) and as a consequence are mirrored in the legal framework surrounding apple breeding and production. These dominating characteristics are not compatible with the perspective of plant genetic resources as a common good and the principles of organic breeding as proposed by the International Federation for Organic Agriculture Movements (IFOAM 2014), in which cooperation and transparency are emphasized (Lammerts van Bueren 2010).

Overall, the problems introduced above – the anthropogenically constructed character of food – and agroecosystems; negative environmental impacts of modern conventional apple breeding and farming; agrobiodiversity loss; privatization-based lack of diversity and limited participation in breeding; and economization trends – culminate in a *lack of resilience* in current apple breeding and cultivation. This corresponds with low capacities for buffering, adaptation, or transformation in the face of external shocks and global environmental changes, such as drought, diseases, or climate change (Bannier 2011; Brown 2012; Byrne

2012). The first step in the transition to a resilient apple cultivation system is the transition to resilient breeding. Thus, the following hypothesis is derived:

> *Commons-based organic apple breeding, including the use of robust cultivars, test-ing in an organic setting, on-farm, participatory breeding methods, and treatment of resources as common goods, is preferable for the social-ecological resilience and long-term sustainability of apple breeding and cultivation as opposed to modern conventional breeding approaches.*

Current State of Research

This book is mainly embedded in four scientific discourses: social-ecological systems, social-ecological resilience, ecosystem services, and commons. The re-spective research gaps concerning apple or fruit breeding and cultivation differ throughout the discourses.

In the field of social-ecological systems, a multitude of theoretical and analyt-ical frameworks with different goals, interdisciplinary approaches, and applica-tions exist (Cumming 2014; Binder et al. 2013; Schlüter et al. 2012). They provide a sound scientific base for conceptualizing and analyzing social-ecological sys-tems. Numerous case studies and basic research have been conducted in various fields of interest (Sakai and Umetsu 2014). In the case of food systems, Norgaard (1984) was the first to describe agriculture as a coupled social-ecological system. Thereafter, the analysis and modeling of fisheries as a social-ecological system by Ostrom (1990) is one of the earliest case studies and pioneering works. Other studies describe maize and coffee systems (Eakin et al. 2017), livestock manage-ment in nomadic systems (Koda and Fujita 2014), or the Californian food system in general (Hodbod and Eakin 2015). Moreover, within this stream, specific re-search concerning plant breeding has been conducted, for example, in the case of potatoes (Pacilly et al. 2016) and soybeans (Achathaler 2015). However, neither a conceptualization nor an analysis of fruit or apple breeding and cultivation as a social-ecological system exists to the knowledge of the author.

As already mentioned previously, the concept of social-ecological resilience is based on the social-ecological systems perspective. Therefore, research in these areas is strongly interlinked. Literature on resilience experienced a tremendous growth over the last years (Folke 2016). A wide range of studies concerning food systems (Fraser 2007; Hodbod and Eakin 2015; Tendall et al. 2015; Walsh-Dilley et al. 2016), seed systems (EUCARPIA 2010; McGuire and Sperling 2013; Global Alliance for the Future of Food 2016), and agricultural systems (Darnhofer 2014; Urruty et al. 2016) already exist, emphasizing different angles of the resilience perspective. Similar to social-ecological systems research, no observations of fruit breeding and cultivation systems have been made applying the conceptual frame-work of social-ecological resilience. However, previous research provides several frameworks, principles, and sets of indicators (Quinlan et al. 2016; Biggs et al. 2015; Cabell and Oelofse 2012). Here, the *Principles for Building Resilience* (Biggs et al. 2015) serve as a central reference point.

Research on ecosystem services is multifaceted, involving different study areas and concepts (Danley and Widmark 2016). Literature already provides several studies on agricultural systems that take a detailed look at the environmental impact of farming practices on ecosystems (Swinton et al. 2007; Zhang et al. 2007; Dale and Polasky 2007). Demestihas et al. (2017) were the first to extensively review on ecosystem services in commercial apple orchards. Albeit they did not include cultural ecosystem services and solely concentrated on provisioning, supporting, and regulating services. Literature on other perennial systems such as meadow orchards (Bieling and Plieninger 2013) or vineyards (Winkler and Nicholas 2016) take a larger focus on cultural aspects. However, an observation of plant breeding in general and concurrently on fruit and apple breeding remain absent in literature on ecosystem services. Based on research already conducted in the fruit and apple context, the inclusion of breeding is valuable to understand and subsequently design resilient breeding systems.

The term commons or common good originated in the field of economics and has been popularized and theorized by Ostrom's work *Governing the Commons* (1990). Recent research uses the conceptualization by Ostrom and colleagues as a foundation for the further development of the concept of commons. Originally constructed for the analysis of natural resource systems, the concept has been expanded to areas of new commons, including digital commons, knowledge commons, cultural commons, or global commons (Hess 2008). In the context of plant breeding and seed production, several studies from the perspective of commons exist, albeit lacking a detailed conceptualization. These include seeds and varieties in general (Zukunftsstiftung Landwirtschaft 2013; Kotschi and Rapf 2016; Wirz et al. 2017), seed exchange systems (Pautasso et al. 2013), and participatory breeding systems (Wilbois 2011; Gali 2013; Chable et al. 2014). Only recently did Sievers-Glotzbach et al. (2020) propose a conceptualization of seed commons. Again, these observations concentrate on grains and vegetables, and a closer look at fruits/apples as well as fruit/apple breeding and cultivation is missing.

Based on this current state of research, four knowledge gaps can be identified: until recently, no scientific observations are available regarding (1) fruit and apple breeding as a social-ecological system, (2) the resilient design of fruit/apple cultivation and breeding systems, (3) ecosystem services connected to fruit/apple breeding, and (4) commons arrangements in fruit or apple breeding. Therefore, there is a lack of understanding regarding the analysis of fruit and apple breeding as a social-ecological system, the potential of commons-based organizations in this system, and the possibilities to enhance the social-ecological resilience of fruit and apple breeding. However, all of these scientific concepts are broadly discussed in the literature, provide large theoretical foundations, and have been already partially implemented to analyze vegetable seeds. These observations and scientific insights are well-suited for the analysis of fruit and apple breeding.

Aims of the Book

This book discusses the following hypothesis: *Commons-based organic apple breeding, including the use of robust cultivars, testing in an organic setting, on-farm, participatory breeding methods, and treatment of resources as common goods, is preferable for the social-ecological resilience and long-term sustainability of apple breeding and cultivation as opposed to modern conventional breeding approaches.* The discussion of this hypothesis contributes to the resilience and commons discourses by applying their conceptual frames to a new and relevant application field.

This leads to the following main question:

> *How can commons-based organic apple breeding improve the social-ecological resilience of current apple breeding and the apple cultivation system, and how can the potential of this approach be applied to praxis?*

Thereby, the focus is on the qualitative potential for resilience and sustainability on ecological-economical, systemic, and organizational levels. While some discussions will refer to fruits more generally, others will concentrate on the empirical context of this book, the apple sector in Germany. Selecting Germany as a case is valuable because it provides an in-depth illustration of socio-economic structures surrounding apple breeding and cultivation and because it is comparable with other apple-producing countries, especially from the EU. The transferability of the findings will be discussed in the course of this book whenever it is appropriate. In particular, the book concentrates on four objectives, derived from the main question.

1. *How is the social-ecological system of fruit breeding and cultivation defined and conceptualized?*

    In a classical sense, social-ecological systems are place-based systems (Ostrom 2009), but breeding also takes place in virtual knowledge systems. Moreover, breeding interacts with fruit cultivation systems, which directly affect the breeding system. Hence, breeding is likely conceptually more complex than classical social-ecological systems. This book aims to integrate fruit breeding and cultivation into contemporary frames and concepts of social-ecological systems research, whereby the general focus on fruit provides further links and research opportunities for fruits other than apples.

2. *What (ideally) characterizes a resilient fruit breeding and cultivation system?*

    Social-ecological resilience refers to the provision, maintenance, and promotion of ecosystem services. Here, it is thus necessary to define, describe, and categorize ecosystem services that are relevant for fruit breeding and cultivation. Thereafter, important characteristics of a resilient fruit breeding and cultivation system are identified by synthesizing existing theoretical and empirical observations. For this, insights from empirical studies regarding agricultural and breeding systems as well as conceptual works from resilience research are used to derive such characteristics. The *Principles for Building Resilience* by Biggs et al. (2015), largely based on empirical data, provides the foundation for the characterization of a resilient system. The book aims to outline ideal characteristics that help to evaluate the resilience potential of different fruit breeding approaches. Again, the discussions offer transferable insights into different fruits and breeding systems.

3. *Which apple breeding approaches exist, what are the differences, and how do they affect apple cultivation and social-ecological resilience?*

    Based on the description of an ideal resilient fruit breeding and cultivation system, current apple breeding approaches in Germany are categorized and analyzed. This provides a basis for the evaluation and comparison of the effects of a commons-based breeding approach on resilience with other models and helps ground the concept of commons-based organic apple breeding in empirical evidence. The analysis further serves as an example of how to apply the general conceptualizations of fruit breeding and cultivation and enables reflections on the previous results.

4. *Which factors inhibit the broad implementation of commons-based apple breeding and how can they be overcome to exploit its full potential?*

    Commons-based organic fruit breeding is currently not widely implemented in Germany but merely a niche. For a better understanding of the reasons behind this status quo, obstacles to the broad implementation of commons-based organic fruit breeding need to be investigated. This calls for an analysis of the driving and inhibiting factors influencing commons-based organic apple breeding to investigate the integration of the approach into practice. Possibilities to overcome the analyzed barriers are described and evaluated.

Overall, this book enhances the understanding of resilience and commons in agroecosystems on conceptual and empirical levels. Beyond the scientific contribution, it derives recommendations for decision-makers and practitioners such as fruit farmers, breeders, and marketeers.

Structure

Part I presents the theoretical (Chapter 1) and methodological (Chapter 2) frameworks and justifies the choice of scientific discourses and the chosen methods. Additionally, this part provides the empirical context (apple breeding and cultivation in Germany) (Chapter 3) and important background knowledge for the following chapters.

Part II investigates the resilience of fruit breeding and cultivation on a conceptual level. It connects fruit breeding and cultivation to the discourse on social-ecological systems (Chapter 4). To understand which ecosystem services have to be maintained and promoted by resilient fruit breeding systems, the current state of research is evaluated and complemented by additional literature on fruit farming, consequently deriving a set of relevant ecosystem services (Chapter 5). Hence, general principles for building resilience are reviewed through the lens of literature on food- and agroecosystems, deducing implications for fruit breeding and cultivation in the form of resilience attributes (Chapter 6). These attributes enable an analysis of the effects of fruit breeding approaches on the system's resilience.

Part III directly connects the conceptual base of the previous chapter to the empirical context by evaluating different breeding approaches. It draws on case studies, including corporate-based apple breeding (Chapter 7), public apple breeding (Chapter 8), and commons-based apple breeding (Chapter 9). Based on these empirical findings, the results are compared and discussed, and the approach of commons-based organic apple breeding is conceptualized (Chapter 10).

Part IV elaborates on how to implement and further propagate commons-based organic apple breeding, based on a market study of organic apple cultivation and breeding in Germany. Results are presented and respectively discussed for market structures and developments (Chapter 11), promoting and inhibiting factors of organic apple cultivation and breeding (Chapter 12), and possible business models that could enhance financial opportunities for organic apple breeding (Chapter 13). Insights from these chapters provide the foundation for discussing research the challenges for a broader implementation of commons-based organic apple breeding (Chapter 14).

The book concludes by reflecting on the central results, discussing possibilities for further research in the respective discourses, and deriving implications for policy and praxis.

Notes

1　For example, Cassman and Wood (2005) identify modern agriculture as the globally greatest threat to biodiversity and the functioning of ecosystems. Bennett et al. (2014) give an overview of the negative impacts of modern agriculture on global resources and processes including land use, biodiversity loss, radiative forcing, freshwater use, and nutrients.

2　In 2018, Monsanto merged with the Bayer AG. Effects of this merger are heavily debated in scientific discourses (e.g., Joseph 2021).

3　Of course, biological, economical, and cultural details differ with other tree fruits (e.g., pears, cherries), small fruits (e.g., blackberries, raspberries), or tree nuts (e.g., almonds, walnuts) (Badenes and Byrne 2012; Hanke and Flachowsky 2017). Nevertheless, apples are the most prominent objects of study in fruit research across scientific disciplines, which makes them an ideal empirical context of this book.

4　Club concepts are also referred to as "club apples" (Legun 2015) and describe both the apple cultivars and their respective business model.

References

Achathaler, L. (2015): Towards a More Sustainable Soybean Production in Austria: A Socio-Ecological Review. Vienna: Master thesis at the Department of Crop Sciences (DNW), Division of Plant Breeding, University of Natural Resources and Life Sciences Vienna (BOKU).

Altieri, M. A.; Nicholls, C. I. (2017): The Adaptation and Mitigation Potential of Traditional Agriculture in a Changing Climate. In *Climatic Change* 140 (1), pp. 33–45. DOI: 10.1007/s10584-013-0909-y.

Badenes, Marisa Luisa; Byrne, David H. (Eds.) (2012): *Fruit Breeding*. Boston, MA: Springer US.

Bannier, H.-J. (2011): Moderne Apfelzüchtung. Genetische Verarmung und Tendenzen zur Inzucht. In *Erwerbs-Obstbau* 52 (3–4), pp. 85–110. DOI: 10.1007/s10341-010-0113-4.

Bennett, E.; Carpenter, S. R.; Gordon, L. J.; Ramankutty, N.; Balvanera, P.; Campbell, B. et al. (2014): Toward a More Resilient Agriculture. In *The Solutions Journal* 5 (5), pp. 65–75.

Berkes, F.; Folke, C. (1998a): Linking Social and Ecological Systems for Resilience and Sustainability. In Fikret Berkes, Carl Folke (Eds.): *Linking Social and Ecological Systems. Management Practices and Social Mechanisms for Building Resilience.* Cambridge, UK: Cambridge University Press, pp. 1–25.

Berkes, F.; Folke, C. (Eds.) (1998b): *Linking Social and Ecological Systems. Management Practices and Social Mechanisms for Building Resilience.* Cambridge, UK: Cambridge University Press.

Bieling, C.; Plieninger, T. (2013): Recording Manifestations of Cultural Ecosystem Services in the Landscape. In *Landscape Research* 38 (5), pp. 649–667. DOI: 10.1080/01426397.2012.691469.

Biggs, R.; Schlüter, M.; Schoon, M. L. (Eds.) (2015): *Principles for Building Resilience. Sustaining Ecosystem Services in Social-Ecological Systems.* Cambridge, UK: Cambridge University Press.

Binder, C. R.; Hinkel, J.; Bots, P. W. G.; Pahl-Wostl, C. (2013): Comparison of Frameworks for Analyzing Social-ecological Systems. In *Ecology and Society* 18 (4). DOI: 10.5751/ES-05551-180426.

Bonny, S. (2017): Corporate Concentration and Technological Change in the Global Seed Industry. In *Sustainability* 9 (9), p. 1632. DOI: 10.3390/su9091632.

Brown, S. (2012): Apple. In Marisa Luisa Badenes, David H. Byrne (Eds.): *Fruit Breeding.* Boston, MA: Springer US, pp. 329–367.

Byrne, D. H. (2012): Trends in Fruit Breeding. In Marisa Luisa Badenes, David H. Byrne (Eds.): *Fruit Breeding.* Boston, MA: Springer US, pp. 3–36.

Cabell, J. F.; Oelofse, M. (2012): An Indicator Framework for Assessing Agroecosystem Resilience. In *Ecology and Society* 17 (1). DOI: 10.5751/ES-04666-170118.

Cassman, K.; Wood, S. (2005): Cultivated Systems. In United Nations Environment Programme (UNEP) (Ed.): *Millenium Ecosystem Assessment. Ecosystems and Human Well-Being: Current State and Trends.* Washington, DC: Island Press, pp. 747–791.

Ceccarelli, S. (2012): *Plant Breeding with Farmers – A Technical Manual.* Aleppo: International Center for Agricultural Research in the Dry Areas (ICARDA).

Chable, V.; Dawson, J.; Bocci, R.; Goldringer, I. (2014): Seeds for Organic Agriculture. Development of Participatory Plant Breeding and Farmers' Networks in France. In Stéphane Bellon, Servane Penvern (Eds.): *Organic Farming, Prototype for Sustainable Agricultures.* Dordrecht: Springer Netherlands, pp. 383–400.

Cleveland, D. A.; Soleri, D. (2002): *Farmers, Scientists, and Plant Breeding. Integrating Knowledge and Practice.* Wallingford, Oxon, New York: CABI.

Cumming, G. (2014): Theoretical Frameworks for the Analysis of Social-Ecological Systems. In Shoko Sakai, Chieko Umetsu (Eds.): *Social-Ecological Systems in Transition.* Tokyo: Springer Japan, pp. 3–26.

Dale, V. H.; Polasky, S. (2007): Measures of the Effects of Agricultural Practices on Ecosystem Services. In *Ecological Economics* 64 (2), pp. 286–296. DOI: 10.1016/j.ecolecon.2007.05.009.

Danley, B.; Widmark, C. (2016): Evaluating Conceptual Definitions of Ecosystem Services and their Implications. In *Ecological Economics* 126, pp. 132–138. DOI: 10.1016/j.ecolecon.2016.04.003.

Darnhofer, I. (2014): Resilience and Why It Matters for Farm Management. In *European Review of Agricultural Economics* 41 (3), pp. 461–484. DOI: 10.1093/erae/jbu012.

Dedeurwaerdere, T. (2012): *Institutionalizing Global Genetic Resource Commons for Food and Agriculture.* Abingdon, Oxon, New York: Earthscan from Routledge.

Demestihas, C.; Plénet, D.; Génard, M.; Raynal, C.; Lescourret, F. (2017): Ecosystem Services in Orchards. A Review. In *Agronomy for Sustainable Development* 37 (2), p. 581. DOI: 10.1007/s13593-017-0422-1.

Eakin, H.; Rueda, X.; Mahanti, A. (2017): Transforming Governance in Telecoupled Food Systems. In *Ecology and Society* 22 (4). DOI: 10.5751/ES-09831-220432.

EUCARPIA (Ed.) (2010): Breeding for Resilience: A Strategy for Organic and Low-Input Farming Systems. *EUCARPIA 2nd Conference of the "Organic and Low-Input Agriculture" Section*. Paris, 1–3 December.

Folke, C. (2016): *Resilience – Oxford Research Encyclopedia of Environmental Science*. Oxford: Oxford University Press.

Folke, C.; Biggs, R.; Norström, A. V.; Reyers, B.; Rockström, J. (2016): Social-Ecological Resilience and Biosphere-Based Sustainability science. In *Ecology and Society* 21 (3). DOI: 10.5751/ES-08748-210341.

Folke, C.; Carpenter, S.; Walker, B.; Scheffer, M.; Elmqvist, T.; Gunderson, L.; Holling, C. S. (2004): Regime Shifts, Resilience, and Biodiversity in Ecosystem Management. In *Annual Review of Ecology and Systematics* 35 (1), pp. 557–581. DOI: 10.1146/annurev.ecolsys.35.021103.105711.

Francis, C. (2015): Community Seed Banks. Origins, Evolution and Prospects. In *Crop Science* 55 (6), p. 2929. DOI: 10.2135/cropsci2015.07.0409br.

Fraser, E. D. G. (2007): Travelling in Antique Lands. Using Past Famines to Develop an Adaptability/Resilience Framework to Identify Food Systems Vulnerable to Climate Change. In *Climatic Change* 83 (4), pp. 495–514. DOI: 10.1007/s10584-007-9240-9.

Gali, A. (2013): Governance of Seed and Food Security through Participatory Plant Breeding. Empirical Evidence and Gender Analysis from Syria. In *Natural Resources Forum* 37 (1), pp. 31–42. DOI: 10.1111/1477-8947.12008.

Global Alliance for the Future of Food (2016): The Future of Food: Seeds of Resilience. A Compendium of Perspectives on Agricultural Biodiversity Form around the World. Available online at https://futureoffood.org/wp-content/uploads/2016/09/Future_of_Food_Seeds_of_Resilience_Report.pdf, checked on 9/11/2017.

Granatstein, D.; Peck, G. (2017): Assessing the Environmental Impact and Sustainability of Apple Cultivation. In Gayle M. Volk, Amit Dhingra, Sally A. Bound, Dugald C. Close, Peter M. Hirst, M. C. Goffinet et al. (Eds.): *Achieving Sustainable Cultivation of Apples*. 1st ed. Cambridge: Burleigh Dodds Science Publishing (Burleigh Dodds Series in Agricultural Science), pp. 523–549.

Halewood, M.; López Noriega, I.; Louafi, S. (2013): Crop Genetic Resources as a Global Commons. Challenges in International Law and Governance. Abingdon, Oxon, New York: Earthscan from Routledge (Issues in agricultural biodiversity).

Hanke, M.-V.; Flachowsky, H. (2017): Obstzüchtung und wissenschaftliche Grundlagen. Berlin, Heidelberg: Springer.

Hess, C. (2008): Mapping the New Commons. In *SSRN Journal*. DOI: 10.2139/ssrn.1356835.

Hodbod, J.; Eakin, H. (2015): Adapting a Social-Ecological Resilience Framework for Food Systems. In *Journal of Environmental Studies and Sciences* 5 (3), pp. 474–484. DOI: 10.1007/s13412-015-0280-6.

IFOAM (2014): The IFOAM Norms for Organic Production and Processing. Version 2014. Available online at http://www.ifoam.bio/sites/default/files/ifoam_norms_version_july_2014.pdf, checked on 3/23/2021.

IPCC (2014): *Climate Change 2014: Synthesis Report. Contribution of Working Groups I, II and III to the Fifth Assessment Report of the Intergovernmental Panel on Climate Change* [Core Writing Team, R.K. Pachauri and L.A. Meyer (eds.)]. Geneva: IPCC.

Janick, J. (2005): The Origins of Fruits, Fruit Growing, and Fruit Breeding. In *Plant Breeding Reviews* 25, pp. 255–321.

Joseph, R. K. (2021): Innovation, Patents, and Competition in Modern Agriculture: A Case Study of Bayer and Monsanto Merger. In *The Antitrust Bulletin* 66 (2), pp. 214–224. DOI: 10.1177/0003603X21997022.

Kloppenburg, J. (2014): Re-Purposing the Master's Tools: The Open Source Seed Initiative and the Struggle for Seed Sovereignty. In *The Journal of Peasant Studies* 41 (6), pp. 1225–1246. DOI: 10.1080/03066150.2013.875897.

Kneen, B. (1999): Restructuring Food for Corporate Profit. The Corporate Genetics of Cargill and Monsanto. In *Agriculture and Human Values* 16 (2), pp. 161–167. DOI: 10.1023/A:1007586710282.

Koda, R.; Fujita, N. (2014): Mongolian Nomadism and the Relationship between Livestock Grazing, Pasture Vegetation, and Soil Alkalization. In Shoko Sakai, Chieko Umetsu (Eds.): *Social-Ecological Systems in Transition*. Tokyo: Springer Japan, pp. 51–70.

Kotschi, J.; Rapf, K. (2016): Befreiung des Saatguts durch open source Lizensierung. Edited by AGRECOL Verein für standortgerechte Landnutzung. Guggenhausen. Available online at http://www.opensourceseeds.org/sites/default/files/downloads/Befreiung_des_Saatguts%20durch_Open_Source_Lizensierung.pdf, checked on 8/6/2019.

Kumar, S.; Volz, R. K.; Alspach, P. A.; Bus, V. G. M. (2010): Development of a Recurrent Apple Breeding Programme in New Zealand: A Synthesis of Results, and a Proposed Revised Breeding Strategy. In *Euphytica* 173 (2), pp. 207–222. DOI: 10.1007/s10681-009-0090-6.

Lammerts van Bueren, E. (2010): Ethics of Plant Breeding: The IFOAM Basic Principles as a Guide for the Evolution of Organic Plant Breeding. In *Ecology & Farming* (February), pp. 7–10, checked on 3/28/2018.

Lammerts van Bueren, E. T.; Jones, S. S.; Tamm, L.; Murphy, K. M.; Myers, J. R.; Leifert, C.; Messmer, M. M. (2011): The Need to Breed Crop Varieties Suitable for Organic Farming, Using Wheat, Tomato and Broccoli as Examples. A Review. In *NJAS – Wageningen Journal of Life Sciences* 58 (3–4), pp. 193–205. DOI: 10.1016/j.njas.2010.04.001.

Lammerts van Bueren, E. T.; Struik, P. C.; van Eekeren, N.; Nuijten, E. (2018): Towards Resilience through Systems-Based Plant Breeding. A Review. In *Agronomy for Sustainable Development* 38 (5), p. 42. DOI: 10.1007/s13593-018-0522-6.

Mace, G. M.; Masundire, J.; Baillie, J. (2005): Biodiversity. In Rashid Hassan, Robert Scholes, Neville Ash (Eds.): *Ecosystems and Human Well-Being: Current State and Trends: Findings of the Condition and Trends Working Group*. Washington, DC: Island Press, pp. 79–115.

McGuire, S.; Sperling, L. (2013): Making Seed Systems More Resilient to Stress. In *Global Environmental Change* 23 (3), pp. 644–653. DOI: 10.1016/j.gloenvcha.2013.02.001.

MEA (2005): *Ecosystems and Human Well-Being. Synthesis; A Report of the Millennium Ecosystem Assessment*. Washington, DC: Island Press.

Mgbeoji, I. (2006): *Global Biopiracy: Patents, Plants, and Indigenous Knowledge*. Ithaca, NY: Cornell University Press.

Niggli, U.; Fließbach, A. (2009): Gut fürs Klima? Ökologische und konventionelle Landwirtschaft im Vergleich. In *Kritischer Agrarbericht*. Konstanz, Germany: AgrarBündnis. pp. 103–109.

Noiton, D. A.; Alspach, P. A. (1996): Founding Clones, Inbreeding, Coancestry, and Status Number of Modern Apple Cultivars. In *Journal of the American Society for Horticultural Science* 121 (5), pp. 773–782. DOI: 10.21273/JASHS.121.5.773.

Norgaard, R. B. (1984): Coevolutionary Agricultural Development. In *Economic Development and Cultural Change* 32 (3), pp. 525–546. DOI: 10.1086/451404.

Ostrom, E. (1990): *Governing the Commons. The Evolution of Institutions for Collective Action*. Cambridge, UK: Cambridge University Press.

Ostrom, E. (2009): A General Framework for Analyzing Sustainability of Social-Ecological Systems. In *Science (New York)* 325 (5939), pp. 419–422. DOI: 10.1126/science.1172133.

Pörtner, H. O.; Scholes, R. J.; Agard, J.; Archer, E.; Arneth, A.; Bai, X.; Barnes, D.; Burrows, M.; Chan, L.; Cheung, W. L. W.; Diamond, S.; Donatti, C.; Duarte, C.; Eisenhauer, N.; Foden, W.; Gasalla, M. A.; Handa, C.; Hickler, T.; Hoegh-Guldberg, O.; ... Ngo, H. (2021): *Scientific outcome of the IPBES-IPCC co-sponsored workshop on biodiversity and climate change.* Intergovernmental Science-Policy Platform on Biodiversity and Ecosystem Services (IPBES). https://zenodo.org/record/5101125

Pacilly, F. C. A.; Groot, J. C. J.; Hofstede, G. J.; Schaap, B. F.; van Bueren, E. T. L. (2016): Analysing Potato Late Blight Control as a Social-Ecological System Using Fuzzy Cognitive Mapping. In *Agronomy for Sustainable Development* 36 (2), p. 25. DOI: 10.1007/s13593-016-0370-1.

Pautasso, M.; Aistara, G.; Barnaud, A.; Caillon, S.; Clouvel, P.; Coomes, O. T. et al. (2013): Seed Exchange Networks for Agrobiodiversity Conservation. A Review. In *Agronomy for Sustainable Development* 33 (1), pp. 151–175. DOI: 10.1007/s13593-012-0089-6.

Peterson, C. A.; Eviner, V. T.; Gaudin, A. C. (2018): Ways Forward for Resilience Research in Agroecosystems. In *Agricultural Systems* 162, pp. 19–27. DOI: 10.1016/j.agsy.2018.01.011.

Power, A. G. (2010): Ecosystem Services and Agriculture. Tradeoffs and Synergies. In *Philosophical Transactions of the Royal Society of London. Series B, Biological Sciences* 365 (1554), pp. 2959–2971. DOI: 10.1098/rstb.2010.0143.

Quinlan, A. E.; Berbés-Blázquez, M.; Haider, L. J.; Peterson, G. D.; Allen, C. (2016): Measuring and Assessing Resilience. Broadening Understanding through Multiple Disciplinary Perspectives. In *Journal of Applied Ecology* 53 (3), pp. 677–687. DOI: 10.1111/1365-2664.12550.

Rasmussen, L. V.; Coolsaet, B.; Martin, A.; Mertz, O.; Pascual, U.; Corbera, E. et al. (2018): Social-Ecological Outcomes of Agricultural Intensification. In *Nature Sustainability* 1 (6), pp. 275–282. DOI: 10.1038/s41893-018-0070-8.

Reganold, J. P.; Glover, J. D.; Andrews, P. K.; Hinman, H. R. (2001): Sustainability of Three Apple Production Systems. In *Nature* 410 (6831), pp. 926–930. DOI: 10.1038/35073574.

Sakai, Shoko; Umetsu, Chieko (Eds.) (2014): *Social-Ecological Systems in Transition.* Tokyo: Springer Japan.

Sandhu, H. S.; Wratten, S. D.; Cullen, R. (2010): Organic Agriculture and Ecosystem Services. In *Environmental Science & Policy* 13 (1), pp. 1–7. DOI: 10.1016/j.envsci.2009.11.002.

Schlüter, M.; McAllister, R.; Arlinghaus, R.; Bunnefeld, N.; Eisenack, K.; Hölkner, F. Milner-Gulland, E. J. et al. (2012): New Horizons for Managing the Environment: A Review of Coupled Social-Ecological Systems Modeling. In *Natural Resource Modeling* 25 (1), pp. 219–272. DOI: 10.1111/j.1939-7445.2011.00108.x.

Shiva, V. (1997): *Biopiracy: The Plunder of Nature and Knowledge.* Boston, MA: South End Press.

Sievers-Glotzbach, S.; Tschersich, J.; Gmeiner, N.; Kliem, L.; Ficiciyan, A. (2020): Diverse Seeds – Shared Practices: Conceptualizing Seed Commons. In *International Journal of the Commons* 14 (1), pp. 418–438. DOI: 10.5334/ijc.1043.

Spengler, R. N. (2019): Origins of the Apple: The Role of Megafaunal Mutualism in the Domestication of Malus and Rosaceous Trees. In *Frontiers in Plant Science* 10, p. 617. DOI: 10.3389/fpls.2019.00617.

Steffen, W.; Crutzen, P. J.; McNeill, J. R. (2007): The Anthropocene. Are Humans Now Overwhelming the Great Forces of Nature. In *AMBIO: A Journal of the Human Environment* 36 (8), pp. 614–621. DOI: 10.1579/0044-7447(2007)36[614:TAAHNO]2.0.CO;2.

Steffen, W.; Richardson, K.; Rockström, J.; Cornell, S. E.; Fetzer, I.; Bennett, E. M. et al. (2015): Sustainability. Planetary Boundaries: Guiding Human Development on a Changing Planet. In *Science* 347 (6223), p. 1259855. DOI: 10.1126/science.1259855.

Swinton, S. M.; Lupi, F.; Robertson, G. P.; Hamilton, S. K. (2007): Ecosystem Services and Agriculture. Cultivating Agricultural Ecosystems for Diverse Benefits. In *Ecological Economics* 64 (2), pp. 245–252. DOI: 10.1016/j.ecolecon.2007.09.020.

Tendall, D. M.; Joerin, J.; Kopainsky, B.; Edwards, P.; Shreck, A.; Le, Q. B. et al. (2015): Food System Resilience. Defining the Concept. In *Global Food Security* 6, pp. 17–23. DOI: 10.1016/j.gfs.2015.08.001.

Urruty, N.; Tailliez-Lefebvre, D.; Huyghe, C. (2016): Stability, Robustness, Vulnerability and Resilience of Agricultural Systems. A Review. In *Agronomy for Sustainable Development* 36 (1), p. 2. DOI: 10.1007/s13593-015-0347-5.

van de Wouw, M.; Kik, C.; van Hintum, T.; van Treuren, R.; Visser, B. (2010): Genetic Erosion in Crops. Concept, Research Results and Challenges. In *Plant Genetic Resources* 8 (01), pp. 1–15. DOI: 10.1017/S1479262109990062.

Walsh-Dilley, M.; Wolford, W.; McCarthy, J. (2016): Rights for Resilience. Food Sovereignty, Power, and Resilience in Development Practice. In *Ecology and Society* 21 (1). DOI: 10.5751/ES-07981-210111.

Wilbois, K.-P. (2011): Ökologisch-partizipative Pflanzenzüchtung. Frankfurt am Main, Bochum: Forschungsanstalt für biologischen Landbau e.V. (FiBL), Zukunftsstiftung Landwirtschaft. Available online at http://orgprints.org/20574/1/1563-oekolog-partizipativ-pflanzenzuechtung.pdf, checked on 9/11/2019.

Winkler, K. J.; Nicholas, K. A. (2016): More than Wine: Cultural Ecosystem Services in Vineyard Landscapes in England and California. In *Ecological Economics* 124, pp. 86–98. DOI: 10.1016/j.ecolecon.2016.01.013.

Winterfeld, U. V.; Biesecker, A.; Katz, C.; Best, B. (2012): *Welche Rolle können Commons in Transformationsprozessen zu Nachhaltigkeit spielen?* Wuppertal: Wuppertal Institut für Klima, Umwelt, Energie GmbH. Available online at https://wupperinst.org/a/wi/a/s/ad/1769/, checked on 9/11/2019.

Wirz, J.; Kunz, P.; Hurter, U. (2017): *Saatgut – Gemeingut. Züchtung als Quelle von Realwirtschaft, Recht und Kultur. Standortbestimmung und Zukunftsperspektiven für gemeinnützige Saatgut- und Züchtungsinitiativen.* Dornach: Goetheanum, Sektion für Landwirtschaft. Available online at https://www.sektion-landwirtschaft.org/fileadmin/landwirtschaft/Saatgut_Gemeingut/saatgut_gemeingut_2_Auflage.pdf, checked on 9/11/2020.

Wolter, H.; Howard, N. P.; Ristel, M.; Sievers-Glotzbach, S.; Albach, D. C.; Sattler, I.; Siebenhüner, B. (2018): Research Project EGON: Development of Organically Bred Fruit Varieties in Commons-Based Initiatives. In FOEKO (Ed.): *Ecofruit. Proceedings of the 18th International Congress on Organic Fruit Growing.* Ecofruit. Hohenheim, Germany. Fördergemeinschaft Ökologischer Obstbau e.V. (FOEKO), pp. 92–95.

Wuppertal Institut (2015): *Apfel ist nicht gleich Apfel. Ökologische & soziale Auswirkungen von Äpfeln, unter Berücksichtigung unterschiedlicher Anbaumethoden.* Wuppertal: Wuppertal Institut für Klima, Umwelt, Energie GmbH. Available online at https://wupperinst.org/fa/redaktion/downloads/projects/Mundraub_Apfel_Factsheet.pdf, checked on 6/8/2019.

Zhang, W.; Ricketts, T. H.; Kremen, C.; Carney, K.; Swinton, S. M. (2007): Ecosystem Services and Dis-Services to Agriculture. In *Ecological Economics* 64 (2), pp. 253–260. DOI: 10.1016/j.ecolecon.2007.02.024.

Zukunftsstiftung Landwirtschaft (Ed.) (2013): Ökologische Pflanzenzüchtung – im Spannungsfeld zwischen Gemeingut und Saatgutwirtschaft. Zusammenfassung der 13. Saatgut-Tagung der Zukunftsstiftung Landwirtschaft am 26.01.2013 in Kassel. Zukunftsstiftung Landwirtschaft. Kassel, 26.01.2013.

Part I

Research Design and Empirical Context

1   Theoretical Framework

The theoretical framework of this book is based on *sustainability science* and therefore links several scientific disciplines and concepts to investigate specific problems (Kates et al. 2001; Kates 2012). Several scholars argue that sustainability science needs to adopt theoretical (and methodological) pluralism to overcome the incommensurability between the natural and the social sciences (Olsson and Ness 2019; Olsson and Jerneck 2018). The theoretical framework thus consists of three scientific concepts: (a) social-ecological resilience, (b) ecosystem services, and (c) commons.

Folke (2016) describes social-ecological resilience as a "subset" of sustainability science, representing a "biosphere-based sustainability science with resilience thinking as a central ingredient" (p. 11). Social-ecological resilience has its conceptual roots in the natural sciences. Resilience thinking aims to reflect on the dynamics and complexity of social-ecological systems by solving sustainability problems with a systems approach (Folke et al. 2010). The current conditions in fruit/apple breeding and cultivation as described in the introduction show that an observation from a systemic point of view is valuable. Therefore, social-ecological resilience provides a fitting lens to analyze and evaluate the lack of sustainability.

Social-ecological resilience is closely intertwined with the concept of ecosystem services, especially in literature on ecosystem stewardship (Chapin et al. 2009, 2010). By presuming that social-ecological systems are constantly confronted with high uncertainty and unpredictable change due to their characterization as complex adaptive systems (Levin et al. 2013), a stewardship of ecosystem services has to respond to these conditions. Therefore, to obtain social-ecological resilience, ecosystem services have to be maintained, promoted, and sustained (Biggs et al. 2015b).

In general, this book adopts an ecological-economic perspective on sustainability. In this view, the economy is defined as a subsystem of society which is in turn embedded in the biosphere as depicted in Figure 1.1 (Folke et al. 2016). Sustainability is understood as sustaining human well-being across current and future generations through the provision, maintenance, and promotion of a desired set of ecosystem services (Folke et al. 2010; Biggs et al. 2015b). Resilient social-ecological systems build capacities to achieve sustainability. A resilient biosphere thus builds the foundation for sustainability – without it, society cannot thrive and human well-being cannot be developed.

DOI: 10.4324/9781003355724-3

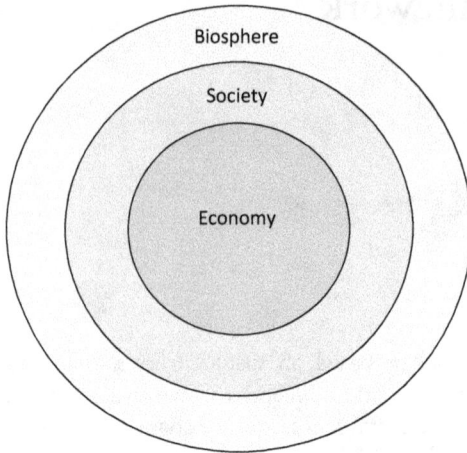

Figure 1.1 Ecological-economic perspective on sustainability (based on Folke et al. 2016).

In addition, the concept of commons is integrated into this framework to further analyze and evaluate fruit and apple breeding as collective action situations. Resilience researchers already partly implemented insights from commons research into their conceptual models (Biggs et al. 2015b). The in-depth analysis of commons-based organic apple breeding aims to further contribute to conceptual linkages of commons and resilience research.

1.1 Resilience and Social-ecological Resilience

Resilience is a concept that is used in several scientific disciplines for analytical, theoretical, or integrative purposes – from ecology to psychology, business management, and engineering (Bhamra et al. 2011; Folke 2016). One main objective of applying this concept to certain problems or systems is to build capacities for responding to changing environments and designing appropriate system structures. The perspective of this book is that of agricultural and food systems, in which resilience provides a fitting lens for analytical observation (Folke 2006; Peterson et al. 2018). As these are social-ecological systems, social-ecological resilience provides the relevant discourse for this book.

Because the perspective of social-ecological resilience is a systemic one, it has its epistemological roots in systems thinking. This form of thinking refers to a "holistic view of the components and the interrelationships among the components of a system" (Berkes and Folke 1998a, p. 8) – in this case among and across social and ecological systems. Holling et al. (1998) describe the intention of this perspective[1] as an alternative approach of doing science:

> The new stream is fundamentally interdisciplinary and combines historical, comparative and experimental approaches at scales appropriate to the issues.

It is a stream of inquiry, that is [...] most relevant for the needs of policy and politics [...] [and] a science of the integration of parts. [...] The premise [...] is that knowledge of the system we deal with is always incomplete. Surprise is inevitable.

(p. 346)

Essentially, the approach by Holling et al. (1998) embraces an alternative way of thinking, for example, in contrast to mechanistic thinking where system dynamics are perceived as linear and uncertainty irrelevant (Chapin et al. 2010). This alternative way is described as *resilience thinking* and lays the conceptual ground for social-ecological resilience.

1.1.1 Social-Ecological Systems and Resilience Thinking

According to Walker and Salt (2012), three central components of resilience thinking exist in the context of social-ecological systems. First, social (humans) and ecological (nature) systems are linked in social-ecological systems. Changes in one (ecological or social) part of the system affect the other part through feedback mechanisms. Second, all social-ecological systems are complex adaptive systems, which means that their overall behavior cannot be fully predicted and a range of different possible system states exist (See Chapter 4). Third, resilience is important and crucial to achieve sustainability in social-ecological systems.

Figure 1.2 illustrates this basic understanding of social-ecological systems as an integrative concept.[2] The dark gray arrows indicate the ecological parts of the system, whereas the light gray arrows indicate the social parts. The social-ecological feedbacks illustrate the influence of both systems on each other. Biggs et al. (2015a) argue that this integrative perspective enables to view social-ecological systems as "cohesive systems in themselves that occur at the interface between social and ecological systems" (p. 8). Albeit social

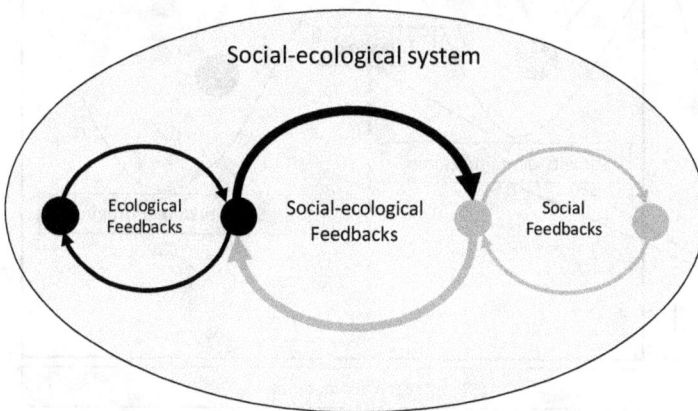

Figure 1.2 Basic model of social-ecological systems (based on Biggs et al. 2015b).

and ecological systems also remain separate entities, their interdependencies and interlinkages in form of social-ecological feedbacks build the core of every social-ecological system.

These reflections indicate that resilience thinking includes several basic assumptions about the way of understanding the world and social-ecological systems, but it is not a specific and detailed theoretical construct. Or, as Darnhofer (2014) puts it: "resilience thinking is not so much a formal theory, than a conceptual framework that helps us think about processes in new ways" (p. 467). However, this conceptual framework also includes a distinct modeling of resilience thinking which was most prominently discussed by Walker and Salt (2012), Folke et al. (2010), and Walker et al. (2004).

A popular way to explain resilience thinking is the metaphor of ball and cup (Walker and Salt 2012; Gunderson 2000). Figure 1.3 provides an illustration of important elements in this metaphor. Each cup illustrates a specific *stability domain* or *regime* of a social-ecological system and has a specific width and depth. The width and depth are defined by controlling variables and their specific thresholds. *Controlling variables* are defined as key processes or conditions inside the system and can both be internal *state variables* (e.g., actors or species) or external variables (e.g., shocks or disturbances). *Thresholds* mark a certain level or condition of one or more controlling variables, symbolized by the edges of the cup. It follows that each cup represents a kind of 'frame' that determines the scope of the respective stability domain. The ball is inside the cup and represents the current state of the system inside this 'frame,' thus the condition of the internal state variables. The ball always tends to the bottom of the cup. The bottom illustrates the equilibrium of the stability domain because the ball would stop moving

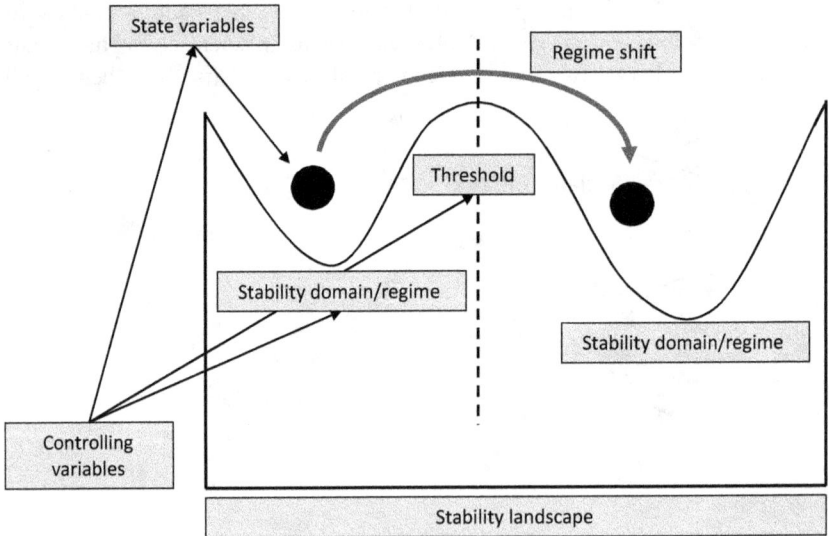

Figure 1.3 Resilience thinking (own figure).

when it reaches there. However, as systems are in a constant mode of change due to changing external and internal conditions – which means that the shape of the cup and hence the location of the ball is constantly changing – the ball never reaches the equilibrium because it is always in a state of movement. However, if controlling variables experience major change, the ball can tip over the threshold of the respective cup and roll over into another cup. This 'rolling over' is called *regime shift*. All cups (stability domains) are located in the *stability landscape* that illustrates the range of possible regimes.

For further illustration of this ball-and-cup-model, Walker and Salt (2012) give the example of a river landscape: one can imagine a river that has two possible system states (stability landscape), symbolized by two cups. In the first system state (regime/stability domain), the river contains clear water. In the second system state, the river contains turbid and muddy water. Several internal state variables, for example, the amount of phosphorus in the water, define which system state is occupied. However, external variables such as rainfall, climate change, or human actions potentially change the system state. For reasons of simplification, we assume that the level of phosphorus is the key controlling variable that defines the current regime and therefore the depth and width of the cup. The level of phosphorus is influenced by internal state variables (e.g., the lake sediments, water that is flowing in through flooding or rain) and external variables (e.g., weather conditions, straightening of the watercourse by humans). If the level of phosphorus tips over a certain threshold, a regime change takes place and the river turns muddy and turbid. The depth and width of the cups in the regime determine how and if the threshold will be crossed, and if recovery eventually takes place. If we imagine the river staying in the second system state, new feedbacks and state variables potentially emerge (e.g., new animal or plant species establish themselves in the river).

The example further illustrates the importance of thresholds as a central element of resilience thinking. This is closely connected to the understanding of sustainability as a major part of the resilience concept:

> A system's resilience can be measured by its distance from these thresholds. […] Sustainability is all about knowing if and where thresholds exist and having the capacity to manage the system in relation to these thresholds
>
> (Walker and Salt 2012, p. 63).

As a result, sustainability is referred to as either trying to avoid crossing specific thresholds in order to remain sustainable or trying to move over specific thresholds to get to a (more) sustainable system as described previously. In addition, resilience is a highly context-dependent concept: "[Resilience thinking] explains why efficiency itself cannot resolve our resource issues, and it offers a constructive alternative that creates options rather than limits them" (Walker and Salt 2012, p. xiv). These options or rather the understanding of these options is perceived differently in the scientific literature on the resilience of social-ecological systems.

1.1.2 Different Understandings of Resilience in Social-Ecological Systems

The concept of resilience was continuously developed and adapted to specific contexts since it was first transferred to an ecological context by Holling (1973). Thereby, resilience scholars assigned the term different meanings. Overall, four different understandings of resilience can be derived from literature[3]: (a) engineering resilience, (b) ecological resilience, (c) evolutionary resilience, and finally (d) social-ecological resilience. The relevance of each understanding differs across specific ecological or social-ecological contexts. Which understanding is adopted depends on the specific goals that are formulated for the system (e.g., conservation, growth, flexibility). The further illustration of the four presented understandings mainly takes place by the description of the desired capabilities a resilient system should have as conceptualized by Darnhofer (2014). She distinguishes between buffer capability, adaptive capability, and transformative capability (See Table 1.1). A capability is defined as "the ability to identify opportunities, to mobilise [sic.] resources, to implement options, to develop processes, to learn as part of an iterative, reflexive process" (ibid., p. 467).

(a) *Engineering resilience* describes the ability of a system to 'bounce back' near an equilibrium state in response to external shocks (Holling 1973, 1986). These shocks can both be natural disturbances (e.g., floods, fire) or social disasters (e.g., financial crisis, riots). The faster the system can return to its desired original state, the more resilient it is. Thresholds are not part of this understanding. Regime changes are perceived as failures in the system's resilience. It follows that the sole focus of this concept is on preservation, recovery, and persistence (Folke et al. 2010). Therefore, in this view, a resilient system has the buffer capability to remain in a specific regime and absorb changing conditions and disturbances (Holling 1973).

(b) *Ecological resilience* also sees persistence as an important ability of a resilient system but adds adaptability as another major characteristic (Holling 1996). To be resilient, a system sometimes needs to adapt to changing conditions to stay outside specific thresholds or move over them to survive. Multiple suitable equilibria can exist instead of just one desired equilibrium. Following Holling (1996), the focus of engineering resilience (see above) lies on maintaining the efficiency of a system's function whereas ecological resilience is concerned with the maintenance of the mere existence of a system's function. As a result, it is sometimes not enough to 'bounce back' but necessary to 'bounce forward' (Davoudi 2012). In the metaphor of ball and cup this means that two options for adaptability exist to react on changing controlling variables: First, the breadth and width of the cup need to change in order to remain in the desired stability domain. Second, a regime change should take place to reach a new regime (bouncing forward), meaning the ball rolls over into another cup. Ecological resilience is hence more focused on identifying and understanding thresholds (Walker and Salt 2012). In this sense, ecological resilience requests buffer capability as well as adaptive capability for a resilient system.

(c) *Evolutionary resilience* is different from both engineering and ecological resilience because it solely rejects buffer capability as a component of resilience. Rather, this approach suggests that change or 'bouncing forward' is always the best response to external disturbances. Equilibria and conditions are always changing and therefore the system has to be in a constant movement of change to be resilient (Davoudi 2012). Evolutionary resilience embraces the concept of adaptive cycles and especially panarchy. Besides adaptive actions, this also includes deliberative actions by actors to create new stability domains and aim at the transformation of the current regime. In this understanding, a system needs adaptive and transformative capability to be resilient (Carpenter et al. 2005; Berkes and Folke 1998b).

(d) *Social-ecological resilience* specifically refers to social-ecological systems and is the most recent understanding of resilience. It includes elements of all other depicted understandings of resilience. For clarification, the definition of this term as introduced in the Introduction is repeated in the following. Social-ecological resilience describes

> the capacity of a social-ecological system to sustain human well-being and a desired set of ecosystem services in the face of disturbance and change, both by buffering shocks and by adapting or transforming in response to change.
>
> (Biggs et al. 2015b, p. 9)

Hence, it acknowledges and integrates buffer capability, adaptive capability, and transformative capability as necessary elements for a resilient system. In this book, social-ecological resilience will be applied as the basic understanding of resilience. Research that adopts this understanding of resilience recently culminated in the formulation of *Principles for Building Resilience* (Biggs et al. 2015b), which will be introduced and discussed in Section 6.1. Table 1.1 gives an overview of all relevant terms which have been introduced so far.

Resilience is not always positively denoted (Gallopín 2006; Walker and Meyers 2004). Especially by adopting the understanding of engineering resilience, a resilient system can also have negative effects on ecosystems or humans. It can be locked into an undesired regime because buffer capability is too high, thus failing to adapt to changing conditions and the approaching of new stability domains. An example is a corrupt governance regime that persists regardless of societal pressure, being resilient in a negative way.

1.1.3 Critical Remarks on the Concept of Resilience

Criticisms on the concept of resilience are mainly expressed by social scientists. Brand and Jax (2007) attest that resilience is a "conceptual vagueness" (p. 1) that has the positive effect of better bridging different scientific disciplines but also

Table 1.1 Glossary of terms in the context of resilience thinking

Stability landscape	The extent of possible system states, characterized by different stability domains and thresholds.
Stability domain/ regime	A possible overall system state, defined by the conditions of controlling variables and their thresholds.
State variables	(Internal) Variables that describe the structure and condition of a social-ecological system.
Controlling variables	Describe specific key state variables and external variables (e.g., shocks, disturbances) with specific thresholds, whose conditions influence the alteration of the stability domain or can lead to regime shifts.
Threshold	Describe a certain level or condition of one or more controlling variables that lead to a regime shift if crossed.
Regime shift	Change from one stability domain to another in the frame of the stability landscape.
Buffer capability	Capacity of a social-ecological system to 'bounce back' and re-organize in response to changing controlling variables to maintain the same functions, structure, and identity as before the change.
Adaptive capability	Capacity of a social-ecological system to adapt to changing controlling variables and 'bounce forward' by staying in the current stability domain or inducing a regime change to another stability domain.
Transformative capability	Capacity of a social-ecological system to alter the stability landscape in response to changing controlling variables and deliberatively create new stability domains.

Based on Walker and Salt (2012), Folke et al. (2010), Gunderson and Holling (2002a, 2002b), Darnhofer (2014).

results in a lack of clarity. This view is also stressed by Brand et al. (2011) and Olsson et al. (2015). Moreover, in their typology of resilience interpretations, Davidson et al. (2016) show that a general confusion exists among resilience scholars because of different understandings of resilience across diverse research fields.[4]

Brand et al. (2011) additionally criticize the low empirical validity of the resilience concept. However, since the publishing of their article, numerous case studies on resilience have been conducted, as for example shown by Biggs et al. (2015b). Another criticism refers to the basic assumption of co-evolutionary social-ecological systems and argues that the linkage of social and ecological systems is not necessarily true. Brand et al. (2011) do not categorize this assumption as primarily scientific. Rather, they see it as a conclusion informed by cultural preconceptions and a specific conservative worldview against homogenization, globalization, industrialization, and urbanization.

Moreover, Olsson et al. (2015) and Sinclair et al. (2017) critically remark on the integration of social-ecological resilience, having its origin in ecology, into social science theories. Sinclair et al. (2017) identify and summarize four common critiques that build on this perspective: First, human agency and heterogeneous perspectives are not adequately recognized in resilience research, and thereby social dynamics are not reflected upon. For example, in studies on local and community resilience, communities are defined as homogenous units. Second, the role and importance of power relations between social actors are not sufficiently stressed, particularly in the case of knowledge production and governance. Third, thresholds are mainly identified as ecological thresholds. Resilience research lacks the identification and evaluation of key social thresholds such as trust. Fourth, the ontological nature of social-ecological systems is not adequately reflected by resilience researchers. Social-ecological systems are essentially social constructs and no definite entities.

The unifying ambition of resilience as an interdisciplinary concept provides several challenges. Olsson et al. (2015) argue that this ambition is indeed fulfilled in ecology and environmental studies, but not in linking these scientific disciplines with social sciences (so far). They even warn of "scientific imperialism" (ibid., p. 9) by failing to address this problem adequately, disregarding diversity in social sciences, and giving resilience as a holistic concept too much credit in overall science. However, Walker and Salt (2012) stress that "resilience is not a panacea for all the world's problems" (p. 151) but, nevertheless, provides an important "foundation for achieving sustainable patterns of resource use" (ibid.). Beichler et al. (2014) particularly emphasize the suitability of the concept for interdisciplinary research:

> [...] the actual process [to find an interdisciplinary definition of resilience] facilitated integration between participating disciplines by encouraging them to explore their own ontological and epistemological fundamentals.
>
> (p. 7)

In their example, researchers from ecology, geography, planning, and communication sciences worked together on a research project in the context of climate change adaptation. Moreover, several researchers aim to further synthesize research on social-ecological resilience with insights from the social sciences. For example, Berkes and Ross (2013) proposed a concept of community resilience that links social-ecological resilience with psychology to emphasize human agency and heterogeneous perspectives in the analysis of social-ecological systems. Recently, Lade et al. (2020) proposed a resilience approach that explicitly links psychological, sociological, and ecological perspectives in a framework they call "pathway diversity" (ibid.). This framework serves as an extension to established resilience approaches and shows the potential of resilience as a conceptual playground for interdisciplinary research.

All criticisms identified above mark important points for reflecting research from the resilience perspective. However, they do not disqualify resilience and specifically social-ecological resilience as fitting concepts for investigating sustainability problems. The conceptual development of the different understandings of resilience shows that resilience is a continuously evolving scientific concept. Most of the critiques by Sinclair et al. (2017) have been already acknowledged in the description of key research gaps in social-ecological resilience research by Biggs et al. (2015b). Although resilience has not yet finally bridged social and environmental sciences, it provides a fitting concept for analyzing sustainability problems in the context of this book. Social-ecological resilience directly aims at bridging these disciplines, which is a necessary condition when investigating fruit cultivation and fruit breeding from a social-ecological perspective. In the course of the application of resilience on this object of investigation, the necessary "clear definitions of system boundaries and normative dimensions of change" (Zanotti et al. 2020, p. 8) will take place in Part II.

1.2 Ecosystem Services

The concept of ecosystem services aims to analyze the relation between nature and humans by capturing the influences of ecosystems and their system components on human well-being (Danley and Widmark 2016). It also analyzes how humans shape natural ecosystems and what kind of effects this has on the provisioning of ecosystem services. As a term, it connects a biological or ecological perspective (ecosystem) with an economic perspective (service). Ecosystems describe specific areas where living organisms like plants, animals, and microorganisms interact with each other and their physical environment that consists of soil and the local atmosphere. In general, the concept assumes that functions and processes in ecosystems provide distinct services for humans – like humans provide economic services for each other in markets. In this way, the (economic) value of nature is integrated into economic thinking.

Most generally defined by the Millennium Ecosystem Assessment (MEA 2005), ecosystem services are "the benefits people obtain from ecosystems" (p. v).[5] As an example, fresh water provided by ecosystems gives humans not only a basic source for living, but also a resource for industrial, personal, or farming activities – depending on its quality and quantity. Another example is pollination services by wild pollinators for home gardens, flower strips, or orchards. These wild pollinators are crucial elements for the functioning of agroecosystems but their service is not part of a commodification.

The valuation of nature's services is not a new idea in economic thinking. Scientific literature provides a range of different definitions for ecosystem services besides the intuitive one by the MEA (2005).

1.2.1 Historical Roots and Conceptions of Ecosystem Services

Different perspectives on ecosystem services have their roots in the historic development of the ecosystem services concept. Gómez-Baggethun et al. (2010) show the origins and developments of ecosystem services both in general economic theory as well as in later *Ecological Economics* as a separate scientific discipline. Inside and around this scientific discipline, the authors divide the development of modern ecosystem services science into three stages. In the first stage, origins and gestation (1970s–1980s), the concept was introduced to explain ecological processes from an economic perspective and hence highlight the benefits of nature for humans, specifically stressing the importance of biodiversity. From its beginning, it was a pedagogic concept with the aim to bridge science and policy discourses. The second stage shows a mainstreaming of the concept (1990s), both in science and policy discourses. The authors highlight the valuation of ecosystem services and natural capital from a global perspective by Costanza et al. (1997) and the MEA (2005) as milestones in this phase. In the third stage, articulation in markets (1990s–2000s), the ongoing research on the commodification of ecosystem services led to the design of market-based instruments to incentivize nature conservation and the internalization of external effects.

Today, the concept of ecosystem services is solidly integrated into scientific and policy discourses, providing the basis for controversial debates about the different values of nature for humanity and understandings of interlinkages between ecological and social systems. A central organization for shaping and moderating this discourse is the Intergovernmental Science-Policy Platform on Biodiversity and Ecosystem Services (IPBES). IPBES was established in 2012 and aims to organize the science-policy-transfer by regularly assessing the state of biodiversity and ecosystem services. Researchers at IPBES recently developed a conceptual framework for the study of human-nature relations (Díaz et al. 2015) and the conception of *Nature's Contributions to People* (NCP) (Pascual et al. 2017). Herewith, they aim to further develop and enhance the ecosystem services concept and focus on the formulation of knowledge-based policies.[6]

Based on these historical roots, different conceptions of ecosystem services have been developed until today.[7] The understanding of ecosystem services from the Ecological Economics perspective is most suitable for the context of this book (resilience and social-ecological systems): ecosystem services are specific services that lie at the interface between ecological and social/economic systems. Paraphrasing Danley and Widmark (2016), the most representative definition from this perspective is the one by Haines-Young and Potschin (2018):

> ecosystem services are defined as the contributions that ecosystems make to human well-being, and distinct from the goods and benefits that people subsequently derive from them. These contributions are framed in terms of 'what ecosystems do' for people.
>
> (p. iii)

This understanding emphasizes the influence of humans in finally shaping or interpreting the outputs of ecosystems to services that meet their needs.

1.2.2 *Approaches for the Study of Ecosystem Services*

Several approaches for the study of ecosystem services have been developed. According to Braat and de Groot (2012), the most prominent basic approaches are the previously mentioned MEA (2005) and the approach by the global initiative The Economics of Ecosystems and Biodiversity (TEEB 2012). Both initiatives conduct their research under the umbrella of the United Nations Environmental Programme (UNEP) and other bodies of the UN. While the MEA is a finished research undertaking, TEEB further investigates ecosystem services and recently developed a framework for the holistic evaluation of eco-agri-food systems (TEEB 2018).

In the MEA (2005), ecosystem services are modeled as flows that influence human well-being (See Figure 1.4). The model shows the different elements of ecosystem services and of human well-being.

According to this model, well-being encompasses security, basic materials for a good life, health, and good social relations. The preferences for these constituents are determined by individual values and freedom of choice. Regarding the flows, the color of the arrows indicates the potential impact on constituents of human well-being, whereas the width of the arrows indicates the intensity of the linkages. For example, provisioning services like food have high impacts and intense linkages to basic materials for a good life. On the other hand, these services have low impacts and weak linkages to good social relations. In comparison to MEA (2005), the TEEB model is more complex and places ecosystem services directly between the ecological and the social system as a separate element (See Figure 1.5). A cascade is modeled with three major elements: ecosystems and biodiversity; human well-being; institutions and human judgments.

Inside ecosystems, the authors differentiate between the overall biophysical structure and processes, and the derived functions as specific subsets. Only these functions consequently lead to services that benefit human well-being and affect (economic) values. Because ecosystem services are not always intentionally shaped, they are – in contrast to the definition by MEA (2005) – defined as the direct and indirect contributions of ecosystems to human well-being (TEEB 2012). Nevertheless, institutions play a major role in this model. Together with human

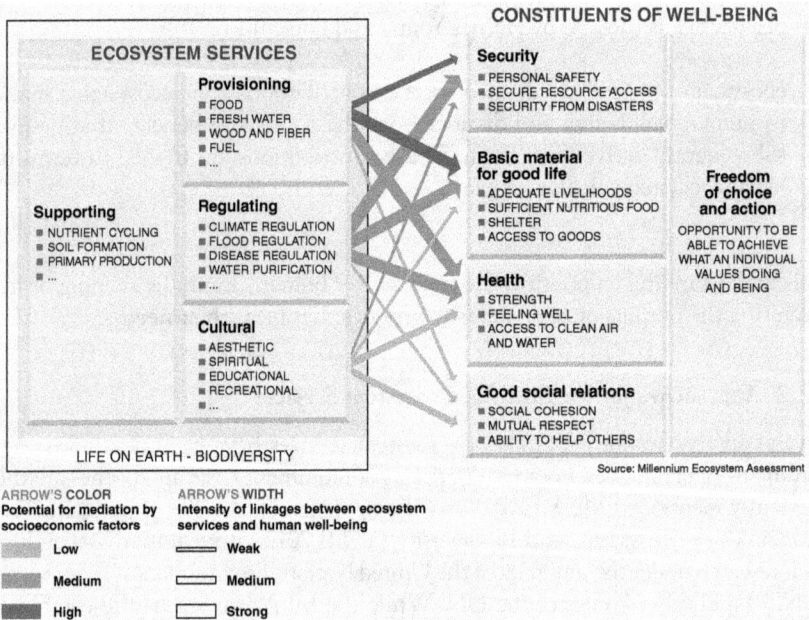

Figure 1.4 Basic ecosystem services approach (MEA 2005, p. vi).

judgments, defined as societal and individual preferences, they influence which ecosystems are managed and/or restored in which way, and subsequently which services have more or less value.

The main objective of TEEB is to communicate the ecosystem services concept to policy and public actors (Danley and Widmark 2016). However, this instrumental understanding of the ecosystem services concept has been harshly criticized for its narrow-minded economic valuation. It would serve the interests of the financial sector, encourage commodification and marketization tendencies, and is difficult to implement in practice (Arsel and Büscher 2012; Spash 2011). In this debate, the authors of TEEB countered these criticisms with the following arguments:

> TEEB does not suggest placing blind faith in the ability of markets to optimize social welfare by privatizing the ecological commons and letting markets discover prices for them. What TEEB offers is both a model for communicating to decision-makers in their own language, dominated by economics, as well as a toolkit for evaluating and integrating good stewardship into their decisions.
> (Sukhdev et al. 2014, p. 8)

Although criticisms on TEEB are valid and have to be considered in a sensitive way, this approach is suitable for research on the resilience of social-ecological systems out of two reasons. First, the approach by TEEB categorizes ecosystem

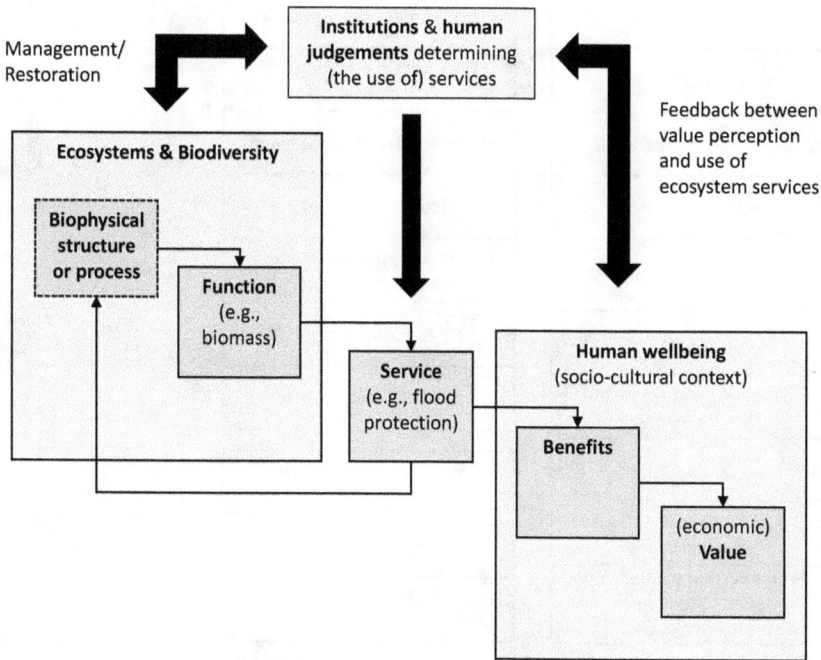

Figure 1.5 Ecosystem services approach by TEEB (based on TEEB 2012, p. 11).

services as contributions and locates them at the interface between ecological and social systems. Second, this approach has communicative benefits which are important in the transdisciplinary framework of this book (See Chapter 2). Overall, this classifies TEEB (2012; 2019) as a useful conceptual perspective for integrating the ecosystem services concept into social-ecological resilience.

1.2.3 Categorization of Ecosystem Services

The first comprehensive categorization of ecosystem services was carried out by the MEA (2005). As shown in Figure 1.4, the approach distinguishes between the following four services that have different influences on human well-being:

- Supporting services: services that are necessary for the presence of other eco-system services.
- Provisioning services: services that represent products obtained from ecosystems.
- Regulating services: services that are necessary for the regulation of processes in ecosystems.
- Cultural services: non-material services provided by ecosystems.

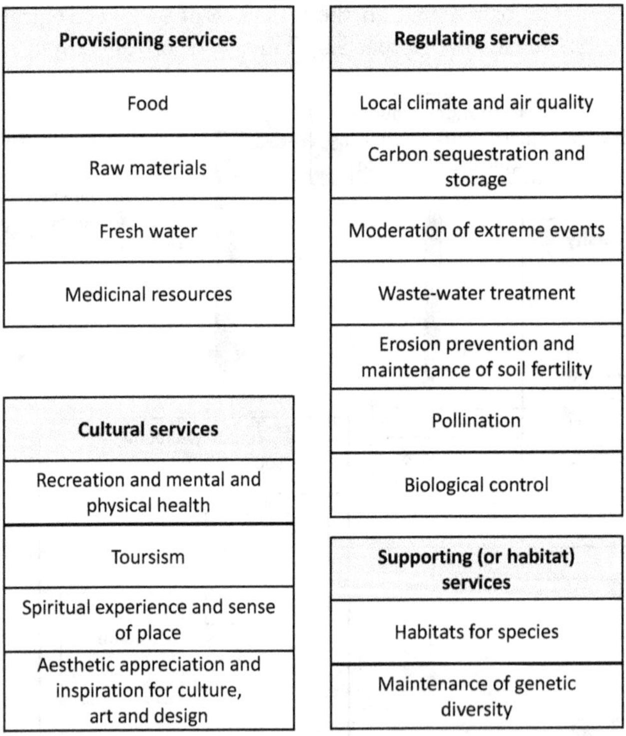

Provisioning services	Regulating services
Food	Local climate and air quality
Raw materials	Carbon sequestration and storage
Fresh water	Moderation of extreme events
Medicinal resources	Waste-water treatment
	Erosion prevention and maintenance of soil fertility

Cultural services	
Recreation and mental and physical health	Pollination
Toursism	Biological control

	Supporting (or habitat) services
Spiritual experience and sense of place	Habitats for species
Aesthetic appreciation and inspiration for culture, art and design	Maintenance of genetic diversity

Figure 1.6 Categorization of ecosystem services by TEEB (based on TEEB 2019).

Biodiversity is an underlying element and influences all types of ecosystem services. The categorization by TEEB (2012) also distinguishes between these four services and is only slightly different regarding the allocation and merging of specific ecosystem services. The particular ecosystem services of each category are depicted in Figure 1.6.[8]

A more detailed categorization is given by the Common International Classification of Ecosystem Services (CICES) provided by the European Environmental Agency (EEA). In this book, the categorizations by MEA (2005) and TEEB (2012) are applied. The main reason is the suitability of these categorizations for communicating ecosystem services to stakeholders and involving them in research processes because the categories are explicitly designed for science-policy interfaces. Further, both categorizations are solidly established in scientific research. Because ecosystem services merely function as a bridging concept, the depth of detail as demanded by the CICES classification is not necessary.

1.2.4 Critical Remarks on the Concept of Ecosystem Services

In their review on criticisms of the ecosystem services concept, Schröter et al. (2014) give a comprehensive overview of critical remarks and respective counterarguments to the criticisms as well as options for further research ("a way forward"). They identify seven major points of critique:

1. The anthropocentric focus on ecosystem services is criticized because it excludes and disregards the intrinsic values of nature. However, intrinsic values are indeed captured in the category of cultural ecosystem services, and the anthropocentric focus tries not to replace a biocentric focus but addresses valid points in a world where most ecosystems are influenced by human activities.
2. The ecosystem services concept may promote an exploitative human-nature relationship and disregards holistic views on nature, such as those of indigenous societies. Others argue that the ecosystem services concept rather challenges exploitative worldviews and practices, and serves to reconnect humans and nature. Non-material values like those of indigenous people can be captured by cultural ecosystem services.
3. Some authors see conflicts between the concepts of ecosystem services and biodiversity. Ecosystem services might divert attention and resources for biodiversity conservation and there is no comprehensive evidence of a win-win situation for both. Counter arguments stress the conceptual overlap of both concepts and the growing body of evidence for this win-win situation.
4. The economic valuation of nature as such is seen critically because it squeezes nature in an economic frame, playing off ecological functions and processes against each other. Other scholars argue that assessments of ecosystem services do not necessarily involve valuation and monetarization, and the use of

an economic frame leads to additional information for political or economic decision-making processes.

5. It is argued that the commodification of nature and the implementation of PES schemes will not ensure the long-term provision of ecosystem services. However, market schemes like PES and ecosystem services are not equivalent and the monetary valuing of nature does not directly lead to the adoption of market schemes.
6. The missing consistency of definitions and conceptions and their vagueness are criticized, hampering comparative research. Some researchers argue against this and state that this imprecision merely fosters the adaptability of the concept and transdisciplinary research.
7. Through optimistic assumptions and normative aims, the concept can be oblivious to ecosystem disservices, which could imply that "ecosystems are benevolent, hence protect them" (ibid., p. 518). However, ecosystem services are just one of many research concepts about environmental problems and thus normative aims are legitimized, especially in the context of sustainability science.

The concept of ecosystem services is thus an issue of controversial debate. Nevertheless, it proves to be an adaptable and beneficial concept for the theoretical framework of this book as it conceptually connects with social-ecological resilience (See Section 1.1). Overall, this book adopts the following understanding of the concept: ecosystem services are the direct and indirect contributions of ecosystems to human well-being (TEEB 2012). They are categorized into provisioning, regulating, supporting, and cultural services (MEA 2005; TEEB 2012).

1.3 Commons

The concept of commons describes in its most general understanding collective forms of management, creation, or control of goods and resources. Helfrich and Stein (2011) identify three dimensions of every commons arrangement that connect to this basic understanding: (1) the material dimension that describes the goods and resources; (2) the social dimension that encompasses the user-community of those goods and resources; and (3) the regulative dimension that refers to rules and norms relevant for their governance. Commons as a term thus describes material, social, and regulative aspects of collective action situations.

In a practical sense, humans have executed those forms of governance for centuries. An example is communal farming on common land owned by the (rural) community where members of the community negotiate rules for land management. This management has been practiced since medieval times (Ostrom 1990). As a scientific concept, commons started to evolve in the midst 20th century when economists reframed economic good categories. In his thoughts about fisheries, Gordon (1954) established the term common good as a certain good that is characterized by a high degree of rivalry and a low degree of exclusivity. This means that many people have access to the good and potentially use it, while it is difficult to exclude people from its use. He thus placed the term beside the classical distinction in public good (low degree of rivalry and exclusivity) versus private goods (high degree of rivalry and exclusivity).

This purely economic and resource-oriented understanding has changed and broadened over the last decades. Faysse and Mustapha (2017) classify these developments in two discourses that partly overlap but follow different epistemological aims: the institutional economics perspective and the sociological-anthropological perspective. Commons studies following the *institutional economics perspective* analyze and explain institutions that try to solve collective action situations and dilemmas. This involves the allocation and use of (mostly natural) resources. The institutional economics perspective is based on methodological individualism[9] and uses distinct frameworks and models to structure and compare conducted research, most prominently the *Institutional Analysis and Development* (IAD) framework (Ostrom 1990). In contrast, studies taking the *sociological-anthropological perspective*[10] aim to analyze and explain power relations, local specificities, negotiations, and social interactions in collective action situations. This perspective is based on social constructivism[11] and uses diverse research subjects, theoretical approaches, and methodologies. Both perspectives have epistemological advantages and disadvantages, and partly criticize the results and approaches of the other perspective (Mosse 1997; Steins and Edwards 1999; Ostrom and Cox 2010).

The sociological-anthropological perspective is especially useful when examining local specificities in case studies and getting a deeper understanding of sociological relationships and the local political dimension. However, this is not particularly relevant to the economic, ecological, and system-oriented focus of this book (See Introduction). Thus, the institutional economics perspective of commons is adopted without solely disregarding the criticisms and objections of the other perspective in the interpretation and reflection of the findings.

1.3.1 Institutional Economics Perspective: Ostrom and the Study of Traditional Commons

The fundamental research that shaped the institutional economics perspective on natural resources was the works by Ostrom (1990, 2005).[12] For decades, she and her colleagues conducted basic research and case studies on collective action situations referring to the management of common-pool resources such as fisheries, irrigation systems, grazing lands, or forests. Common-pool resources describe resources in settings where it is difficult to exclude potential beneficiaries from resource use and subtractability[13] of use is high (Ostrom 2005). She concluded that shaping the social institutions (rules, norms, shared understandings) that govern those resources decide on the effectiveness and sustainability of the commons and calls these social institutions *common-property regimes* (CPR). Those objects of study are referred to as *traditional commons* (Hess 2008; Ostrom and Hess 2007b).

The findings of Ostrom (1990) directly addressed the false assumptions and conclusions in Hardin's broadly reviewed article *The Tragedy of the Commons* (1968) that served as an economic paradigm for many decades. In his article,

Hardin argued that common goods ultimately have to be privatized because they are managed and used inefficiently and, in the end, will be degraded. His nowadays famous example is a pasture open to all where every livestock farmer wants to maximize his or her individual profit and thus the meadow becomes overgrazed and degraded. However, Hardin's thesis was heavily criticized because he did not describe a common good but an open-access good, thus an open access-regime (Feeny et al. 1990; Ostrom 1990), which he acknowledged in a later article (Hardin 1998). CPRs are not open-access but transform open-access goods into common goods, restricting access and managing the usage of goods with social institutions.

With the establishment of a new empirically grounded paradigm in commons research, Ostrom (1990) derived eight *design principles* for the effective long-term and sustainable management of CPRs. Cox et al. (2010) prove the general validity of the eight design principles with a review of over 90 empirical studies and further identify possible additional design aspects to consider. There are more critical remarks on the design principles, for example, the problem of scaling-up (e.g., Berkes 2006) or the perceived universality of the principles (e.g., Young 2002a).[14] However, all of the design principles are only guidelines that likely benefit the success of CPRs and no determining absolute factors (Ostrom 1990). The relevance of each design principle depends on the specific context and scale of the collective action situation (Baggio et al. 2016; Schlager 2016).

Beyond the design principles, case study research from the institutional perspective uses established analytical frameworks such as the IAD (Ostrom 1990, 2005) and the *Social-ecological Systems Framework* (SESF) (Anderies et al. 2004; Ostrom et al. 2007; Ostrom 2009). These frameworks enable comparable analyses of diverse collective action situations and institutions and represent the most prominent frameworks used in the study of traditional commons.

To this date, the prime studied subjects in research on traditional commons are the 'big five' of pastures, fisheries, forests, irrigation systems, and water management (van Laerhoven and Ostrom 2007; van Laerhoven et al. 2020). However, additional research endeavors slowly move to other topics and subjects like digital platforms, biodiversity, or climate change. These new objects of studies have been described as new commons (Hess 2008).

1.3.2 Institutional Perspective: Global and Knowledge Commons as Representations of New Commons

In general, the term *new commons* refers to commons arrangements where material and non-material goods beyond the sector of natural resources are managed. Unlike traditional commons that look at given natural resource systems and their management, here an institutionalized community is actively created and aspects of trust, participation, and normative goals play a significant role (Hess 2008). This understanding conceptionally opens the commons approach for a "seemingly limitless diversity" (ibid., p. 3) of research subjects – from cultural topics, knowledge creation, and infrastructure management to global sectors and social movements. In this universe of potential fields of study, refinements and adaptations of established frameworks and research results are necessary. Global commons and knowledge commons are suitable to exemplify these necessities.

When analyzing *global commons*,[15] it is to consider that resource systems on a global scale are different from local resource systems in many ways. Stern (2011) gives an overview of these different characteristics in the context of natural resources. He concludes that besides the geographic scale and the number of resource users, local and global resource systems differ in the aspects of salience, distribution of interests and power, cultural and institutional homogeneity, and feasibility of learning. Despite these differences, McGinnis and Ostrom (1996) argue that an upscaling of insights from local commons research is beneficial for three reasons: (1) the analytical structure of global problems is similar to the structure of local problems; (2) adapting the 'lessons learned' accelerates theory and model building for global commons; and (3) many global problems are the result of inadequate solutions on the local level. Upscaling insights from commons research to a global scale is thus possible but not easy, and provides a range of challenges. Hence, Ostrom et al. (1999) identify the following challenges when lessons learned from research on traditional commons are transferred to global commons:

- the very large number of potential participants in global resource systems
- cultural diversity of relevant actors challenges the identification of shared understandings and interests
- the complexity of interlinked common-pool resources such as biodiversity or climate and their accompanied interlinked social connections, when actors are simultaneously distanced from each other and environmental problems are not directly visible
- accelerating rates of change in technological, economic, and social sectors, which makes it difficult to identify points and causalities of the crossing of environmental thresholds
- monitoring and sanctions are difficult to implement in international voluntary agreements
- on a global scale, no experimental resource systems or spots exist because there is only one globe to experiment with

These challenges of upscaling became a prominent research topic since the turn of the millennium, starting with Ostrom (2002) and temporarily leading to the proposal of polycentric governance systems (Ostrom 2010) as fitting institutional structures for global commons. Beyond natural resources as the main research topic of global commons, literature provides a multitude of other topics (Hess 2008), including knowledge resources.

Knowledge commons are most comprehensively defined by Frischmann et al. (2014, p. 3) as "institutionalized community governance of the sharing and, in some cases, creation, of information, science, knowledge, data, and other types of intellectual and cultural resources" (p. 3). Knowledge is thus defined as a collectively created or managed resource that opens up a range of questions and challenges (Hess and Ostrom 2007). A first effort to bring order in this diverse field of study was the adaptation of the IAD (see above) as an analytical framework by Ostrom and Hess (2007a) to study knowledge commons. They applied all components of the IAD to the context of knowledge, described them subsequently, and concluded that the basic framework proves a fitting tool for the analysis of knowledge commons. Further research would adapt this framework even more to the uniqueness of this field of study and could possibly formulate design principles for the design of long-term and robust knowledge commons (ibid.). This endeavor was taken further by Frischmann et al. (2014) who extended and adapted the IAD for studying knowledge commons. Besides education, libraries, or indigenous and local knowledge commons, digital commons such as the development of open-source software seem to be one of the most prominent branches of knowledge commons (Hess 2008; Schweik and English 2012).

1.3.3 Commons and Resilience Research

The scientific discourses on commons and social-ecological resilience (See Section 1.1) show several interlinkages, and scholars from both discourses aim to integrate insights into the design of commons and resilient social-ecological systems. This interlinkage is in line with the two general challenges of managing social-ecological systems: solving collective action problems and dealing with uncertainty and change (Schlüter et al. 2015). Both challenges represent the key aspects of the social-ecological resilience discourse while solving collective action problems is the main focus of commons research.

Armitage (2008) argues that resilience thinking complements commons research in the sense that it enables the precise recognition of system dynamics and complexity in the governance system. By synthesizing both research strings, he derives nine attributes for the governance of common-pool resources on a multilevel scale, representing a set of "normative governance values or principles" (ibid., p. 18). This set includes, for example, leadership, knowledge pluralism, learning, and trust. Resilience thinking is conceptually beneficial for understanding how actors in collective action situations respond and adapt to changes in the social-ecological system and in response to external disturbances and shocks (Folke 2006; Duit et al. 2010). Berkes (2017) thus proposes to use the social-ecological systems concept (See Sections 1.1 and 4.1) as a conceptual foundation for analyzing local, regional, or global commons, and social-ecological resilience as the framework to design and foster adaptive governance for managing commons. The other way around, resilience scholars increasingly adopt insights from commons research to analyze and evaluate social-ecological resilience.

In general, the design principles for the effective long-term and sustainable management of CPRs (Ostrom 1990) provided by commons research serve as important connecting points for the design of resilient governance structures in social-ecological systems (Schlüter et al. 2015). Several case studies analyze the influence of commons-based governance solutions on the resilience of social-ecological systems. This includes, for example, fishery management in indigenous communities (Galappaththi et al. 2021), small-scale agriculture in Guatemala (Hellin et al. 2018), pasture and forest management in Ethiopia (Dessalegn 2016), and perhaps most prominently urban commons in different settings (Feinberg et al. 2020; Esopi 2018; Schauppenlehner-Kloyber and Penker 2016; Petrescu et al. 2016; Colding and Barthel 2013). The case studies reveal that commons-based solutions foster resilience in social-ecological systems throughout different contexts.

Barnes (2013) concludes that CPRs as property arrangements are generally better suited to deal with complexity and change, compared to private property institutions. His main argument is that they show a "greater sensitivity to the natural constraints on resource use" (ibid., p. 14). Overall, the current state of research on the interlinkages between commons and resilience emphasizes the value of this book's aim to evaluate the resilience of commons-based organic apple breeding.

1.3.4 Criticisms on the Institutional Economics Perspective

Several criticisms on the institutional economics perspective exist, which are largely stated by representatives of the sociological-anthropological perspective. In their synthesis article, Hall et al. (2014) summarize three major critical points on what they depict as "mainstream institutionalism" (ibid., p. 72), which includes the concepts and perspectives described previously:

(1) The assumption of a homogenous community and insufficient consideration of the heterogeneity of actors involved in CPRs is criticized. Actors are necessarily involved in more than one social or ecological system such as family, job practice, etc. Thus, local practices of those actors in CPRs are not only a result of internal actions but evolve out of a complex interplay with external actors and systems. Further, the whole conception of the community as a distinct entity is challenged because the community as a concept and term is vague and elusive. This is also described as the "myth of community" (ibid., p. 74).

(2) Mainstream institutionalism would 'avoid politics' because the analytical focus is on rules and system characteristics. The aspect of power is disregarded, although power relations on the local level significantly influence the practices of actors in CPRs and outcomes of social processes. This 'avoidance' also includes ideational politics such as overarching norms and narratives that are present in social-ecological systems.

(3) Institutional analyses are socially inadequate, which means that they miss the depth of social relations and structures that influence outcomes and rules of CPRs. This includes personal interests and values of involved actors, local politics and competition between subgroups, and the cultural and symbolic value of resources. Additionally, the institutional perspective misses to distinguish between the biophysical aspects and the socially constructed aspects of resource scarcity and overexploitation. Resource scarcity can be socially constructed when local leaders aim to express their power with the overexploitation of a resource.

The overview of Hall et al. (2014) shows the main differences and conflicts between the sociological-anthropological and the institutional economics perspective. Faysse and Mustapha (2017) argue that although there have been attempts to bridge both perspectives, and institutional economists appreciate sociological and anthropological in-depth-studies, the general differences and positions remain firm. The main problem seems to be that the main protagonists of the institutional economics perspective refrain from including the aspects described above, because their frameworks would become too complex. This could hamper the epistemological and methodological connection to the discipline of institutional economics.

In this book, it is aimed to reflect on these valid criticisms when it is applicable and possible. However, the focus still remains on institutional economic aspects of commons as the aim of this book is to identify general patterns and principles and not the in-depth analysis of a specific CPR. Commons are thus understood as collective action situations where a distinct community of actors collectively own, share, manage, and/or develop a potentially diverse range of goods and resources in a particular institutional setting. Commons encompass a material (goods and resources), social (user community), and regulative (rules and norms) dimension.

Notes

1 Holling et al. (1998) call this perspective the 'new stream', dichotomous to the 'conventional stream' that is describes disciplinary science based on a reductionist and mechanistic worldview.
2 A further review on more detailed theoretical and empirical conceptualizations of social-ecological systems takes place in Section 4.1.
3 These four understandings represent the most common and broadly used resilience perspectives in relevant literature on social-ecological systems. Other scholars differentiate even further. Brand and Jax (2007), for example, distinguish between ten definitions of resilience according to their degree of normativity. However, the differentiation of these four understandings of resilience is sufficient for the context of this book.
4 By developing a typology of ideal resilience interpretations that show core and discipline-specific conceptual elements, their research aims to enhance conceptual clarity and offer guidance for resilience scholars (Davidson et al. 2016). This point is also stressed by Zanotti et al. (2020) who describe different tensions and problems with the application of different conceptual understandings in resilience research.
5 The MEA was an initiative by the United Nations (UN) with the goal to systematically categorize ecosystem services. This categorization provides an analytical framework for their study and put the concept on the political agenda.
6 Recently, IPBES featured a global assessment of ecosystem services that suggests enhanced and collaborative science-policy processes for tackling the identified challenges (Ruckelshaus et al. 2020). With a co-sponsored workshop with the IPCC, it is also aimed to further connect research approaches and insights on biodiversity and climate change (Otto-Portner et al. 2021).
7 In their literature review on definitions of ecosystem services, Danley and Widmark 2016 show this multitude of different perspectives.
8 A short and intuitive description of each specific category is depicted in TEEB (2019).
9 The principle of methodological individualism argues that actors in social situations behave in a rational way according to their subjective individual motivations and preferences (Schumpeter 1909).

10 This perspective does also include the side discourse on *commoning* (Euler 2018), that focuses on the social relations and processes in commons, largely disregarding any material and resource aspects. The concept of commoning evolved out of an activist background and often promotes alternative societal concepts (Gibson-Graham et al. 2016; Euler 2016).

11 Social constructivism is a social theory claiming that individuals develop and change their motivations and preferences through social interactions, especially when constructing knowledge (Burr 1997).

12 Especially Ostrom (1990) is still the core of this discourse as van Laerhoven et al. (2020) show in their trend study on citations of commons research.

13 Subtractability describes the gradual nature of goods that is not captured by the original term of rivalry. High subtractability means that other actors can use the resource at the same time the resource is consumed by another actor. Low subtractability means that other actors cannot use the resource at the same time it is consumed by another actor, like it is the case for club or public goods (Ostrom 2005).

14 See Cox et al. (2010) for a comprehensive overview of the critical remarks regarding Ostrom's design principles.

15 Besides commons research, global natural resource systems have been and are still the object of study in international governance and political economy research (Young 1996, 2002b, 2002a, 2010; Buck 1998). Here, commons are understood as "resource domains in which common pool resources are found. [...] The very large resource domains that do not fall within the jurisdiction of any one country are termed international commons or global commons" (Buck 1998, pp. 5). Building regimes for the governance of those resource domains mostly address problems of overexploitation of natural resources (Grafton 2004).

References

Anderies, J. M.; Janssen, M. A.; Ostrom, E. (2004): A Framework to Analyze the Robustness of Social-Ecological Systems from an Institutional Perspective. In *Ecology and Society* 9 (1). Available online at https://www.ecologyandsociety.org/vol9/iss1/art18/.

Armitage, D. (2008): Governance and the Commons in a Multi-level World. In *International Journal of the Commons* 2 (1), pp. 7–32.

Arsel, M.; Büscher, B. (2012): Nature™ Inc.: Changes and Continuities in Neoliberal Conservation and Market-based Environmental Policy. In *Development and Change* 43 (1), pp. 53–78. DOI: 10.1111/j.1467-7660.2012.01752.x.

Baggio, J. A.; Barnett, A. J.; Perez-Ibarra, I.; Brady, U.; Ratajczyk, E.; Rollins, N. et al. (2016): Explaining Success and Failure in the Commons: The Configural Nature of Ostrom's Institutional Design Principles. In *International Journal of the Commons* 10 (2), p. 417. DOI: 10.18352/ijc.634.

Barnes, R. A. (2013): The Capacity of Property Rights to Accommodate Social-Ecological Resilience. In *Ecology and Society* 18 (1). DOI: 10.5751/ES-05292–180106.

Beichler, S. A.; Hasibovic, S.; Davidse, B. J.; Deppisch, S. (2014): The Role Played by Social-ecological Resilience as a Method of Integration in Interdisciplinary Research. In *Ecology and Society* 19 (3). DOI: 10.5751/ES-06583-190304.

Berkes, F. (2006): From Community-Based Resource Management to Complex Systems: The Scale Issue and Marine Commons. In *Ecology and Society* 11 (1), p. 45.

Berkes, F. (2017): Environmental Governance for the Anthropocene? Social-Ecological Systems, Resilience, and Collaborative Learning. In *Sustainability* 9 (7), pp. 12–32. DOI: 10.3390/su9071232.

Berkes, F.; Folke, C. (1998a): Linking Social and Ecological Systems for Resilience and Sustainability. In Fikret Berkes, Carl Folke (Eds.): *Linking Social and Ecological Systems.*

Management Practices and Social Mechanisms for Building Resilience. Cambridge, UK: Cambridge University Press, pp. 1–25.

Berkes, F.; Folke, C. (Eds.) (1998b): *Linking Social and Ecological Systems. Management Practices and Social Mechanisms for Building Resilience*. Cambridge, UK: Cambridge University Press.

Berkes, F.; Ross, H. (2013): Community Resilience: Toward an Integrated Approach. In *Society & Natural Resources* 26 (1), pp. 5–20. DOI: 10.1080/08941920.2012.736605.

Bhamra, R.; Dani, S.; Burnard, K. (2011): Resilience. The Concept, a Literature Review and Future Directions. In *International Journal of Production Research* 49 (18), pp. 5375–5393. DOI: 10.1080/00207543.2011.563826.

Biggs, R.; Schlüter, M.; Schoon, M. L. (2015a): An Introduction to the Resilience Approach and Principles to Sustain Ecosystem Services in Social-ecological Systems. In Reinette Biggs, Maja Schlüter, Michael L. Schoon (Eds.): *Principles for Building Resilience. Sustaining Ecosystem Services in Social-Ecological Systems*. Cambridge, UK: Cambridge University Press, pp. 1–31.

Biggs, R.; Schlüter, M.; Schoon, M. L. (Eds.) (2015b): *Principles for Building Resilience. Sustaining Ecosystem Services in Social-Ecological Systems*. Cambridge, UK: Cambridge University Press.

Braat, L. C.; Groot, R. de (2012): The Ecosystem Services Agenda. Bridging the Worlds of Natural Science and Economics, Conservation and Development, and Public and Private Policy. In *Ecosystem Services* 1 (1), pp. 4–15. DOI: 10.1016/j.ecoser.2012.07.011.

Brand, F.; Hoheisel, D.; Kirchhoff, T. (2011): Der Resilienz-Ansatz auf dem Prüfstand: Herausforderungen, Probleme, Perspektiven. In *Laufener Spezialbeiträge*, pp. 78–83.

Brand, F.; Jax, K. (2007): Focusing the Meaning(s) of Resilience: Resilience as a Descriptive Concept and a Boundary Object. In *Ecology and Society* 12 (1), p. 23.

Buck, S. J. (1998): *The Global Commons. An Introduction*. Washington, DC: Island Press.

Burr, V. (1997): *An Introduction to Social Constructionism*. Reprinted. London: Routledge.

Carpenter, S. R.; Westley, F.; Turner, M. G. (2005): Surrogates for Resilience of Social–Ecological Systems. In *Ecosystems* 8 (8), pp. 941–944. DOI: 10.1007/s10021-005-0170-y.

Chapin, F. S.; Carpenter, S. R.; Kofinas, G. P.; Folke, C.; Abel, N.; Clark, W. C. et al. (2010): Ecosystem Stewardship. Sustainability Strategies for a Rapidly Changing Planet. In *Trends in Ecology & Evolution* 25 (4), pp. 241–249. DOI: 10.1016/j.tree.2009.10.008.

Chapin, F. S.; Kofinas, G. P.; Folke, C. (Eds.) (2009): *Principles of Ecosystem Stewardship. Resilience-based Natural Resource Management in a Changing World*. New York: Springer.

Colding, J.; Barthel, S. (2013): The potential of 'Urban Green Commons' in the resilience building of cities. In *Ecological Economics* 86, pp. 156–166. DOI: 10.1016/j.ecolecon.2012.10.016.

Costanza, R.; d'Arge, R.; de Groot, R.; Farber, S.; Grasso, M.; Hannon, B. et al. (1997): The Value of the World's Ecosystem Services and Natural Capital. In *Nature* 387 (6630), pp. 253–260. DOI: 10.1038/387253a0.

Cox, M.; Arnold, G.; Villamayor Tomás, S. (2010): A Review of Design Principles for Community-Based Natural Resource Management. In *Ecology and Society* 15 (4). DOI: 10.5751/ES-03704-150438.

Danley, B.; Widmark, C. (2016): Evaluating Conceptual Definitions of Ecosystem Services and Their Implications. In *Ecological Economics* 126, pp. 132–138. DOI: 10.1016/j.ecolecon.2016.04.003.

Darnhofer, I. (2014): Resilience and Why It Matters for Farm Management. In *European Review of Agricultural Economics* 41 (3), pp. 461–484. DOI: 10.1093/erae/jbu012.

Davidson, J. L.; Jacobson, C.; Lyth, A.; Dedekorkut-Howes, A.; Baldwin, C. L.; Ellison, J. C. et al. (2016): Interrogating Resilience: Toward a Typology to Improve its Operationalization. In *Ecology and Society* 21 (2). DOI: 10.5751/ES-08450-210227.

Davoudi, S. (2012): Resilience: A Bridging Concept or a Dead End? In *Planning Theory & Practice* 13 (2), pp. 299–307. DOI: 10.1080/14649357.2012.677124.

Dessalegn, M. (2016): Threatened Common Property Resource System and Factors for Resilience: Lessons Drawn from Serege-Commons in Muhur, Ethiopia. In *Ecology and Society* 21 (4). DOI: 10.5751/ES-08768-210422.

Díaz, S.; Demissew, S.; Carabias, J.; Joly, C.; Lonsdale, M.; Ash, N. et al. (2015): The IPBES Conceptual Framework—Connecting Nature and People. In *Current Opinion in Environmental Sustainability* 14, pp. 1–16. DOI: 10.1016/j.cosust.2014.11.002.

Duit, A.; Galaz, V.; Eckerberg, K.; Ebbesson, J. (2010): Governance, Complexity, and Resilience. In *Global Environmental Change* 20 (3), pp. 363–368. DOI: 10.1016/j.gloenvcha.2010.04.006.

Esopi, G. (2018): Urban Commons: Social Resilience Experiences to Increase the Quality of Urban System. In *Tema. Journal of Land Use, Mobility and Environment* 11 (2), pp. 173–194. DOI: 10.6092/1970-9870/5532.

Euler, J. (2016): Commons-creating Society. In *Review of Radical Political Economics* 48 (1), pp. 93–110. DOI: 10.1177/0486613415586988.

Euler, J. (2018): Conceptualizing the Commons: Moving Beyond the Goods-based Definition by Introducing the Social Practices of Commoning as Vital Determinant. In *Ecological Economics* 143, pp. 10–16. DOI: 10.1016/j.ecolecon.2017.06.020.

Faysse, N.; Mustapha, A. B. (2017): Finding Common Ground Between Theories of Collective Action: The Potential of Analyses at a Meso-scale. In *International Journal of the Commons* 11 (2), pp. 928–949. DOI: 10.18352/ijc.776.

Feeny, D.; Berkes, F.; Mccay, B. J.; Acheson, J. M. (1990): The Tragedy of the Commons: Twenty-Two Years Later. In *Human Ecology* 18 (1), pp. 1–19. DOI: 10.1007/bf00889070.

Feinberg, A.; Ghorbani, A.; Herder, P. M. (2020): Commoning Toward Urban Resilience: The Role of Trust, Social Cohesion, and Involvement IN A Simulated Urban Commons Setting. In *Journal of Urban Affairs*, pp. 1–26. DOI: 10.1080/07352166.2020.1851139.

Folke, C. (2006): Resilience. The Emergence of a Perspective for Social–Ecological Systems Analyses. In *Global Environmental Change* 16 (3), pp. 253–267. DOI: 10.1016/j.gloenvcha.2006.04.002.

Folke, C. (2016): *Resilience – Oxford Research Encyclopedia of Environmental Science.* Oxford: Oxford University Press, p. 1.

Folke, C.; Biggs, R.; Norström, A. V.; Reyers, B.; Rockström, J. (2016): Social-ecological Resilience and Biosphere-based Sustainability Science. In *Ecology and Society* 21 (3). DOI: 10.5751/ES-08748-210341.

Folke, C.; Carpenter, S.; Walker, B.; Scheffer, M.; Chapin, T.; Rockström, J. (2010): Resilience Thinking: Integrating Resilience, Adaptability and Transformability. In *Ecology and Society* 15 (4). Available online at http://www.ecologyandsociety.org/vol15/iss4/art20/.

Frischmann, Brett M.; Madison, Michael J.; Strandburg, Katherine J. (Eds.) (2014): *Governing Knowledge Commons.* Oxford: Oxford Univ. Press.

Galappaththi, E. K.; Ford, J. D.; Bennett, E. M.; Berkes, F. (2021): Adapting to Climate Change in Small-Scale Fisheries: Insights from Indigenous Communities in the Global North and South. In *Environmental Science & Policy* 116, pp. 160–170. DOI: 10.1016/j.envsci.2020.11.009.

Gallopín, G. C. (2006): Linkages between Vulnerability, Resilience, and Adaptive Capacity. In *Global Environmental Change* 16 (3), pp. 293–303. DOI: 10.1016/j. gloenvcha.2006.02.004.

Gibson-Graham, J. K.; Healy, S.; Cameron, J. (2016): Commoning as a Postcapitalist Politics. In Ash Amin, Philip Howell (Eds.): *Releasing the Commons.* Abingdon, Oxon, New York: Routledge, 2016. | Series: Routledge Studies in Human Geography: Routledge, pp. 192–212.

Gómez-Baggethun, E.; de Groot, R.; Lomas, P. L.; Montes, C. (2010): The History of Ecosystem Services in Economic Theory and Practice. From Early Notions to Markets and Payment Schemes. In *Ecological Economics* 69 (6), pp. 1209–1218. DOI: 10.1016/j. ecolecon.2009.11.007.

Gordon, H. S. (1954): The Economic Theory of a Common-Property Resource: The Fishery. In *Journal of Political Economy* 62 (2), pp. 124–142. DOI: 10.1057/9780230523210_10.

Grafton, R. Q. (2004): *The Economics of the Environment and Natural Resources.* Malden, MA: Blackwell Publ.

Gunderson, L. H. (2000): Ecological Resilience—In Theory and Application. In *Annual Review of Ecology and Systematics* 31 (1), pp. 425–439. DOI: 10.1146/annurev. ecolsys.31.1.425.

Gunderson, L. H.; Holling, C. S. (Eds.) (2002a): *Panarchy. Understanding Transformations in Human and Natural Systems.* Washington, DC: Island Press.

Gunderson, L. H.; Holling, C. S. (Eds.) (2002b): *Panarchy: Understanding Transformations in Human and Natural Systems.* Washington, DC: Island Press.

Haines-Young, R.; Potschin, M. B. (2018): Common International Classification of Ecosystem Services (CICES) V5.1 and Guidance on the Application of the Revised Structure.

Hall, K.; Cleaver, F.; Franks, T.; Maganga, F. (2014): Capturing Critical Institutionalism: A Synthesis of Key Themes and Debates. In *The European Journal of Development Research* 26 (1), pp. 71–86. DOI: 10.1057/ejdr.2013.48.

Hardin, G. (1968): The Tragedy of the Commons. In *Science* 162 (3859), pp. 1243–1248. DOI: 10.1126/science.162.3859.1243.

Hardin, G. (1998): Extensions of "The Tragedy of the Commons". In *Science* 280 (5364), pp. 682–683. DOI: 10.1126/science.280.5364.682.

Helfrich, S.; Stein, F. (2011): Was sind Gemeingüter. In Bundeszentrale für politische Bildung (Ed.): *Aus Politik und Zeitgeschichte.* Gemeingüter, vol. 61 (61), pp. 9–15.

Hellin, J.; Ratner, B. D.; Meinzen-Dick, R.; Lopez-Ridaura, S. (2018): Increasing Social-Ecological Resilience within Small-Scale Agriculture in Conflict-Affected Guatemala. In *Ecology and Society* 23 (3). DOI: 10.5751/ES-10250–230305.

Hess, C. (2008): Mapping the New Commons. In *SSRN Journal.* DOI: 10.2139/ssrn.1356835.

Hess, C.; Ostrom, E. (Eds.) (2007): *Understanding Knowledge as a Commons. From Theory to Practice.* Cambridge: MIT Press.

Holling, C. S. (1973): Resilience and Stability of Ecological Systems. In *Annual Review of Ecology and Systematics* 4, pp. 1–23.

Holling, C. S. (1986): The Resilience of Terrestrial Ecosystems: Local Surprise and Global Change. In William C. Clark, R. E. Munn (Eds.): *Sustainable Development of the Biosphere.* Cambridge, UK: Cambridge University Press.

Holling, C. S. (1996): Engineering Resilience versus Ecological Resilience. In *Engineering Within Ecological Constraints* 31, pp. 31–43.

Holling, C. S.; Berkes, F.; Folke, C. (1998): Science, Sustainability and Resource Management. In Fikret Berkes, Carl Folke (Eds.): *Linking Social and Ecological Systems. Management Practices and Social Mechanisms for Building Resilience.* Cambridge, UK: Cambridge University Press, pp. 342–362.

Kates, R. W. (2012): From the Unity of Nature to Sustainability Science: Ideas and Practice. In Michael P. Weinstein, R. Eugene Turner (Eds.): *Sustainability Science. The Emerging Paradigm and the Urban Environment.* New York: Springer New York, pp. 3–19.

Kates, R. W.; Clark, W. C.; Corell, R.; Hall, M.; Jaeger, C.; Lowe, I. et al. (2001): Sustainability Science. In *Science* 292 (5517), pp. 641–642. DOI: 10.1126/science.1059386.

Lade, S. J.; Walker, B. H.; Haider, L. J. (2020): Resilience as pathway diversity: linking systems, individual, and temporal perspectives on resilience. In *Ecology and Society* 25 (3). DOI: 10.5751/ES-11760-250319.

Levin, S.; Xepapadeas, T.; Crépin, A.-S.; Norberg, J.; Zeeuw, A. de; Folke, C. et al. (2013): Social-ecological Systems as Complex Adaptive Systems. Modeling and Policy Implications. In *Environment and Development Economics* 18 (2), pp. 111–132. DOI: 10.1017/ S1355770X12000460.

McGinnis, M. D.; Ostrom, E. (1996): Design Principles for Local and Global Commons. In Oran R. Young (Ed.): *The International Political Economy and International Institutions.* Cheltenham: Edward Elgar, pp. 465–493.

MEA (2005): *Ecosystems and Human Well-being. Synthesis; a Report of the Millennium Ecosystem Assessment.* Washington, DC: Island Press.

Mosse, D. (1997): The Symbolic Making of a Common Property Resource: History, Ecology and Locality in a Tank-irrigated Landscape in South India. In *Development & Change* 28 (3), pp. 467–504. DOI: 10.1111/1467-7660.00051.

Olsson, L.; Jerneck, A. (2018): Social Fields and Natural Systems: Integrating Knowledge about Society and Nature. In *Ecology and Society* 23 (3). DOI: 10.5751/ES-10333-230326.

Olsson, L.; Jerneck, A.; Thoren, H.; Persson, J.; O'Byrne, D. (2015): Why Resilience Is Unappealing to Social Science. Theoretical and Empirical Investigations of the Scientific Use of Resilience. In *Science Advances* 1 (4), e1400217. DOI: 10.1126/sciadv.1400217.

Olsson, L.; Ness, B. (2019): Better Balancing the Social and Natural Dimensions in Sustainability Research. In *Ecology and Society* 24 (4). DOI: 10.5751/ES-11224-240407.

Ostrom, E. (1990): *Governing the Commons. The Evolution of Institutions for Collective Action.* Cambridge: Cambridge University Press.

Ostrom, E. (2005): *Understanding Institutional Diversity.* Princeton, NJ, Oxford: Princeton University Press (Princeton paperbacks).

Ostrom, E. (2009): A General Framework for Analyzing Sustainability of Social-ecological Systems. In *Science (New York)* 325 (5939), pp. 419–422. DOI: 10.1126/science.1172133.

Ostrom, E. (2010): Beyond Markets and States: Polycentric Governance of Complex Economic Systems. In *American Economic Review* 100 (3), pp. 641–672. DOI: 10.1257/aer.100.3.641.

Ostrom, E.; Burger, J.; Field, C. B.; Norgaard, R. B.; Policansky, D. (1999): Revisiting the Commons: Local Lessons, Global Challenges. In *Science* 284 (5412), pp. 278–282. DOI: 10.1126/science.284.5412.278.

Ostrom, E.; Cox, M. (2010): Moving beyond Panaceas: A Multi-Tiered Diagnostic Approach for Social-Ecological Analysis. In *Environmental Conservation* 37 (4), pp. 451–463. DOI: 10.1017/S0376892910000834.

Ostrom, E.; Hess, C. (2007a): A Framework for Analyzing the Knowledge Commons. In Charlotte Hess, Elinor Ostrom (Eds.): *Understanding Knowledge as a Commons. From Theory to Practice.* Cambridge: MIT Press, pp. 41–82.

Ostrom, E.; Hess, C. (2007b): Private and Common Property Rights. In *SSRN Journal.* DOI: 10.2139/ssrn.1936062.

Ostrom, E.; Janssen, M. A.; Anderies, J. M. (2007): Going Beyond Panaceas. In *Proceedings of the National Academy of Sciences of the United States of America* 104 (39), pp. 15176–15178. DOI: 10.1073/pnas.0701886104.

Ostrom, E. (Ed.) (2002): *The Drama of the Commons. National Research Council (U.S.).* Washington, DC: National Academy Press.

Pörtner, H. O.; Scholes, R. J.; Agard, J.; Archer, E.; Arneth, A.; Bai, X.; Barnes, D.; Burrows, M.; Chan, L.; Cheung, W. L. W.; Diamond, S.; Donatti, C.; Duarte, C.; Eisenhauer, N.; Foden, W.; Gasalla, M. A.; Handa, C.; Hickler, T.; Hoegh-Guldberg, O.; ... Ngo, H. (2021). Scientific outcome of the IPBES-IPCC co-sponsored workshop on biodiversity and climate change. Intergovernmental Science-Policy Platform on Biodiversity and Ecosystem Services (IPBES). https://zenodo.org/record/5101125

Pascual, U.; Balvanera, P.; Díaz, S.; Pataki, G.; Roth, E.; Stenseke, M. et al. (2017): Valuing Nature's Contributions to People. The IPBES Approach. In *Current Opinion in Environmental Sustainability* 26–27, pp. 7–16. DOI: 10.1016/j.cosust.2016.12.006.

Peterson, C. A.; Eviner, V. T.; Gaudin, A. C. (2018): Ways Forward for Resilience Research in Agroecosystems. In *Agricultural Systems* 162, pp. 19–27. DOI: 10.1016/j.agsy.2018.01.011.

Petrescu, D.; Petcou, C.; Baibarac, C. (2016): Co-producing Commons-based Resilience: Lessons from R-Urban. In *Building Research & Information* 44 (7), pp. 717–736. DOI: 10.1080/09613218.2016.1214891.

Ruckelshaus, M. H.; Jackson, S. T.; Mooney, H. A.; Jacobs, K. L.; Kassam, K.-A. S.; Arroyo, M. T. et al. (2020): The IPBES Global Assessment: Pathways to Action. In *Trends in Ecology & Evolution.* DOI: 10.1016/j.tree.2020.01.009.

Schauppenlehner-Kloyber, E.; Penker, M. (2016): Between Participation and Collective Action—From Occasional Liaisons towards Long-Term Co-Management for Urban Resilience. In *Sustainability* 8 (7), p. 664. DOI: 10.3390/su8070664.

Schlager, E. (2016): Introducing the "The Importance of Context, Scale, and Interdependencies in Understanding and Applying Ostrom's Design Principles for Successful Governance of the Commons". In *International Journal of the Commons* 10 (2), p. 405. DOI: 10.18352/ijc.767.

Schlüter, M.; Biggs, R.; Schoon, M. L.; Robards, M. D.; Anderies, J. M. (2015): Reflections on Building Resilience – Interactions among Principles and Implications for Governance. In Reinette Biggs, Maja Schlüter, Michael L. Schoon (Eds.): *Principles for Building Resilience. Sustaining Ecosystem Services in Social-Ecological Systems.* Cambridge, UK: Cambridge University Press, pp. 251–282.

Schröter, M.; van der Zanden, E. H.; van Oudenhoven, A. P.; Remme, R. P.; Serna-Chavez, H. M.; de Groot, R. S.; Opdam, P. (2014): Ecosystem Services as a Contested Concept. A Synthesis of Critique and Counter-Arguments. In *Conservation Letters* 7 (6), pp. 514–523. DOI: 10.1111/conl.12091.

Schumpeter, J. (1909): On the Concept of Social Value. In *Quarterly Journal of Economics* 23, pp. 213–232.

Schweik, C. M.; English, R. C. (2012): *Internet Success. A Study of Open-source Software Commons.* Cambridge: MIT Press.

Sinclair, K.; Rawluk, A.; Kumar, S.; Curtis, A. (2017): Ways Forward for Resilience Thinking: Lessons from the Field for Those Exploring Social-ecological Systems in Agriculture and Natural Resource Management. In *Ecology and Society* 22 (4), p. 21.

Spash, C. L. (2011): Editorial: Terrible Economics, Ecosystems and Banking. In *Environmental Values* 20 (2), pp. 141–145. DOI: 10.3197/096327111X12997574391562.

Steins, N. A.; Edwards, V. M. (1999): Collective Action in Common-Pool Resource Management: The Contribution of a Social Constructivist Perspective to Existing Theory. In *Society & Natural Resources* 12 (6), pp. 539–557. DOI: 10.1080/089419299279434.

Stern, P. C. (2011): Design Principles for Global Commons: Natural Resources and Emerging Technologies. In *International Journal of the Commons* 5 (2), p. 213. DOI: 10.18352/ijc.305.

Sukhdev, P.; Wittmer, H.; Miller, D. (2014): The Economics of Ecosystems and Biodiversity (TEEB): Challenges and Responses. In Dieter Helm, Cameron Hepburn (Eds.): *Nature in the Balance. The Economics of Biodiversity*. 1st edition. Oxford: Oxford University Press.

TEEB (Ed.) (2012): The Economics of Ecosystems and Biodiversity. Ecological and Economic Foundations. In *The Economics of Ecosystems and Biodiversity (TEEB)*. London: Routledge.

TEEB (2018): TEEB for Agriculture & Food: Scientific and Economic Foundations. Geneva.

TEEB (2019): Ecosystem Services. The Economics of Ecosystems and Biodiversity (TEEB). Available online at http://www.teebweb.org/resources/ecosystem-services/, checked on 8/7/2019.

van Laerhoven, F.; Ostrom, E. (2007): Traditions and Trends in the Study of the Commons. In *International Journal of the Commons* 1 (1), p. 3. DOI: 10.18352/ijc.76.

van Laerhoven, F.; Schoon, M.; Villamayor-Tomas, S. (2020): Celebrating the 30th Anniversary of Ostrom's Governing the Commons: Traditions and Trends in the Study of the Commons, Revisited. In *International Journal of the Commons* 14 (1), pp. 208–224. DOI: 10.5334/ijc.1030.

Walker, B.; Holling, C. S.; Carpenter, S.; Kinzig, A. (2004): Resilience, Adaptability and Transformability in Social-Ecological Systems. In *Ecology and Society* 9 (2), p. 5.

Walker, B.; Meyers, J. A. (2004): Thresholds in Ecological and Social–Ecological Systems: A Developing Database. In *Ecology and Society* 9 (2). DOI: 10.5751/ES-00664-090203.

Walker, B.; Salt, D. (2012): *Resilience Thinking. Sustaining Ecosystems and People in a Changing World*. Washington, DC: Island Press.

Young, O. R. (2002a): *The Institutional Dimensions of Environmental Change. Fit, Interplay, and Scale*. Cambridge: MIT Press.

Young, O. R. (2002b): *The Institutional Dimensions of Environmental Change. Fit, Interplay, and Scale*. Cambridge, London, England: The MIT Press (Global environmental accord).

Young, O. R. (2010): Institutional Dynamics: Resilience, Vulnerability and Adaptation in Environmental and Resource Regimes. In *Global Environmental Change* 20 (3), pp. 378–385. DOI: 10.1016/j.gloenvcha.2009.10.001.

Young, Oran R. (Ed.) (1996): *The International Political Economy and International Institutions*. Cheltenham: Edward Elgar.

Zanotti, L.; Ma, Z.; Johnson, J. L.; Johnson, D. R.; Yu, D. J.; Burnham, M.; Carothers, C. (2020): Sustainability, Resilience, Adaptation, and Transformation: Tensions and Plural Approaches. In *E&S* 25 (3). DOI: 10.5751/ES-11642-250304.

2 Methodological Framework

Qualitative Research in a Transdisciplinary Context

The methodological and epistemological approach of this book is embedded in transdisciplinary sustainability research. This form of research aims at generating and integrating knowledge through (a) the collaboration of different scientific disciplines, and (b) science-practice interactions where participating actors collaborate on equal levels (Brandt et al. 2013; Lang et al. 2012). It is a solution- and action-oriented research approach that aims to strengthen the science-society interface, specifically targeted at solving complex real-world problems (ibid.). Hereby, transdisciplinary research activates and connects scientific knowledge from various disciplines with real-world or experiential knowledge which is particularly suitable for sustainability research (Hirsch Hadorn et al. 2006).

The findings of this book are embedded in the transdisciplinary setting of the research project EGON (See Preface). In this vein, the framing, problem definition, and research focus were developed as part of the transdisciplinary process of EGON. The transdisciplinary embedding allowed to integrate different forms of knowledge into the research process. Literature on transdisciplinary research typically differentiates between three types of knowledge: system knowledge, target knowledge, and transformative knowledge (Adler et al. 2018; Brandt et al. 2013; Mauser et al. 2013). *System knowledge* describes current basic understandings of elements and dynamics of social-ecological systems. *Target knowledge* refers to normative goals and values about how a social-ecological system ought to be, thus focusing on solutions and visions. *Transformative knowledge* includes knowledge about how those goals and values can be achieved and how this change could be managed, for example, with specific competencies or measures. This book aims to include the whole diversity of knowledge types.

In the first section of this chapter, the general qualitative research design is presented and justified (See Section 2.1). Afterward, applied methods for data collection and analysis are explained and classified in the research design (See Section 2.2). Finally, a reflection on the benefits as well as the limitations and challenges of the methodological design takes place (See Section 2.3).

2.1 Qualitative Research

Besides literature research and analysis, empirical social research was applied for pursuing the objectives of this book in the context of the object of investigation:

DOI: 10.4324/9781003355724-4

apple breeding and cultivation in Germany. The empirical investigation of this case aimed to deduce insights for resilient fruit breeding and cultivation in general, but also to inductively show conceptual shortcomings of used theories and concepts (See Chapter 1).

For this book, a qualitative research design was chosen for three reasons. In line with the goals and leading questions, there was first the need to conduct a deep and differentiated analysis of the object of investigation. It is the aim to comprehensively understand the phenomenon of apple breeding and cultivation. Secondly, the research design had to give credit to the explorative character of this undertaking because there is a large research gap on the resilience of fruit breeding and cultivation. Qualitative research is suitable to understand and analyze components and linkages of social-ecological systems and collect appropriate data (Biggs et al. 2021). Thirdly, qualitative research is especially useful in transdisciplinary settings (Pohl and Hirsch Hadorn 2008). It enables the access to and the collection of different forms of knowledge (See Section 2.2) and acknowledges diverse perspectives.

Within the field of qualitative research, two empirical research methods have been identified as valuable for this undertaking: case studies and a Delphi study. Both methods are particularly suited for explorative research designs as it is the case in this research project. *Case studies* are a well-established method for the indepth explanation and comparison of contemporary social phenomena (Yin 2017, 2011). *Delphi studies* give the possibility to involve a diverse scope of experts in the collection of in-depth insights about certain phenomena in several questioning rounds (Häder 2014). Figure 2.1 shows the general methodological framework which includes the methods for data collection and analysis.

For the case studies, qualitative interviews and a focus group have been chosen for data collection. For the Delphi study, two online questionnaires have been used. In both studies, literature research and analysis were also of importance. Interview data was analyzed with a qualitative content analysis. The following section further explains the empirical study design and applied methods.

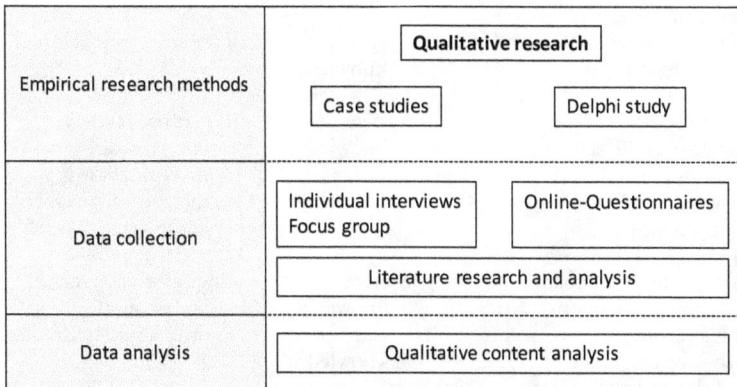

Figure 2.1 General methodological framework (own figure).

2.2 Methods for Data Collection and Analysis

Overall, three qualitative interviews, one focus group, and two online question-naires within a two-step Delphi study have been carried out to collect empiri-cal data. Table 2.1 gives an overview of the leading questions of this book, their knowledge contribution in light of the above-explained knowledge types, and the applied methods for data collection and analysis.

The evaluations of the first two questions are solely based on literature research and analysis. The broad literature base enables the conceptualization of fruit breeding and cultivation as social-ecological systems (system knowledge) and the discussion of how those systems need to be designed for achieving resilience (target knowledge).

The evaluations of apple breeding approaches (third question) take place in form of case studies, using a holistic multiple-case study design (Yin 2011). This means that multiple (or to be more specific: three) case studies are conducted with the same analytical focus, enabling to contrast and compare the results coher-ently (ibid.). The case studies aim to shed light on how different approaches are designed (system knowledge), and which solutions are most suitable for achieving resilience in apple breeding and cultivation (target knowledge).

A variety of data has been generated for the case studies with three quali-tative interviews and one focus group. Qualitative interviews are particularly recommended for conducting multiple-case study research (Yin 2011; Bryman 2016; Pahl-Wostl et al. 2021) and collecting data on components and interlink-ages of social-ecological systems (Shackleton et al. 2021). All interviews were

Table 2.1 Overview of applied methods (own table)

Leading question	Relevant chapters	Knowledge contribution	Methods for data collection & analysis
How is the social-ecological system of fruit breeding and cultivation defined and conceptualized?	4–6	System knowledge	Literature research and analysis
What (ideally) characterizes a resilient fruit breeding and cultivation system?	4–6	Target knowledge	Literature research and analysis
Which apple breeding approaches exist, what are the differences, and how do they affect apple cultivation and social-ecological resilience?	7–10	System knowledge, target knowledge	Literature research and analysis, three qualitative interviews, one focus group, qualitative content analysis
Which factors inhibit the broad implementation of commons-based apple breeding and how can they be overcome to exploit its full potential?	11–14	System knowledge, transformative knowledge	Literature research and analysis, online-questionnaires

conceptualized as *semi-standardized interviews* to allow a focused but flexible interview process (Misoch 2019). These characteristics of semi-standardized interviews allowed to put the emphasis on the interviewees' own perspectives and on what they perceive as relevant and important. This is necessary when it is aimed at understanding their breeding approaches in detail.

Focus groups are "moderated discourses where a small group discusses a specific topic on the base of a certain information input" (Schulz et al. 2012, p. 9, own translation). As such, this method connects two methods from qualitative research with each other: the focused interview and the group interview (Bryman 2016; Dürrenberger and Behringer 1999). The main goal of focus groups is to identify different perspectives on a thematic focus and accompanied collective dynamics that emerge in the group discussion. They are suitable for explorative research designs applied to social-ecological systems research (Shackleton et al. 2021). A focus group was carried out with members of *apfel:gut e.V.* for the case study on commons-based apple breeding (See Chapter 9). This novel breeding approach involves collective and participatory elements that cannot be captured with individual interviews. This required the necessity to generate additional qualitative data that is better able to explore these collective dynamics and "study the ways in which individuals collectively make sense of a phenomenon and construct meanings around it" (Bryman 2016, p. 502).

All semi-standardized interviews and the focus group have been analyzed with *qualitative content analysis* (Mayring 2015). This method of analysis is well established in qualitative research and a standard tool to code and categorize qualitative data generated through semi-standardized interviews (Preiser et al. 2021; Flick 2017; Gläser and Laudel 2009). Hereby, the *structuring content analysis* was applied because it allowed to analyze the interview data in a comparative way by a filtration of the most important aspects that were relevant for the conduction of the case studies.

For the last leading question, a Delphi study was carried out with the application of two online questionnaires. A universally agreed-upon definition of Delphi studies does not exist, but any Delphi study at least consists of two central elements: (1) aggregating opinions and prognoses centered around a specific thematic focus or problem (2) in several rounds, so-called waves (Häder 2014). This study involved mostly qualitative elements but also some quantitative ones to reveal insights about the status quo of the German apple market (system knowledge) and about how to induce change toward resilience (transformative knowledge).

Questionnaires are the usual method for data collection in Delphi studies. Here, *self-administered online questionnaires* were used to collect data. In the first wave, qualitative data on market structures and developments was collected (system knowledge). After aggregating the data, identified structures and developments were assessed by the participants with quantitative ratings.[1] Additional qualitative elements were added to the second wave to collect new information, specifically transformative knowledge. Online questionnaires of both waves have been designed following the recommendations by Häder (2014) and Bryman (2016), with the aim to create well-structured surveys with diverse question types and an appropriate length.

2.3 Reflection of the Methodological Design

Qualitative research has some general limitations. Following Bryman (2016), there are essentially four criticisms:

1. Qualitative research is perceived as too subjective because it relies on observations of the researcher who takes subjective decisions on what is (not) important and significant.
2. Results of qualitative research are difficult to replicate as they are highly context- and time-dependent.
3. Qualitative findings are difficult to generalize as often specific cases are assessed.
4. Some critics see a lack of transparency in the data collection process.

A possible solution to address these criticisms is to implement *forms of triangulation* that mark an established method to include different perspectives and verify research insights (Flick 2011). Triangulation strengthens the quality of qualitative research. Flick (2017) names four different types of triangulation: data triangulation (using different data sources); investigator triangulation (employing different observers/interviewers); theory triangulation (involving various theoretical points of view); and methodological triangulation (using different methodological elements within and between methods). Here, investigator and methodological triangulation are implemented to certain extents.

As the research covered by this book was embedded in the transdisciplinary research project EGON (see Preface), the author developed close professional relationships with the consortium practitioners that were included in the case study on commons-based apple breeding (See Chapter 9). However, this fact also explains the chosen research area and justifies the practical relevance of this book. All interim research results and/or the research process were discussed with the consortium practitioners, with the research team of EGON, and at conferences with practitioners. This gives the aim and scope of this book a sufficient level of credibility. Especially the regular in-depth discussions within the research team created a high level of *investigator triangulation.*

To tackle the challenge of replication, different perspectives and data sources were used. This data was collected at different points in time with different research subjects and analyzed from various theoretical points of view (See Chapter 1). Within the case studies and the Delphi study, a moderate form of *methodological triangulation* took place because both studies involved different methodological elements to create valid insights. Regarding the case studies, this involved individual interviews and one focus group, which proved challenging because different methods were used for different case studies. Regarding the Delphi study this involved qualitative and quantitative elements in the online questionnaires. Although the Delphi study proved a sound method for data collection, some participants might better respond to another method. In Appendix B, it is reflected upon those aspects in detail.

Moreover, choosing the apple sector in Germany as an empirical context for this book (See Chapter 3) addresses time- and context-dependency. Market structures and norms are generally slow to change in the apple sector (Zander 2011) which increases the replication potential of conducted research. As apples are one of the most economically and culturally important fruits in the temperate zone (See Chapter 3), apple-specific findings can be generalized in a moderate way. This moderate generalization is not only valid across countries but also for fruit breeding and cultivation in general (See Introduction). Hence, the conceptualization of resilience is carried out on the general fruit level (See Part II) and then applied to the apple sector (See Part III).

As a result, the above-described four criticisms of qualitative research have been addressed as far as possible. First, investigator triangulation lowers the level of subjectivity throughout the research process. Second, methodological triangulation and the chosen empirical context increase the replication potential of conducted research. Third, the empirical context allows for a moderate generalization. Fourth, the data collection and analysis process for both the case studies and the Delphi study are documented extensively in Appendix B and C to create full transparency.

Note

1 It has to be emphasized that the prime direction of the Delphi study was a qualitative one, thus causal reasons and mechanisms are analyzed qualitatively. All quantitative ratings were analyzed with basic statistic methods, whereby only mean values have been of relevance to illustrate the distribution of specific characteristics.

References

Adler, C.; Hirsch Hadorn, G.; Breu, T.; Wiesmann, U.; Pohl, C. (2018): Conceptualizing the Transfer of Knowledge across Cases in Transdisciplinary Research. In *Sustainability Science* 13 (1), pp. 179–190. DOI: 10.1007/s11625-017-0444-2.

Biggs, R.; de Vos, A.; Preiser, R.; Clements, H.; Maciejewski, K.; Schlüter, M. (Eds.) (2021): *The Routledge Handbook of Research Methods for Social-Ecological Systems.* New York: Routledge (Routledge International Handbooks).

Brandt, P.; Ernst, A.; Gralla, F.; Luederitz, C.; Lang, D. J.; Newig, J. et al. (2013): A Review of Transdisciplinary Research in Sustainability Science. In *Ecological Economics* 92, pp. 1–15. DOI: 10.1016/j.ecolecon.2013.04.008.

Bryman, A. (2016): Social Research Methods. 5th ed. Oxford: Oxford University Press.

Dürrenberger, G.; Behringer, J. (1999): *Die Fokusgruppe in Theorie und Anwendung.* Stuttgart: Akademie für Technikfolgenabschätzung in Baden-Württemberg.

Flick, U. (2011): *Triangulation.* Wiesbaden: Springer Fachmedien (Qualitative Sozialforschung).

Flick, U. (2017): *Qualitative Sozialforschung. Eine Einführung.* Originalausgabe, 8. Auflage. Reinbek bei Hamburg: rowohlts enzyklopädie im Rowohlt Taschenbuch Verlag (Rororo Rowohlts Enzyklopädie, 55694).

Gläser, J.; Laudel, G. (2009): *Experteninterviews und qualitative Inhaltsanalyse als Instrumente rekonstruierender Untersuchungen.* 3, überarb. Aufl. Wiesbaden: VS Verlag für Sozialwissenschaften (Lehrbuch).

Häder, M. (2014): *Delphi-Befragungen. Ein Arbeitsbuch.* 3, Aufl. Wiesbaden: Springer VS (Lehrbuch).

Hirsch Hadorn, G.; Bradley, D.; Pohl, C.; Rist, S.; Wiesmann, U. (2006): Implications of Transdisciplinarity for Sustainability Research. In *Ecological Economics* 60 (1), pp. 119–128. DOI: 10.1016/j.ecolecon.2005.12.002.

Lang, D. J.; Wiek, A.; Bergmann, M.; Stauffacher, M.; Martens, P.; Moll, P. et al. (2012): Transdisciplinary Research in Sustainability Science: Practice, Principles, and Challenges. In *Sustainability Science* 7 (S1), pp. 25–43. DOI: 10.1007/s11625-011-0149-x.

Mauser, W.; Klepper, G.; Rice, M.; Schmalzbauer, B. S.; Hackmann, H.; Leemans, R.; Moore, H. (2013): Transdisciplinary Global Change Research: The Co-creation of Knowledge for Sustainability. In *Current Opinion in Environmental Sustainability* 5 (3–4), pp. 420–431. DOI: 10.1016/j.cosust.2013.07.001.

Mayring, P. (2015): *Qualitative Inhaltsanalyse. Grundlagen und Techniken.* 12th ed. Weinheim: Beltz.

Misoch, S. (2019): *Qualitative Interviews. 2., erweiterte und aktualisierte Auflage.* Berlin: De Gruyter Oldenbourg.

Pahl-Wostl, C.; Basurto, X.; Villamayor-Tomas, S. (2021): Comparative Case Study Analysis. In Reinette Biggs, Alta de Vos, Rika Preiser, Hayley Clements, Kristine Maciejewski, Maja Schlüter (Eds.): *The Routledge Handbook of Research Methods for Social-ecological Systems.* New York: Routledge (Routledge International Handbooks), pp. 282–294.

Pohl, C.; Hirsch Hadorn, G. (2008): Methodological Challenges of Transdisciplinary Research. In *Natures Sciences Sociétés* 16 (2), pp. 111–121. DOI: 10.1051/nss:2008035.

Preiser, R.; Garcia, M. M.; Hill, L.; Klein, L. (2021): Qualitative Content Analysis. In Reinette Biggs, Alta de Vos, Rika Preiser, Hayley Clements, Kristine Maciejewski, Maja Schlüter (Eds.): *The Routledge Handbook of Research Methods for Social-ecological Systems.* New York: Routledge (Routledge International Handbooks), pp. 270–281.

Schulz, M.; Mack, B.; Renn, O. (2012): *Fokusgruppen in der empirischen Sozialwissenschaft.* Wiesbaden: VS Verlag für Sozialwissenschaften.

Shackleton, S.; Bezerra, J. C.; Cockburn, J.; Reed, M. G.; Abu, R. (2021): Interviews and Surveys. In Reinette Biggs, Alta de Vos, Rika Preiser, Hayley Clements, Kristine Maciejewski, Maja Schlüter (Eds.): *The Routledge Handbook of Research Methods for Social-Ecological Systems.* New York: Routledge (Routledge International Handbooks), pp. 107–118.

Yin, R. K. (2011): *Applications of Case Study Research.* 3rd ed. Thousand Oaks, CA: Sage.

Yin, R. K. (2017): *Case study Research and Applications. Design and Methods.* 6th ed. Thousand Oaks, CA: Sage.

Zander, K. (2011): Ausländisches Angebot an ökologischen Äpfeln: Bedeutung für deutsche Öko-Apfelerzeuger. Universität Kassel, Fachgebiet Agrar- und Lebensmittelmarketing. Witzenhausen.

3 Empirical Context
Apple Breeding and Cultivation in Germany

This chapter provides an overview of the empirical context of this book – the apple sector in Germany – from a political-legal, economic, and ecological perspective. Apples are the most consumed fruits in Germany and are hence of particular importance in German fruit cultivation (BMELV 2013). The description of the empirical context focuses on apples for direct consumption, not apples for processing, as these apples are the main subject of apple breeding efforts. Scientific and gray literature provides the basis for the description of the empirical context. Additionally, a qualitative semi-standardized interview with the director from the Federal Plant Variety Office *Bundessortenamt* (BS) has been carried out in 2018 to get background knowledge on the political-legal framework. References to the interview are marked with 'Interview BS 2018.'

An in-depth look is taken at the German political-legal framework (See Section 3.1) and the value chain in the apple sector with specific references to conditions in Germany (See Section 3.2). For a better understanding of the value of the German case, it is then placed into the global and European context of apple breeding and cultivation (See Section 3.3).

3.1 Political-Legal Framework in Germany

This section provides a compact overview of the political-legal framework for apple breeding[1] in Germany. Several international and supranational agreements and conventions are relevant to seed law. Those "institutional complexes" (Stokke and Oberthür 2011, p. 326) are well researched and mapped. Tschersich (2021) provides an overview of the landscape of seed governance. However, as this international level is not of particular relevance for the scope of this book, the depiction of the political-legal framework focuses on the national level.

3.1.1 Basic German Regulatory Framework

The distinct process of apple breeding is not directly regulated but rather indirectly influenced by certain legislations (Interview BS 2018). The specific regulation starts with the introduction of a variety into the market and its propagation. A compact description of these generally relevant legislations is given by Hanke

DOI: 10.4324/9781003355724-5

and Flachowksy (2017), which serves as the major source for the following elaborations. Several EU directives[2] provide the legal basis for German legislation and
encompass quality criteria regarding the commercialization of plants and plant
health. The following German norms are relevant for fruits in general:

- German Seed Marketing Act (*Saatgutverkehrsgesetz*, SaatG)
- German Plant Protection Act (*Pflanzenschutzgesetz*, PflSchG)
- German Growing Material Act (*Verordnung über das Inverkehrbringen von
  Anbaumaterial von Gemüse-, Obst- und Zierpflanzenarten*, AGOZV)
- German Plant Inspection Ordinance Act (*Pflanzenbeschauverordnung*,
  PBVO)
- German Plant Breeders' Rights Act (*Sortenschutzgesetz*, SortG)
- Conservation varieties ordinance relating to the certification of conservation
  varieties and marketing of conservation varieties of seed and plant material
  (*Erhaltungssortenverordnung*, ErhaltungsV)

The SaatG defines general quality and variety criteria for the market placement of
seeds (which is not relevant in the context of fruits) and propagation material. In
addition, the PflSchG determines basic plant health criteria for the market placement of propagation material. Based on SaatG and PflSchG, the AGOZV adds
more detailed norms on the market placement of propagation and cultivation material, for example, budwood, fruit trees, or rootstocks.[3] Organizations that trade
with propagation and cultivation material have to be registered by the norms of
the PBVO and are yearly controlled by regional authorities. For placing cultivation material on the market, the cultivar either has to hold variety protection (see
below), be officially registered in the national list, or be commonly known.[4]

The BS is responsible for the official registration process and national listing of
new varieties. An official registration is a necessary obligation for the commercial
distribution of a variety as regulated in the SaatG. The process of national listing
is described by the BS (2020) as follows:

> A pre-requisite for a variety to be admitted to the national list includes dis-
> tinctiveness from other varieties, uniformity and stability, which are tested in
> open field or greenhouse testing procedures (DUS testing) and a suitable va-
> riety denomination. For agricultural plant varieties value for cultivation and
> use is also required. A variety is considered to have value for cultivation and
> use if its qualities taken as a whole offer a clear improvement for cultivation,
> for use of the harvest or use of products derived from the harvest compared
> to comparable listed varieties. The value features of a variety are determined
> by properties shown in cultivation testing and laboratory testing relating to
> cultivation, resistance, yield, quality and use (VCU testing).[5]

The admission to be included in the national list is granted for 20 years and can
be extended upon application. At the moment, more than 10.000 fruit varieties
are listed as (nationally) marketable varieties including more than 2.000 apple

varieties (Interview BS 2018). Because the registration process is similar to the demands of applying for variety protection, most applicants simultaneously apply for it and national listing. An overview of this general process is given by Messmer et al. (2015) as depicted in Figure 3.1.

Variety protection is the most common instrument to place certified cultivation material on the market. The SortG defines the norms for receiving variety protection for a newly bred cultivar. It can be applied for at the BSA and is granted for 25 years. Decision criteria for a successful application are the DUS criteria as explained above and as defined in §1, SortG. The holder of variety protection gets the sole right to (a) place propagation material of the protected variety on the market; (b) produce, import, or store this material; and (c) issue licenses to third parties to propagate and cultivate the variety. Third parties have to pay tree royalties to the holder of variety protection. However, the protection rights have to be enforced by the holder itself.[6] According to the interview partner from the BS, applications for variety protection rose up in the past few years but now experiences a certain "saturation" (Interview BS 2018, own translation) because it is not economically feasible[7] to apply for protection rights for every newly developed variety (ibid.). It is worth noting that despite this protection possibility, access of breeders to any variety is not restricted if it is used for breeding new varieties. This

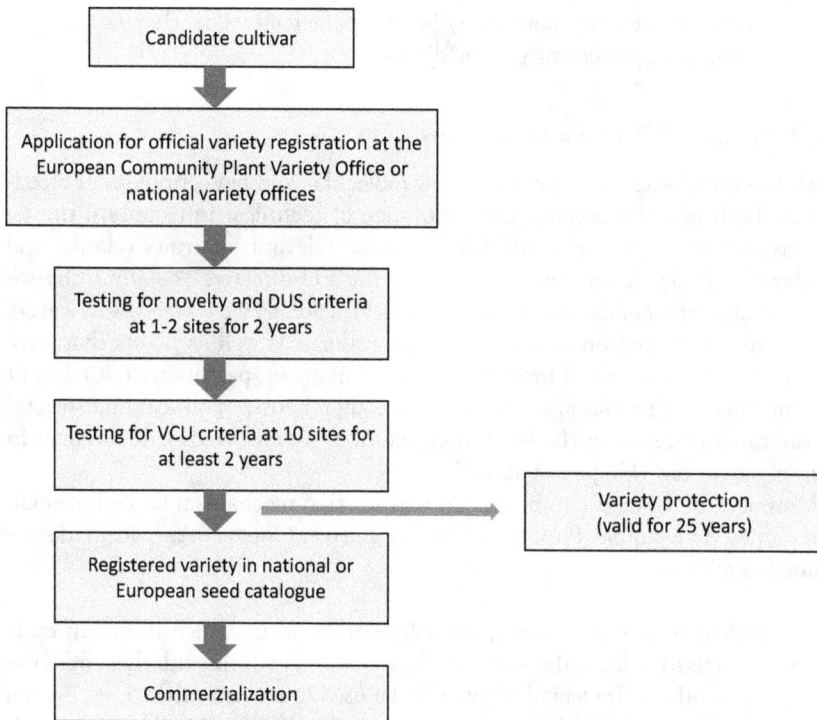

Figure 3.1 Basic registration process for a new plant variety (based on Messmer et al. 2015, p. 8).

is called the *breeder's exemption*. A holder of variety protection rights can thus not restrict access to his or her variety if it is used for breeding activities.

There have been some recent changes in relevant legislation to further specify the mutual recognition of fruit varieties throughout the EU:

> National listing is obligatory for fruit varieties since 2017. Plant material can be sold throughout the EU if the variety, in addition to plant health requirements, is protected or listed in one member state or the variety was already in circulation prior to 30/09/2012. There must be at least an officially recognised [sic.] variety description. Varieties that may only be marketed in Germany include so-called amateur varieties and varieties for the preservation of genetic diversity.
>
> (BS 2020)

For amateur varieties that are not of commercial value (§57a, 1, No. 6, SaatG), a simplified national listing process exists.[8] Amateur varieties are of interest for regional production and/ or hobby gardeners and can be marketed in Germany without a quantitative limit. The interview partner sees amateur fruit varieties as an "excellent instrument" (Interview BS 2018, own translation) to promote marketable varieties without much effort. Similarly, conservation varieties as regulated in the ErhaltungsV also apply to a simplified listing process. Although conservation varieties can potentially be of commercial value, they rather have the sole purpose of preserving genetic diversity.

3.1.2 Additional Relevant Regulations

With the increasing development of new molecular and biotechnological breeding methods and the accompanied relevance of technical innovations, the legal instrument of patents could become more relevant for fruits (Hanke and Flachowsky 2017). *Patents* are regulated by the EU directive 98/44/EG, the so-called *Biopatentrichtlinie*, and nationally in the Patent Act (*Patentgesetz*, PatG). They can be granted on new technical procedures as well as plants that were developed with a technical innovation. At the moment, patents are not relevant for fruit breeding in Germany. However, a comprehensive political and societal debate takes place about the legal interpretation of new breeding techniques in light of patent law (Nuijten et al. 2017).

Moreover, *trademarks* can be used as a protection mechanism for commercialized names, for example, Pink Lady® for the cultivar *Cripps Pink*. Trademarks are defined as follows:

> A trademark is a word, name, symbol, or other device which is used in trade with goods to indicate the source of the goods and to distinguish them from the goods of others. Trademark rights may be used to prevent others from using a confusingly similar mark, but not to prevent others from making the same goods or from selling the same goods or services under a completely different mark.
>
> (Clark and Jondle 2008, p. 445)

Trademarks can be applied for at the German Patent and Trademark Office *Deutsches Patent- und Markenamt* as a national protection instrument or EU-wide at the European Union Intellectual Property Office (EUIPO). Because the trademark can be continuously extended, no temporal restriction exists and the holder of the trademark has the sole right to use it and/or transfer the usage right to third parties upon license fees.[9]

The combination of variety protection, trademarks, and (in the future) possibly patents is seen as critical for the freedom of breeding. The interview partner states that patents on fruit's gene sequences in combination with variety protection could be "the end of the breeder's exemption" (Interview BS, own translation) as it restricts access to the variety. As Hanke and Flachowksy (2017) put it:

> The significance of this possibility for protecting a variety [patents] becomes clear when plants incorporate variety and patent protection. In this case, the breeder's exemption is restricted. Indeed, it is allowed to breed with this variety, but a resulting new variety with a patented attribute needs the approval of the patent holder for its market introduction.
>
> (p. 152, own translation)

Overall, additional relevant regulations play an important role in the apple sector and discussions on patents will probably further continue.

3.2 Value Chain in the Apple Sector

Most examinations of value chains in the fruit sector focus on cultivation, marketing, and trade (Garming et al. 2018; Dirksmeyer 2009). This section adds an often-neglected part to the value chain: breeding and multiplication. Because apples are perennial crops, their biological character dictates several requirements and conditions along the value chain. For instance, cultivating apples needs special production techniques and know-how. Any planted apple tree is a long-term investment and it is economically not feasible for farmers to make a short-term switch to another variety (Zander 2011).

Figure 3.2 shows a simplified illustration of the value chain in the apple sector. It involves the steps for apple breeding, distribution of apple varieties, apple cultivation, marketing and trade of apples, and lastly consumption.

In the following, major aspects of every part of the value chain are explained to get an overview of the value cycle of apple varieties. All explanations focus on Germany while further descriptions from actors out of other countries are included when it proves valuable. All following descriptions of biological processes are generally valid for other tree fruits (Badenes and Byrne 2012) which is important for the elaborations on the general fruit level in Part II.

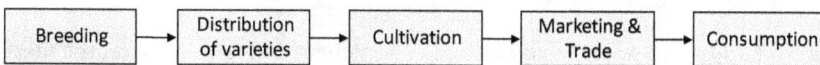

Breeding → Distribution of varieties → Cultivation → Marketing & Trade → Consumption

Figure 3.2 Value chain in the apple sector (own figure).

3.2.1 *Processes in Apple Breeding*

Fruit breeding in general is very different from other forms of plant breeding. The major distinction is the long-term breeding process of 15–20 years compared with 5–7 years for most vegetables. Following basic works by Hanke and Flachowsky (2017) or Badeners and Byrne (2012), there are several characteristics that make fruit/apple breeding particularly challenging, including two major aspects:

1. *Heterozygosity:* the genome of apples is heterozygous as they are obligatory cross-pollinated and self-incompatible. This means that any pollen is genetically unique. In comparison, tomato genes of hybrid varieties are for example largely homozygous – all seeds inside one tomato are genetically uniform and, when planted, potentially show the same phenotype under the same environmental conditions (Cheema and Dhaliwal 2005).[10] All apple seeds inside one apple on the other hand are different and always show different phenotypes when planted. Open-pollinated apple seedlings thus consist of a highly heterozygous mix of apples. When crossing two apple cultivars with each other, the outcome is always genetically unique.

2. *Vegetative propagation:* because of the heterozygous character of apples, they cannot be multiplied by seeds (generative propagation) if their original characteristics shall be maintained. Instead, they have to be propagated asexually in a vegetative way by cutting the budwood of grown apple trees. This budwood is grafted on rootstocks and cultivated as new apple trees. Consequently, unique phenotypes based on the original apple tree are cloned with grafting. Vegetative propagation is thus the only way to propagate specific apple varieties.

The general process of apple breeding follows the classical crossbreeding approach.[11] This process follows four steps: (a) defining breeding goals; (b) planning and doing crossings; (c) cultivation and selection of seedlings; and (d) registration and assessment of a new variety. In a nutshell:

> The traditional apple breeding strategy has involved crossings among a few top commercial cultivars and elite selections, and planting full-sib families in order to forward-select (based on phenotypic performance) individuals for further clonal testing. Selection of parents with complementary characteristics is essential in order to produce seedlings that have the desired attribute inherited from each parent. Crossing of two commercial cultivars produces progeny exhibiting a wide variation for any one quality trait, with only a very small proportion showing trait improvements over the parents. In apple, selected plants are used as potential cultivars and/or as breeding parents, so that frequencies of desirable alleles[12] are increased in successive populations.
>
> (Isik et al. 2015, p. 104)

In standard literature on apple breeding, researchers generally agree on the most important *breeding goals* for apples (Hanke and Flachowksy 2017, Brown 2012): fruit quality (e.g., juiciness and crispness), disease and pest resistance (e.g., apple scab), nutritional components and postharvest traits (e.g., storage characteristics, freshness), and yield. Breeding apples with multiple resistances against several pests and diseases is perceived as one of the main challenges because with increased resistances, fruit quality with regard to commercially acceptable fruits often decreases (Brown 2012). This exemplifies that there are always trade-offs between the pursuit of the different breeding goals.[13]

Crossing plans are developed with cultivars that fit to the identified breeding goals. Breeders take phenotypical characteristics as well as genotypes into account. For example, if a new apple cultivar with resistance to fire blight shall be developed, cultivars with resistant traits against fire blight (phenotype) and cultivars with fitting genetic traits (genotype) enter the crossing plan as parents. As research on pedigree relations and genotyping is growing ever faster, genetic information plays a more and more important role in the development of crossing plans (Howard et al. 2021).[14] However, most pedigree relations of apples are still largely unknown. Based on the crossing plan, crossings are carried out manually with *controlled pollination*. Hereby, pollen is harvested from the male parent and painted on the flower of the female parent, fertilizing the egg cells. Nets are hung around the pollinated trees to create a controlled environment and prevent wild pollination. When the pollinated flowers bear fruits, the apple seeds (progeny) are harvested and stratified.[15] Afterward, they are planted in fitting environments, mostly in greenhouses.

From those seeds, seedlings are grown and the *selection process* starts. The number and duration of selection steps depend on the breeding goals and approach as well as used methods. For an illustration, Isik et al. (2015) explain a typical breeding scheme[16]: in the first selection stage, the performance of planted seedlings is continuously evaluated and the seedlings that perform satisfactorily either again enter the crossing stage (forward-selection) and/or the second selection stage. It takes about seven years until good-performing cultivars are identified. In the second selection stage (year 7–11), those promising cultivars are cloned, grafted, and planted at two or more different sites so that multiple copies can be further evaluated on different conditions. All seedlings that perform satisfyingly enter the third and last selection stage (year 11–15), in which, again, multiple copies are made and planted at more than two sites. Finally, potential cultivars are identified that are suitable for registration and assessment (see the previous section).

Fruit breeding in Germany has a long tradition with different scientists, pomologists, and private breeders developing a great diversity of cultivars and breeding techniques. An extensive overview of the historic development is given by Hanke and Flachowsky (2017). Today, apple breeding in Germany is carried out by public institutions but also by private actors. An overview and a categorization are given in Part III.

3.2.2 Distribution of Apple Varieties

Plant material of apple varieties is propagated and distributed by nurseries, regulated by AGOZV and PBVO (see the previous section). *Nurseries* serve as a 'bottleneck' between breeders and farmers that are interested in cultivating specific varieties. They have a particular form of market power because if they do not offer certain varieties, those varieties get lost in the commercial sector. However, the business of nurseries is confronted with different challenges and forms of competition:

> Nurseries have played a significant role in defining and marketing varieties, and yet their economic power has been historically tempered by a lack of control over their products: once sold, varieties could easily and exactly be reproduced through grafting, and other nurseries could reproduce the same trees at a cheaper rate.
>
> (Legun 2015, p. 302)

Many nurseries are organized in nursery consortia such as the Associated International Group of Nurseries (AIGN) or, in Germany, Artevos and the *Consortium Deutscher Baumschulen* (CDB). Those consortia often serve as cooperation and coordination platforms for connecting breeders, farmers, and marketeers (Hanke and Flachowsky 2017). An example is the apple breeding program by the French National Research Institute for Agriculture, Food and Environment (INRAE) which is cooperating with French nurseries organized by the company NOVADI. Breeders and nurserymen from INRAE and NOVADI work together by sharing responsibilities and knowledge, for example, about breeding methodologies, apple genetic diversity, and marketing (Laurens and Pitiot 2003).

3.2.3 Apple Cultivation: Concepts and Status Quo

Apples are generally an expensive commodity to produce and the planting of apple trees on dwarfing rootstocks has become a standard in apple cultivation (Granatstein and Peck 2017; Gallardo and Garming 2017). In Germany, apples are cultivated with integrated production, organic farming, or in meadow orchards.[17]

Apple cultivation in a conventional and industrial sense – maximizing crop productivity and yield through the farming of monocultures combined with the intensive use of pesticides and chemical-synthetic fertilizers – plays no role in Germany (See also Part IV). Due to legal regulations on a European level, integrated fruit production as a modern concept of conventional fruit cultivation was widely adopted across Europe.[18] *Integrated fruit production* aims to "[...] produce[s] high quality fruits by using natural resources and regulating mechanisms to replace polluting inputs and to secure sustainable farming" (Damos et al. 2015, p. 636). It thus tries to maintain high productivity and yield while replacing chemical-synthetic inputs with mechanisms of biological control. The International Organization of Biological Control (IOBC) provides general guidelines for integrated fruit production regarding professional trainings, farming practices, methods, and treatments. However, integrated fruit production is criticized for missing transparency as standards "are not precisely defined or legally specified" (ibid., p. 638).

Organic apple production generally follows the norms for organic production and processing as defined by the IFOAM (2014), adopting their principles of Health, Ecology, Fairness, and Care. This production approach aims to preserve and promote soil health as well as the genetic and functional diversity of plants, animals, and other organisms in agroecosystems. Natural resources are treated with care and organic apple farmers have a special responsibility for protecting the environment and livelihoods for current and future generations. Organic apple production is strictly regulated by EU norms and follows labeling and certification schemes (see the previous section).

Apple cultivation in Germany is concentrated in regional clusters, particularly in the regions of Lake Constance in Baden-Wuerttemberg and *Altes Land* in Lower Saxony, especially in the district of Stade (Garming et al. 2018). The most recent data of 2017 illustrated by the *Bundesinformationszentrum Landwirtschaft* (BZL 2020) shows that the overall cultivation area of apples is 34.000 hectares. Eighteen percent of this area is cultivated in an organic way. Of the overall cultivation area, 83% is used for the production of fresh apples and 17% for the cultivation of apples used for processing. The most cultivated variety is *Elstar* (24% of fresh apple area), followed by *Braeburn* (10%), *Gala* (8.4%), and *Jonagold* (8.1%). Standard varieties in organic cultivation are also *Elstar* and *Jonagold*, but additionally *Idared, Boskoop, Piros, Ingrid Marie*, and *Holsteiner Cox*.

3.2.4 Distribution and Marketing Channels

Apple farmers usually have three options to market their harvest: self-promotion to wholesale or food retail, direct marketing to consumers, or the bundling of harvested apples by production organizations (in German *Erzeugerorganisationen* or *Erzeugergemeinschaften*). According to the study by Garming et al. (2018) which refers to data from 2014, 43% of fresh fruit produce is bundled by production organizations.[19] Thus, those actors play a significant role in the market. Hanke and Flachowksy (2017) describe this role as follows:

> Production organizations play an important part in the utilization of varieties in Germany and Europe because only via them, new varieties can be established in a specific area in a fast way and on a large scale. They are able to execute a variety-specific quality management and can convince marketeers on food retail level.
>
> (p. 157, own translation)

Important production organizations in Germany are the *Marktgemeinschaft Altes Land* (MAL), *Marktgemeinschaft Bodenseeobst* (MaBo), *Württembergische Obstgenossenschaft Raiffeisen* (WOG), and the *Deutsches Obstsorten Konsortium* (DOSK). In most cases, they simultaneously function as distribution or marketing organizations. Thus, they are responsible for sorting, storage, packaging, and logistics – pooling all tasks that take place in the stage of marketing and trade (See Figure 3.2).

The distribution and trade of apples take place in a highly complex net of actors and institutions. In their commodity flow analyses, Dirksmeyer et al. (2009) and Garming et al. (2018) make this complexity visible. Besides weekly markets, direct marketing, and specialist trade, food retail is the most important distribution channel. In 2014, 83% of fresh fruit produce was marketed via food retail, of which discounter had the biggest share with 46% (Garming et al. 2018).

3.2.5 Consumption of Apples

At the end of the value chain are the consumers that buy fresh apples. Their demand has an influence on several stages of the value chain (See Figure 3.2): the selection and quantity of propagated varieties (distribution of varieties stage); the selection and quantity of cultivated varieties (cultivation stage); and the breeding goals (breeding stage). In line with that, sensory and consumer testing is getting more and more important for breeding to meet the consumer's demands (Brown 2012).

For fruits and vegetables, Garming et al. (2018) generally identify a rising demand on regional produce. A representative study of the SINUS Institute (SINUS 2017) gives specific insights into popular traits of apples for consumers. Most consumers (38%) like multicolored apples, followed by red apples (36%) and green apples (21%). The majority also prefers sweet apples (46%), whereas 29% like sweet-sour and 24% like sour apples.

Although this work focuses on breeding, the description of the whole value chain makes clear that all later stages are intricately linked to it. Breeders take the demands of those later stages into account. Further knowledge on the interrelations (See Parts II–IV) helps to understand the breeder's motives and the challenges they are confronted with.

3.3 German Apple Breeding and Cultivation in the Global Context

While this book focuses on apple breeding and cultivation in Germany, modern crop breeding is always subject to global influences and developments. Apple breeders across the world face the same sustainability challenges (See Introduction): the lack of robust cultivars, inbreeding tendencies, as well as an ongoing mechanization, economization, and privatization of breeding. Overall, the chosen empirical context of this book has particular value because it is comparable with other apple-producing countries, especially from the EU, as they have (a) similar economic structures along the apple value chain, and (b) similar jurisdictions relevant for breeding as part of a global political-legal framework.

The globally most important countries in apple cultivation regarding overall production are China (about 40.5 million tons in 2020/21), the EU (about 12.2 million tons), the United States (US) (about 4.5 million tons), and Turkey (about 4.3 million tons) (United States Department of Agriculture 2021). The leading role of China, the EU, and the US in worldwide production remained largely unchanged for the last few decades (Brown 2012). In the EU, Germany is regularly the fourth largest apple producer (0.7 million tons in 2020), following Italy (2.4 million tons), Poland (2.3 million tons), and France (1.2 million tons) (European Commission 2022). Germany is thus not a major player in apple cultivation on an international scale but an important one in the EU.

Because Germany is a member state of the EU, the legislative framework for apple cultivation is consistent with all other EU countries.[20] Every country translated the relevant EU directives mentioned in Section 3.2 into national law, which means that standards for integrated and organic production of apples are uniform across the union. European countries were specifically supported by the EU to expand organic apple production (Delate et al. 2008) and experienced similar developments regarding the rise of organic fruit production:

> The first few pioneer fruit growers in several European regions started [organic farming] with little or no scientific support in the 1970s. They experienced losses and took risks that nowadays are difficult to imagine. The first specific research activities were started in close collaboration with local producers in the 1980s. Since then, the sector has continuously increased and now at least 10% of the fruit-growing area in several important fruit-growing regions in Europe, such as Northern Italy, Germany and Austria, is under organic cultivation.
>
> (Kienzle and Kelderer 2017, p. 2)

On the contrary, apples cultivated in organic farming systems in the US currently have a share of 7% (Mordor Intelligence 2022). In China, organic apple production plays no role at all as conventional production is the dominant standard (Zhu et al. 2018). The EU thus acts as a forerunner in growing apples under organic standards and shows a different production setting than the other globally important countries.

Similar to the conditions in Germany, apple cultivation in other European countries also takes place in regional clusters such as Southern Tyrol in Italy, central, southwestern and southeastern Poland, or Pays de Lore and Provence-Alpes-Côte d'Azur in France. Equivalent to MAL or MaBo in Germany, production organizations or cooperatives that have an important share in apple distribution and marketing are also present in other European countries such as Melinda (Italy), Appolonia (Poland), Blue Whale (France), or Fruit Masters (Netherlands). Germany is thus a representative case for many apple-producing European countries with a 10–20% share of organic production, regional production centers, and a patchwork of important production organizations.

Relevant legislations that concern apple breeding such as DUS and VCU testing are also similar across all EU countries, thus nationally registered and protected varieties may be marketed EU-wide (See Section 3.1). However, these legislations are also comparable worldwide because their objective – combining DUS criteria with intellectual property rights – has its origin in the international convention by the International Union for the Protection of New Plant Varieties of Plants (UPOV). UPOV specifies the DUS testing procedure for plant breeding and is one of the most important global agreements that are also relevant for apple breeding. Any country that ratified the UPOV convention commits its national seed law to these DUS guidelines and resulting intellectual property rights (such as variety protection in Germany).

Today's most important centers of professional and systematic apple breeding are in the US, China, New Zealand, Australia, and Europe (Brown 2012). All of these countries, including Germany via the EU, ratified the UPOV convention which makes their seed laws and variety protection rules highly comparable. The same is true for trademarking which is applied in all major countries where apple breeding is conducted. All insights of this book that refer to Germany but also concern the global political-legal framework may thus be applied to other countries as well.

However, the specific legal construction of those intellectual property rights following UPOV may differ as Clark et al. (2012) show for several of the above mentioned and other countries. Another difference in legislations lies in the relevance of patent law. While patent law currently plays no role in apple breeding in the EU (See Section 3.1) and also in China or New Zealand, "plant patents" are an established instrument in the US (Clark et al. 2012). This opened up possibilities for the application of genetic engineering in apple breeding, as the example of the *Arctic Apple* shows: Using the method of gene editing, breeders developed this an apple variety with a non-browning trait, which means that the apple flesh's oxidation is eliminated, and patented this particular trait or rather the respective gene sequence (Stowe and Dhingra 2021).[21] There is a large debate about the application of gene editing in apple breeding and its concurring impacts on apple cultivation, trade relations, and other economic aspects. Fritsche et al. (2018) reflect this debate in the context of New Zealand and conclude that "a global consensus on regulation of gene editing [is] impossible in the immediate future" (ibid., p. 6). While the EU remains skeptical, other countries such as China or Australia are more open for not regulating gene-edited plants.

In the global apple breeding centers, a wide range of different organizations engage in apple breeding, from public to private breeding initiatives. Some short examples illustrate parts of the organizational structure of breeding in different countries to get a better impression on organizational similarities and the globalization of the fruit market:

- **New Zealand**: Following a long tradition of apple breeding, the New Zealand Institute for Plant and Food Research Ltd. (in short: Plant & Food Research) and its predecessor institutes developed several globally important apple varieties such as *Braeburn* or *Royal Gala* and conduct research on apple breeding (Plant & Food Research 2022). Plant & Food Research is a private research institute with historical roots in former public institutes. A collaborator of this institute is New Zealand Apples & Pears Inc. that represents the country's fruit cultivation industry and coordinates breeding programs between fruit growers and researchers (New Zealand Apples & Pears 2021).

- **Switzerland**: Agroscope, the public center of excellence for agricultural research, carries out research along the entire value chain of agriculture and the food sector. One strategic research division is particularly responsible for plant and particularly fruit breeding (Agroscope 2022). Agroscope bred several economically important apple varieties, for example, *Diwa*. For market introduction, Agroscope cooperates with the private company VariCom. In addition, private breeding initiatives exist such as the association Poma Culta e.V., which concentrates on apple breeding specifically for organic production. Agroscope and Poma Culta initiated the project NAGBA in 2016, with the goal to "revive apple genetic resources preserved in Switzerland through breeding for organic production" (Kellerhals et al. 2018, p. 12).

- **The Netherlands**: At Wageningen University, the expert group on plant breeding is one of Europe's most prominent breeding actors and has been researching fruit breeding and breeding apple varieties for decades, for example, *Natyra* or *Wellant* (WUR 2022). The university works together with the private company Fresh Forward Breeding & Marketing, which cooperates in breeding activities and is responsible for the market introduction of bred varieties.

As the illustrative cases show, apple breeding often has its roots in public centers or universities, whereby the level of sophistication and breeding success differs. However, the level of how further actors such as private companies or associations/NGOs are involved in breeding activities differs from a strong public-private nexus (Netherlands, New Zealand) to a patchwork of initiatives and companies (Switzerland). Germany is embedded into this breeding landscape, particularly in the European one because of geographical proximity. The evaluation of apple breeding approaches (See Part III) will show how other countries may also profit from these insights.

In sum, analyzing and discussing the case of Germany provide value in the global context of apple breeding and cultivation; concurring insights on the resilience and sustainability of apple breeding/cultivation may be transferable to some degree. This is particularly true for the EU context because Germany shares its relevant political-legal framework with other member states. Limitations occur in the comparability of other national political-legal developments, as the case of gene editing shows, but also regarding the role and importance of organic standards.

Notes

1 As the focus of this book is on breeding, other regulatory frameworks that concern cultivation or marketing are excluded. Norms for agriculture and fruit cultivation in Germany are highly complex, on which Norer (2018), for example, provides an overview. An overview of marketing norms is given by the Federal Office for Agriculture and Food *Bundesanstalt für Landwirtschaft und Ernährung* (BLE 2021).
2 This concerns the EU directives 2014/98/EU, 2014/96/EU, 2014/97/EU, and 2008/90/EU. All directives can be accessed via https://eur-lex.europa.eu/.

3 In more detail, the AGOZV distinguishes between the following types of material: *Standardmaterial, Vorstufenmaterial, Basismaterial, zertifiziertes Material* (§2).

4 Commonly known means that a variety has been introduced into the market before September 30, 2012. This is particularly relevant for traditional and heirloom cultivars.

5 It is interesting to note that in the registration process no formal differences exist between conventionally and organically bred cultivars (Interview BS). The testing procedure is the same for every cultivar. The cultivation procedure it is targeted on is irrelevant.

6 Because of this legal classification, it is also possible for the holder to not claim license fees and only use variety protection to protect the developed variety from misuse (Interview BS).

7 In addition to one-time payments for the registration process, one-time and yearly fees have to be paid to the BS for variety protection. As the enforcement of protection rights is usually outsourced to specialized companies (Interview BS 2018), further additional costs occur.

8 It is sufficient to deliver a further description of the cultivar, tree growth aspects, and other basic information.

9 This is of particular relevance if variety protection expired but the variety is still successful on the market. The only possibility to further restrict the use of this variety is with its trademark (Interview BS 2018).

10 The genetic code of organisms, in this case plants, is called *genotype* whereas the visible traits of an organism is called *phenotype*. Not all genetic characteristics (genotype) are expressed in the visible traits (phenotype). Although the phenotype largely depends upon the genotype, environmental factors also influence the phenotype (Miedaner 2017).

11 For an in-depth overview of breeding methods, their systematization, and the creation and development of breeding programs, see Hanke and Flachowsky (2017).

12 An allele is a specific variation or condition of a gene. Every gene has several alleles, which means it has different characteristic values, for example, red or white flowers. Because of their heterozygosity, apple genes have always different alleles.

13 The evaluations of apple breeding approaches in Part III will shed some light on specific goals in organic and conventional breeding programs.

14 Literature provides several overviews on past global apple breeding programs (Brown and Maloney 2003; Knight et al. 2005; Laurens 1998), which highlight important knowledge resources for further breeding programs. Research communities have formed that give detailed access to breeding data and results by databases like the AppleBreedDatabase, HIDRAS, DARE, ISAFRUIT, or RosBREED. However, with regard to methodologies used in breeding programs, a knowledge gap exists because many methodologies are not published or hard to find in many different scientific journals (Brown 2012).

15 For breaking the dormancy of apple seeds, they are stratified for several weeks. This cold stratification forces an earlier start of the germination process to better control the start and time period of the growing process.

16 Although the authors refer their scheme to New Zealand conditions, it can be classified as generally valid.

17 Because meadow orchards have no significant commercial relevance, they will not be further explained. A coherent overview of the history, characteristics, challenges, and potential of meadow orchards is given by Barde and Hochmann (2019).

18 This primarily refers to the legislation of pesticide use. With the implementation of EU directive 128/2009, member states of the EU were obliged to reduce risks and negative impacts of pesticide use.

19 Self-promotion takes place for 27%, and 13% of fresh fruits that are marketed directly. Interestingly, the trend is reverse for vegetables, that show a share of 22% for production organizations and 40 % for self-promotion (Garming et al. 2018).

20 International guidelines also provide standards in apple cultivation, for example, IOBC and IFOAM for integrated and organic fruit production (See Section 3.2).
21 In general, it is differentiated between genetic modification (GM) and gene editing (GE): "GM plants are created by inserting another organism's DNA (transgene insertion) into the original plant's DNA. [...] GE events are created by editing a part of the plant's DNA (transgene or non-transgene insertion). [...] Unlike GM, which largely depends on trial and error, GE technologies have greater precision" (Bullock et al. 2021). In the case of the Arctic Apple, "gene-silencing" took place by suppressing the genes that are responsible for the browning of the apple (Stowe and Dhingra 2021).

References

Agroscope (2022): About us. Available online at https://www.agroscope.admin.ch/agroscope/en/home/about-us/agroscope.html, checked on 8/14/2022.

Badenes, M. L.; Byrne, D. H. (Eds.) (2012): *Fruit Breeding*. Boston, MA: Springer US.

Barde, M.; Hochmann, L.; Barde, M. (2019): Streuobstwirtschaft. Aufbruch zu einem neuen sozialökologischen Unternehmertum. München: Oekom-Verl. Ges. für Ökologische Kommunikation.

BLE (2021): Vermarktungsnormen für Obst und Gemüse. Bundesanstalt für Landwirtschaft und Ernährung. Available online at https://www.ble.de/DE/Themen/Ernaehrung-Lebensmittel/Vermarktungsnormen/Obst-Gemuese/obst-gemuese_node.html, updated on 1/13/2021.

BMELV (2013): Zahl der Woche. Available online at http://www.bmel.de/SharedDocs/Pressemitteilungen/2013/232-Zahl-der-Woche-Obstverbrauch.html?nn=312878, checked on 9/11/2017.

Brown, S. (2012): Apple. In Marisa Luisa Badenes, David H. Byrne (Eds.): *Fruit Breeding*. Boston, MA: Springer US, pp. 329–367.

Brown, S. K.; Maloney, K. E. (2003): Genetic Improvement of Apple. Breeding, Markers, Mapping and Biotechnology. In David Curtis Ferree, Ian J. Warrington (Eds.): *Apples. Botany, production, and Uses*. New York: CABI Publisher, pp. 31–59.

BS (2020): National Listing. Admission to the National List. Bundessortenamt. Available online at https://www.bundessortenamt.de/bsa/en/variety-testing/national-listing, updated on 12/12/2020.

Bullock, D. W.; Wilson, W. W.; Neadeau, J. (2021): Gene Editing Versus Genetic Modification in the Research and Development of New Crop Traits: An Economic Comparison. In *American Journal of Agricultural Economics* 103 (5), pp. 1700–1719. DOI: 10.1111/ajae.12201.

BZL (2020): Äpfel. Bundesinformationszentrum Landwirtschaft. Available online at https://www.landwirtschaft.de/landwirtschaftliche-produkte/wie-werden-unsere-lebensmittel-erzeugt/pflanzliche-produkte/aepfel, updated on 8/27/2020, checked on 5/11/2021.

Cheema, D. S.; Dhaliwal, M. S. (2005): Hybrid Tomato Breeding. In *Journal of New Seeds* 6 (2–3), pp. 1–14. DOI: 10.1300/J153v06n02_01.

Clark, J. R.; Aust, A. B.; Jondle, R. J. (2012): Intellectual Property Protection and Marketing of New Fruit Cultivars. In Marisa Luisa Badenes, David H. Byrne (Eds.): *Fruit Breeding*. Boston, MA: Springer US, pp. 69–96.

Clark, J. R.; Jondle, R. J. (2008): Intellectual Property Rights for Fruit Crops. In James F. Hancock (Ed.): *Temperate Fruit Crop Breeding*. Dordrecht: Springer Netherlands, pp. 439–455.

Damos, P.; Colomar, L.-A. E.; Ioriatti, C. (2015): Integrated Fruit Production and Pest Management in Europe: The Apple Case Study and How Far We Are from the Original Concept? In *Insects* 6 (3), pp. 626–657. DOI: 10.3390/insects6030626.

Delate, K.; McKern, A.; Turnbull, R.; Walker, J. T.; Volz, R.; White, A. et al. (2008): Organic Apple Systems: Constraints and Opportunities for Producers in Local and Global Markets: Introduction to the Colloquium. In *HortScience* 43 (1), pp. 6–11. DOI: 10.21273/HORTSCI.43.1.6.

Dirksmeyer, W. (Ed.) (2009): Status quo und Perspektiven des deutschen Produktionsgartenbaus. Johann Heinrich von Thünen-Institut. Braunschweig: vTI (Landbauforschung Sonderheft, 330).

European Commission (2022): The apple market in the EU. Vol. 1: production, areas and yields. Available online at https://agriculture.ec.europa.eu/data-and-analysis/markets/overviews/market-observatories/fruit-and-vegetables/pip-fruit-statistics_en, checked on 8/14/2022.

Fritsche, S.; Poovaiah, C.; MacRae, E.; Thorlby, G. (2018): A New Zealand Perspective on the Application and Regulation of Gene Editing. In *Frontiers in Plant Science* 9, p. 1323. DOI: 10.3389/fpls.2018.01323.

Gallardo, R. K.; Garming, H. (2017): The Economics of Apple Production. In Gayle M. Volk, Amit Dhingra, Sally A. Bound, Dugald C. Close, Peter M. Hirst, M. C. Goffinet et al. (Eds.): *Achieving Sustainable Cultivation of Apples*. 1st ed. Cambridge: Burleigh Dodds Science Publishing (Burleigh Dodds Series in Agricultural Science), pp. 485–510.

Garming, H.; Dirksmeyer, W.; Bork, L. (2018): Entwicklungen des Obstbaus in Deutschland von 2005 bis 2017: Obstarten, Anbauregionen, Betriebsstrukturen und Handel. Braunschweig, Germany (Thünen working paper).

Granatstein, D.; Peck, G. (2017): Assessing the Environmental Impact and Sustainability of Apple Cultivation. In Gayle M. Volk, Amit Dhingra, Sally A. Bound, Dugald C. Close, Peter M. Hirst, M. C. Goffinet et al. (Eds.): *Achieving Sustainable Cultivation of Apples*. 1st ed. Cambridge: Burleigh Dodds Science Publishing (Burleigh Dodds Series in Agricultural Science), pp. 523–549.

Hanke, M.-V.; Flachowsky, H. (2017): *Obstzüchtung und wissenschaftliche Grundlagen*. Berlin, Heidelberg: Springer Berlin Heidelberg.

Howard, N. P.; Luby, J. J.; van de Weg, E.; Durel, C.-E.; Denancé, C.; Muranty, H. et al. (2021): Applications of SNP-based Apple Pedigree Identification to Regionally Specific Germplasm Collections and Breeding Programs. In *Acta Horticulturae* (1307), pp. 231–238. DOI: 10.17660/ActaHortic.2021.1307.36.

IFOAM (2014): The IFOAM Norms for Organic Production and Processing. Version 2014. Available online at http://www.ifoam.bio/sites/default/files/ifoam_norms_version_july_2014.pdf, checked on 3/23/2021.

Interview BS (2018): Qualitative telephone interview with a director from the Bundessortenamt. Transcription from the original language, Germany.

Isik, F.; Kumar, S.; Martínez-García, P. J.; Iwata, H.; Yamamoto, T. (2015): Acceleration of Forest and Fruit Tree Domestication by Genomic Selection. In *Advances in Botanical Research* 74, pp. 93–124. DOI: 10.1016/bs.abr.2015.05.002.

Kellerhals, M.; Schütz, S.; Baumgartner, I. O.; Andreoli, R.; Gassmann, J.; Bolliger, N. et al. (2018): Broaden the genetic basis in apple breeding by using genetic resources. In FOEKO (Ed.): *Ecofruit. Proceedings of the 18th International Congress on Organic Fruit Growing*. Ecofruit. Hohenheim, Germany: Fördergemeinschaft Ökologischer Obstbau e.V. (FOEKO), pp. 12–18.

Kienzle, J.; Kelderer, M. (2017): Growing organic apples in Europe. In Gayle M. Volk, Amit Dhingra, Sally A. Bound, Dugald C. Close, Peter M. Hirst, M. C. Goffinet et al. (Eds.):

Achieving Sustainable Cultivation of Apples. 1st ed. Cambridge: Burleigh Dodds Science Publishing (Burleigh Dodds Series in Agricultural Science), pp. 551–578.

Knight, V. H.; Evans, K. M.; Simpson, D. W.; Tobutt, K. R. (2005): Report on a desktop study to investigate the current world resources in Rosaceous fruit breeding programmes. East Malling Research, New Road East Malling Kent ME19 6BJ.

Laurens, F. (1998): Review of the Current Apple Breeding Programmes in the World. Objectives for Scion Cultivar Improvement. In *Acta Hortic.* (484), pp. 163–170. DOI: 10.17660/ActaHortic.1998.484.26.

Laurens, F.; Pitiot, C. (2003): French Apple Breeding Program. A New Partnership Between INRA and the Nurserymen of NOVADI. In *Acta Hortic.* (622), pp. 575–582. DOI: 10.17660/ActaHortic.2003.622.61.

Legun, K. A. (2015): Club apples: a biology of markets built on the social life of variety. In *Economy and Society* 44 (2), pp. 293–315. DOI: 10.1080/03085147.2015.1013743.

Messmer, M.; Wilbois, K.-P.; Baier, C.; Schäfer, F.; Arncken, C.; Drexler, D.; Hildermann, I. (2015): *Plant Breeding Techniques. An Assessment for Organic Farming.* Frick: Forschungsinstitut für biologischen Landbau FIBL.

Miedaner, T. (2017): *Grundlagen der Pflanzenzüchtung.* Frankfurt am Main: DLG-Verlag.

Mordor Intelligence (2022): Fresh Apple Market - Growth, Trends, COVID-19 Impact, and Forecasts (2022–2027).

New Zealand Apples & Pears (2021): Industry Scheme. Available online at https://www.applesandpears.nz/About_Us/About_Us/Industry_Scheme, checked on 8/14/2022.

Norer, R. (Ed.) (2018): *Handbuch zum Agrarrecht.* 1. Auflage. Bern: Stämpfli Verlag.

Nuijten, E.; Messmer, M.; Lammerts van Bueren, E. (2017): Concepts and Strategies of Organic Plant Breeding in Light of Novel Breeding Techniques. In *Sustainability* 9 (1), p. 18. DOI: 10.3390/su9010018.

Plant & Food Research (2022): Our history. Available online at https://www.plantand-food.com/en-nz/our-history, checked on 8/14/2022.

SINUS (2017): Studie zum Apfel: Die Hälfte hat schon einmal Äpfel vom Nachbarsbaum gepflückt. SINUS-Institut. Available online at https://www.sinus-institut.de/veroeffentli-chungen/meldungen/detail/news/studie-zum-apfel-die-haelfte-hat-schon-einmal-aepfel-vom-nachbarsbaum-gepflueckt/news-a/show/news-c/NewsItem/, updated on 4/8/2021.

Stokke, O. S.; Oberthür, S. (2011): *Managing Institutional Complexity. Regime Interplay and Global Environmental Change.* Cambridge, MA: MIT Press (Institutional dimensions of global environmental change).

Stowe, E.; Dhingra, A. (2021): Development of the Arctic® Apple. In Irwin Goldman (Ed.): *Plant Breeding Reviews.* Hoboken, NJ: Wiley, pp. 273–296.

Tschersich, J. (2021): Norm Conflicts as Governance Challenges for Seed Commons: Comparing Cases from Germany and the Philippines. In *Earth System Governance* 7, p. 100097. DOI: 10.1016/j.esg.2021.100097.

United States Department of Agriculture (2021): Fresh Apples, Grapes, and Pears: World Market and Trade. Available online at https://downloads.usda.library.cornell.edu/usda-esmis/files/1z40ks800/nc581j510/zw130269m/fruit.pdf, checked on 8/14/2022.

WUR (2022): About us. Wageningen University & Research. Available online at https://www.wur.nl/en/Research-Results/Research-Institutes/plant-research/Plant-Breeding/About-us.htm, checked on 8/14/2022.

Zhu, Z.; Jia, Z.; Peng, L.; Chen, Q.; He, L.; Jiang, Y.; Ge, S. (2018): Life Cycle Assessment of Conventional and Organic Apple Production Systems in China. In *Journal of Cleaner Production* 201, pp. 156–168. DOI: 10.1016/j.jclepro.2018.08.032.

Part II

Resilience in Fruit Breeding and Cultivation

Part II

Introduction

In this part, an in-depth look is taken at the social-ecological resilience of fruit breeding and cultivation. The aim is to understand and conceptualize fruit breeding and cultivation as social-ecological systems and elaborate on how to design these systems in a resilient way.

This chapter especially draws on literature from the research communities on social-ecological resilience (Folke et al. 2010; Walker and Salt 2012; Biggs et al. 2015), ecosystem services in the context of agriculture and fruit cultivation (Zhang et al. 2007; Dale and Polasky 2007; Kremen and Miles 2012; Demestihas et al. 2017), and food systems (Cabell and Oelofse 2012; Tendall et al. 2015). Resilient agricultural systems have the following overarching goals:

> A resilient agriculture is one that meets both food and development needs over both the short- and very long-terms, from local to global scales, without destabilizing the Earth system. [...] [It] explicitly allows for adaptive changes or transformations to meet evolving environmental conditions and human needs.
>
> (Bennett et al. 2014)

This is also valid for fruit breeding and cultivation. With the inclusion of these systems into this debate – being largely neglected so far – general and specific challenges and promising system attributes can be identified and discussed. This offers the research communities mentioned above and practitioners further learning opportunities for designing resilient agricultural systems.

First, fruit breeding and cultivation are described with the logic of social-ecological systems research (See Chapter 4). This conceptualization enables the uncovering of interrelations and feedback mechanisms between both systems. Subsequently, general ecosystem services in agricultural systems as well as in fruit breeding and cultivation systems are evaluated (See Chapter 5). Because resilient systems aim to sustain a desired set of ecosystem services, this chapter aims to define this particular set in the context of fruit breeding and cultivation. Finally, characteristics of resilient fruit breeding and cultivation systems are discussed (See Chapter 6) based on the *Principles for Building Resilience* (Biggs et al. 2015).

DOI: 10.4324/9781003355724-7

This provides the basis for the development of an analytical framework to evaluate the general resilience of fruit breeding systems.

References

Bennett, E.; Carpenter, S. R.; Gordon, L. J.; Ramankutty, N.; Balvanera, P.; Campbell, B. et al. (2014): Toward a More Resilient Agriculture. In *The Solutions Journal* 5 (5), pp. 65–75.

Biggs, R.; Schlüter, M.; Schoon, M. L. (Eds.) (2015): *Principles for Building Resilience. Sustaining Ecosystem Services in Social-Ecological Systems.* Cambridge, UK: Cambridge University Press.

Cabell, J. F.; Oelofse, M. (2012): An Indicator Framework for Assessing Agroecosystem Resilience. In *Ecology and Society* 17 (1). DOI: 10.5751/ES-04666–170118.

Dale, V. H.; Polasky, S. (2007): Measures of the Effects of Agricultural Practices on Ecosystem Services. In *Ecological Economics* 64 (2), pp. 286–296. DOI: 10.1016/j.ecolecon.2007.05.009.

Demestihas, C.; Plénet, D.; Génard, M.; Raynal, C.; Lescourret, F. (2017): Ecosystem Services in Orchards. A Review. In *Agronomy for Sustainable Development* 37 (2), p. 581. DOI: 10.1007/s13593-017-0422-1.

Folke, C.; Carpenter, S.; Walker, B.; Scheffer, M.; Chapin, T.; Rockström, J. (2010): Resilience Thinking: Integrating Resilience, Adaptability and Transformability. In *Ecology and Society* 15 (4). Available online at http://www.ecologyandsociety.org/vol15/iss4/art20/.

Kremen, C.; Miles, A. (2012): Ecosystem Services in Biologically Diversified Versus Conventional Farming Systems. Benefits, Externalities, and Trade-Offs. In *Ecology and Society* 17 (4). DOI: 10.5751/ES-05035–170440.

Tendall, D. M.; Joerin, J.; Kopainsky, B.; Edwards, P.; Shreck, A.; Le, Q. B. et al. (2015): Food System Resilience. Defining the Concept. In *Global Food Security* 6, pp. 17–23. DOI: 10.1016/j.gfs.2015.08.001.

Walker, B.; Salt, D. (2012): Resilience Thinking. Sustaining Ecosystems and People in a Changing World. Washington, DC: Island Press.

Zhang, W.; Ricketts, T. H.; Kremen, C.; Carney, K.; Swinton, S. M. (2007): Ecosystem Services and Dis-services to Agriculture. In *Ecological Economics* 64 (2), pp. 253–260. DOI: 10.1016/j.ecolecon.2007.02.024.

4 Fruit Breeding and Cultivation as a Social-Ecological System

The perspective of social-ecological systems is an established concept in scientific research and crucial for understanding and shaping complex and interlinked dynamics in the interaction of ecosystems and humans (Herrero-Jáuregui et al. 2018; Fischer et al. 2015; Young et al. 2006). For a better understanding, Figure 4.1 shows a recap of the basic model of social-ecological systems as introduced in Chapter 1.

Fischer et al. (2015) identify four major advances of the perspective's integration in sustainability research. First, it recognizes the dependence of human well-being on ecosystem services. Second, it fosters interdisciplinary research as a necessary element to understand the multiple characteristics of sustainability problems. Third, it additionally promotes pluralism in conceptuality and methodology by combining qualitative and quantitative approaches, which leads to a more holistic view of systems. Fourth, it has the potential to influence policymakers by clearly recognizing social-ecological interactions rather than making isolated policy decisions for social or ecological aspects.

Because the social-ecological systems' perspective lays the conceptual groundwork for social-ecological resilience (See Section 1.1), it is necessary to integrate fruit breeding and cultivation into this perspective. Hence this section provides

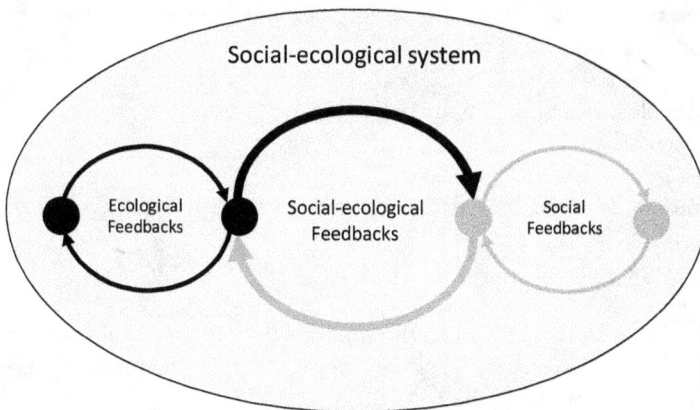

Figure 4.1 Recap. Basic model of social-ecological systems (based on Biggs et al. 2015).

DOI: 10.4324/9781003355724-8

an overview of general theoretical and empirical conceptualizations of social-ecological systems and, subsequently, a conceptualization of fruit breeding and cultivation as social-ecological systems.

4.1 Basic Theoretical and Empirical Conceptualizations

The general concept of social-ecological systems adopts an anthropocentric perspective. One of the most acknowledged general definitions of social-ecological systems (Herrero-Jáuregui et al. 2018, p. 5) is the one by Berkes and Folke (1998b) who describe social-ecological systems as nested multilevel systems that provide essential ecosystem services for human well-being. Research on social-ecological systems is growing fast since 2006[1] and the concept as well as its theoretical foundation is still broadly discussed: a general definition of social-ecological systems does not exist (Herrero-Jáuregui et al. 2018). This could partly be due to the context-specificity of social-ecological systems (Walker and Salt 2012), which makes it challenging to define them coherently. As a term, the social-ecological system is synonymous with the socio-ecological system or socio-ecosystem, which are also applied in research on this topic (Herrero-Jáuregui et al. 2018).

However, the basic understanding of social-ecological systems in the context of resilience was already introduced in Section 1.1. Social and ecological systems are linked through feedback mechanisms. Beyond this basic understanding, it is important to note that the study of these systems is in most cases combined with the study of ecosystem services (Herrero-Jáuregui et al. 2018), as it is also the case for social-ecological resilience (Biggs et al. 2015; Folke 2016). Social-ecological systems provide or influence ecosystem services through reciprocal feedbacks between social and ecological variables in the social-ecological core of the respective system. Probably the most illustrating and simple example is agriculture (as the social-ecological system) where humans create provisioning services (food) in their interaction with the ecosystem (land). Agriculture is carried out at the social-ecological core.

Based on that, the basic definition of social-ecological systems in the context of this book includes the following points:

1. Social and ecological systems are linked and
2. influence each other
3. through feedback mechanisms,
4. that culminate in complex, dynamic, and reciprocal social-ecological interactions,
5. which create and influence ecosystem services.

This perspective lies in stark contrast to the traditional view of natural resource systems, in which linkages and feedbacks are neglected and dynamics are perceived as linear and monotonic (Schlüter et al. 2012, p. 222).

Numerous frameworks have been developed to study social-ecological systems. According to Binder et al. (2013), these frameworks can be categorized with the following three aspects: their conceptualization of the relationship between the social and ecological system (bi- or unidirectional); their perspective on the ecological system (ecocentric or anthropocentric); and the purpose of the framework (action- or analysis-oriented). For a better understanding, two established frameworks will be shortly described: The SESF by Ostrom (2009) and Turners Vulnerability Framework (TVUL) by Turner et al. (2003).

The SESF is a unidirectional (reciprocity between the social and ecological system), anthropocentric, and analysis-oriented framework. This detailed framework explains certain variables in social-ecological systems (e.g., resource system, governance system, actors) to "provide a common language for case comparison" (Binder et al. 2013, p. 6) in the field of natural resource systems. All variables are ordered in a multitier hierarchy. By selecting and describing relevant variables for a specific social-ecological system, (un-)sustainable outcomes in the human management of this system (e.g., fishery or forestry) can be analyzed and explained. On the contrary, TVUL is a bidirectional (the social system influences the ecological system), anthropocentric, and action-oriented framework. In short, TVUL serves to "analyze[s] who and what are vulnerable to multiple environmental and human changes, and what can be done to reduce these vulnerabilities" (ibid.), always with regard to a specific location. It is specifically designed for the vulnerability analysis of communities in developing countries. The comparison of TVUL and the SESF illustrates the general differences between frameworks for the study of social-ecological systems very well: both investigate these systems, but with different conceptualizations and purposes.

Binder et al. (2013) conclude that not one single framework has the potential to be used for every social-ecological system. Rather, a framework should be chosen based on the problem context, the specific conceptualization of the social-ecological system, and the research aim (Binder et al. 2013). This corresponds to the above-described context-specificity of social-ecological systems.[2]

It follows that although a full range of different frameworks for the study of social-ecological systems exist, they remain partly isolated from one another (Cumming 2014) and concurrently cannot be integrated under one overarching theoretical framework. This makes the selection of a specific framework for the conceptualization of fruit breeding and cultivation as social-ecological systems challenging. Moreover, the purpose of this conceptualization primarily is (a) to get a better understanding of interlinkages between breeding and cultivation, and (b) to conceptually integrate these insights into research on social-ecological systems as a necessary step to evaluate resilience. The frameworks described above are either too context-specific or detailed for an application. A simplified conceptualization using a more general, analysis-oriented framework proves sensible.

One of the first analysis-oriented frameworks in the study of social-ecological systems is the analytical framework for analyzing interlinkages of social-ecological systems in the context of resilience and sustainability by Berkes and Folke (1998b). It is illustrated in Figure 4.2.

This framework is anthropocentric and unidirectional. The authors define four sets of elements that describe the linked system characteristics: ecosystem,

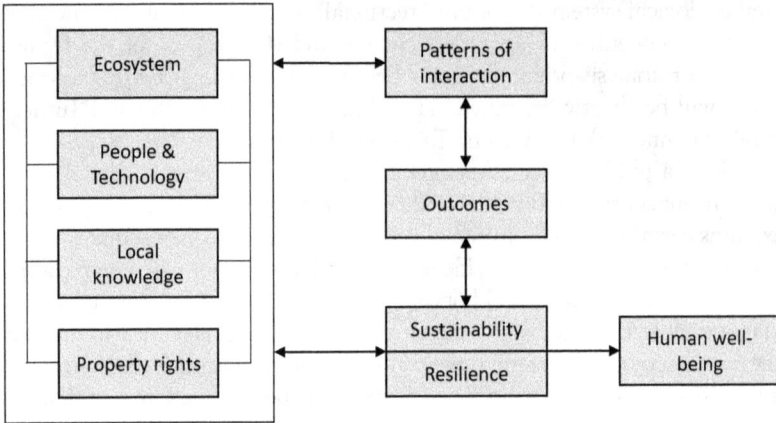

Figure 4.2 Basic analytical framework for social-ecological systems (based on Berkes and Folke 1998b).

people and technology, local knowledge, and property rights. *Ecosystem* refers to the physical and biological characteristics and processes of the ecological system, whereby their significances vary. *People and technology* describe the relevant user community and the technology they use for managing the resources of the ecological system. *Local knowledge* is especially important for indigenous groups and historically continued communities as it describes specific (local) knowledge about the environmental management of resources. *Property rights* deal with the type(s) of property regime(s) – private, state, or common property – that are present in the social-ecological system. Similar to the IAD or SESF (See Section 1.3), the set of elements is shaped by dynamic patterns of interaction. This leads to outcomes that ultimately define the sustainability and resilience of the social-ecological system and, hence, human well-being.

However, although these elements may be able to categorize fruit breeding and cultivation as local social-ecological systems, they miss to capture an important characteristic: although agricultural systems such as fruit cultivation are per se social-ecological systems (Herrero-Jáuregui et al. 2018), breeding and cultivation are not isolated activities or purely spatial-dependent social-ecological systems.[3] Agricultural systems are embedded parts of larger food systems (Ericksen 2008; Urruty et al. 2016). *Food systems* as social-ecological systems are comprehensively defined by Tendall et al. (2015):

> They comprise, at a minimum, the activities involved in food production, processing and packaging, distribution and retail, and consumption. These activities encompass social, economic, political, institutional and environmental processes and dimensions, referred to as scales. The processes play out at different [...] positions on a scale. [...] [A]ctivities lead to a number

of social and environmental outcomes [...]. Food system activities and outcomes eventually result in processes that [...] may lead to unintended consequences.

(p. 18)

This food system approach addresses the complexity of food systems.[4] Food systems are exposed to global change processes and unexpected shocks. In many aspects, some parts of food systems are influenced by other parts although there are no connections at first sight. For example, local farmers producing food on their fields can be affected by national or international policies pushing biofuels, for example, through subsidies. These policy changes on a totally different scale could influence the local farmers to grow maize or rapeseed on their fields as inputs for biogas plants. Eventually, this complexity is described as telecoupling.

Telecoupling is a rather new concept in the social sciences and refers to social or environmental processes between spatially, socially, or institutionally distant social-ecological systems (Hull and Liu 2018). Telecoupling processes involve five elements: systems, agents, flows, causes, and effects (Liu et al. 2013; Hull and Liu 2018). Figure 4.3 illustrates this conceptual model.

Systems interact by receiving or sending *flows*, for example, of information or goods. Within and across these systems, *agents* potentially direct, influence, and observe these flows. Understanding the *causes* and *effects* of specific flows provides valuable insights into the sustainability and resilience of telecoupled systems. For example, Gordon et al. (2017) conceptualize global food systems as coupled social-ecological systems,[5] focusing on feedbacks between important variables that concern production and consumption. They conclude that while in the 1960s strong links between local production and consumption existed (flow of matter, specifically food), today these links are weak and processes on a larger, global scale are more important. These changes in key feedback mechanisms, mainly in trade and consolidation, and the followed dissociation of producers and consumers

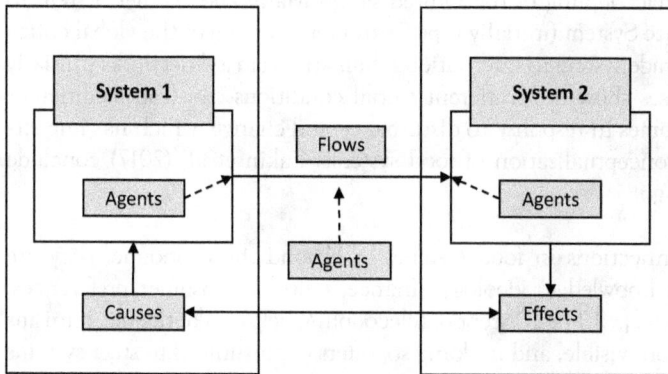

Figure 4.3 Conceptual model of telecoupling (based on Liu et al. 2013).

reduce the resilience of the biosphere as a whole. This connects with the findings of Krausmann and Langthaler (2019) who describe the current regime (since 1986/1995) as the corporate food regime. In this regime, the rising land use and intensification of agriculture in some areas contradict greening developments in other areas.

Eakin et al. (2017) specifically conducted research on telecoupled food systems with a focus on governance challenges. They redefine the boundaries of food systems by looking at problem-determined systems (e.g., food waste) rather than system-determined problems (e.g., focus on certain value chains). Taking this perspective allows for the identification of telecoupling processes between systems. Consequently, they propose a typology of four initial conditions under which distant processes and actors become telecoupled. The purpose of this typology is to introduce analytical categories for the definition of boundaries in social-ecological systems and the relationships between actors and resources:

- Type 1 describes two or more physically distant systems that share no social or institutional ties, meaning that no flows between actors and resources of the systems exist.
- Type 2 describes two or more physically distant systems that share some social or institutional ties, for example, through trade agreements. Here, institutional arrangements lead to flows of resources or information.
- Type 3 describes two overlapping social-ecological systems that share the same resource base but have no or few social or institutional ties.
- Type 4 describes two overlapping social-ecological systems in which no physical distance between actors and a common resource base exist, but very few institutional ties are in place.

The effects of telecoupling on these initial states can be observed when change or disturbances happen. Depending on the initial condition, telecoupling processes activate, change, or emerge. In their article, Eakin et al. (2017) analyze two cases of telecoupling: telecoupling of the United States Maize and Biofuel System to the Mexican Maize System (initially type 2) and telecoupling of the global coffee production and trade system to international humanitarian aid networks (initially type 3). Both cases show how different initial conditions can lead to different governance outcomes in response to disturbance and change, which has impacts on the general conceptualization of food systems. Eakin et al. (2017) conclude with the following:

> The interconnections in food systems go beyond the economic: they are grounded in knowledge, ideology, finance, culture, consumer preferences, and ways of life. […] The concept of telecoupling helps make the mechanisms of change more visible, and in doing so, offers opportunities to steer systems to more sustainable outcomes.
>
> (p. 14)

Following these insights, the definition of social-ecological systems is extended for the context of food systems (added phrases):

1. Social and ecological systems are linked **and coupled across scales and distances,**
2. **adjust and change in response to global changes or disturbances and**
3. influence each other
4. through feedback mechanisms **or flows,**
5. that culminate in complex, dynamic, and reciprocal social-ecological interactions
6. that create and influence ecosystem services.

As a result, both the place-based character of fruit breeding and cultivation as well as its integration into telecoupled food systems have to be recognized in their conceptualization as social-ecological systems. For this undertaking, the analytical framework by Berkes and Folke (1998a) will serve as a conceptual grounding. The variable *property rights* will be replaced by the variable *institutions*, as these usually determine property rights and subsequently include them in the terminology. Because social-ecological resilience is the analytical dimension of sustainability in the context of this book, both terms are not understood as separate entities in this frame. Additionally, above-described insights about telecoupled food systems need to be included in the conceptualization. Figure 4.4 illustrates the adjusted analytical framework.

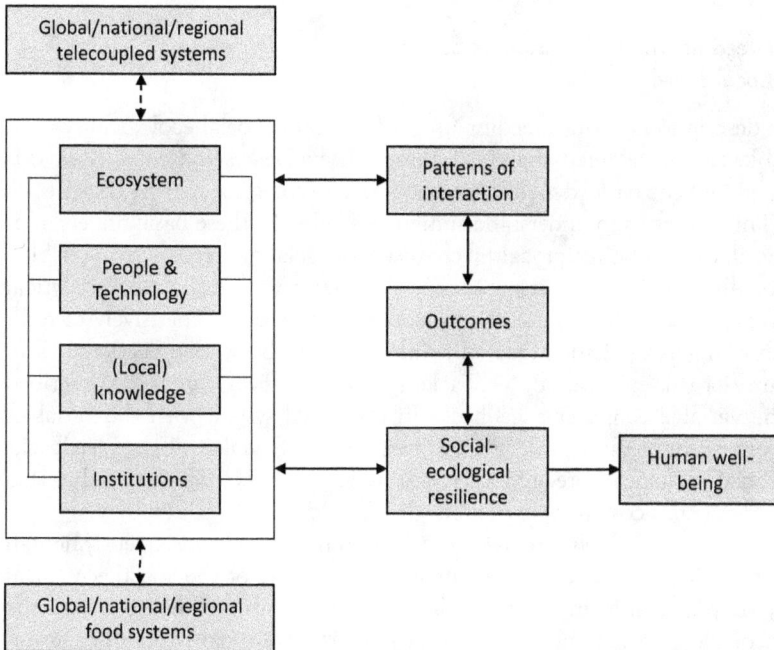

Figure 4.4 Adjusted basic analytical framework for social-ecological systems (own figure).

Fruit breeding and cultivation can be categorized as type 3 (Eakin et al. 2017) regarding their initial condition: Both social-ecological systems have common resource bases (e.g., fruit cultivars or budwood), but only some social and institutional ties. The following section will dive deeper into these issues.

4.2 Application of Fruit Breeding and Cultivation

The conceptualization of fruit breeding and cultivation as social-ecological systems is a theoretical one, based on the insights of the previous section. Although it periodically draws on practical background information on apples (the empirical context of this book, See Chapter 3) for illustrative reasons, it is not an empirical analysis of fruit breeding and cultivation in Germany. Rather, the conceptualization provides a general integration of this topic into the conceptual frame of social-ecological systems research. Thus, the interlinkages between breeding and cultivation as well as their contextualization in a broader multilevel system will become visible.

A stepwise approach is chosen for a consistent description and conceptualization of fruit breeding and cultivation as social-ecological systems. Each foregoing step lays the base for the next step. Breeding and cultivation are described (1) as social-ecological systems at local levels; (2) as overlapping activities at the local level (special case); (3) as telecoupled social-ecological systems in a multilevel system[6]; and (4) as telecoupled social-ecological systems in a multilevel system that is embedded in a multilevel food system.

4.2.1 Breeding and Cultivation as Social-Ecological Systems at Local Levels

For the description of fruit breeding and cultivation as social-ecological systems at local levels, the adapted analytical framework by Berkes and Folke (1998a) is integrated into the basic, dynamic model of social-ecological systems (Biggs et al. 2015). This allows us to understand similarities between these basic understandings and illustrate the reciprocal mechanisms of social and ecological variables. Additionally, this integration puts the rather static variables of the analytical framework in a dynamic context, connecting it more comprehensively to resilience thinking as the basic model of thinking in this book (See Section 1.1). In the course of this integration, feedbacks in the ecological system are described with the variable ecosystem; feedbacks in the social system with the variables people and technology, knowledge, and institutions. Social-ecological feedbacks in the social-ecological core are illustrated by the patterns of interaction that lead to outcomes (e.g., ecosystem services).

These outcomes influence social-ecological resilience and consequently human well-being. Whereas the outcomes themselves are part of the social-ecological system, they not only influence resilience and human well-being inside the social-ecological system but also outside the social-ecological system. For example, the reduction of chemical or biological plant protection treatments in the cultivation of robust cultivars does not only affect the resilience of the agricultural social-ecological system. It also affects the resilience of agriculture itself, as this particular social-ecological system is part of a larger multilevel system (See Sections 4.2.3 and 4.2.4) and has a role model function. Hence, both variables lie at the intersection of the social-ecological system and the sphere outside. Figure 4.5 illustrates this conceptual integration.

4.2.1.1 Fruit Cultivation as a Social-Ecological System

In the ecological system, fruit cultivation takes place in orchards, in which the long-term perennial trees are typically grown on dwarfing rootstocks for an efficient management.[7] Orchards additionally function as ecological habitats for a wide range of flora (e.g., weeds, herbs) and fauna (e.g., insects) as well as different benign and beneficial pests, for example, certain species of fungi. In the social system, the cultivation of fruit trees in this ecosystem is carried out by farmers with certain technical resources relevant to farming practice. Usually, manual, chemical, or biological plant protection treatments are carried out in the cultivation process. Farmers manage their orchards with their farming knowledge of environmental processes and respective measures for cultivation. By practicing agriculture, farmers are bound to certain institutions. These include the political-legal framework[8] and possibly rules or norms that result from memberships in certain associations, such as organic agriculture certification schemes. While the land on which the orchard is grown as well as the used technology is usually private property of the respective farmer, property rights are sometimes limited concerning the grown cultivars.[9] The interaction of the social and the ecological system results in a range of different outcomes, mainly ecosystem services or disservices, which are described in detail in Chapter 5. Depending on the relation between these services and disservices,

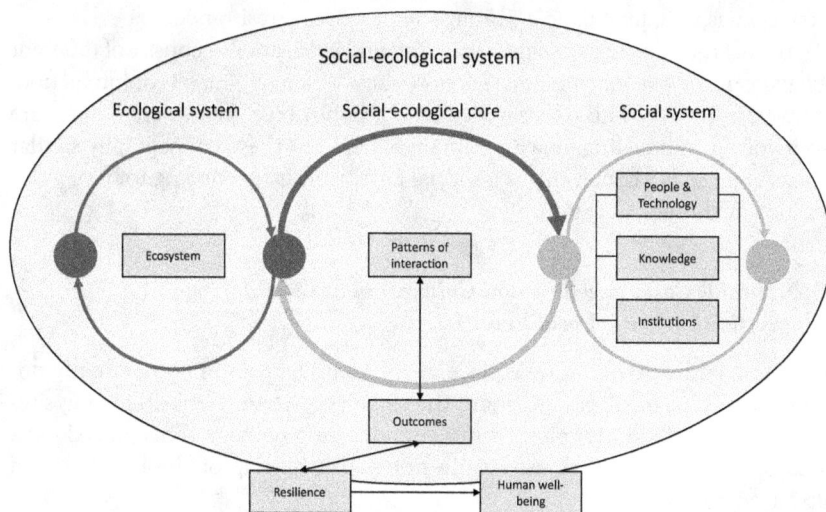

Figure 4.5 Fruit breeding and cultivation at local levels (own figure).

social-ecological resilience is influenced accordingly, which finally increases or decreases human well-being. Overall, fruit cultivation at the local level takes place in a classical place-based social-ecological system.

4.2.1.2 Fruit Breeding as a Social-Ecological System

Breeding of fruits typically takes place in natural ecosystems such as open fields, or in controlled ecosystems, for example, greenhouses. In these environments, different fruit cultivars are crossed based on distinct breeding goals. The resulting seeds are harvested and planted in the form of seedlings. All seedlings go through several selection processes and the most promising ones are then tested in trial areas.[10] Plant breeders who hold the necessary knowledge of the breeding process carry out the crossings and selections. Here, they use several manual tools (e.g., brushes for pollination) as well as possibly plant protection treatments. New technological instruments for the conduction of plant breeding, such as marker-assisted breeding technologies, are becoming more and more relevant (See Section 3.2). In comparison to cultivation, few rules and norms exist for fruit breeding in the political-legal framework. With the mechanism of the breeder's exemption (See Section 3.1), breeders have (at least theoretically) legal access to all existing fruit cultivars – whether or not they are private property. Different organizations provide access to these cultivars, for example, gene banks. In the interaction of the social and ecological system, it becomes very clear that humans are directly shaping and developing natural resources according to their needs and purposes. Just as in fruit cultivation, the outcomes of these interactions are ecosystem services and disservices. The most obvious desired outcome is a new fruit cultivar that can be introduced to the market and provides economic returns. The outcome potentially improves the social-ecological resilience of fruit breeding and cultivation as well, for example, through better disease resistance.

Both social-ecological systems – breeding and cultivation – consist of different ecological and social systems and therefore show different patterns of interaction. However, in cases of PPB (Ceccarelli 2012) or hobby breeding, where farmers are also involved in breeding, new interlinkages between these conceptually similar systems emerge. Although this is a special case, it seems promising to have a further look at this constellation.

4.2.2 Special Case: Breeding and Cultivation as Overlapping Activities at the Local Level

If breeding and cultivation take place at the same location, two nominally different social-ecological systems share the same ecosystem. Hence, growing sites for breeding are built and planted inside an existing orchard. This procedure is called on-farm breeding. Figure 4.6 illustrates this overlap of the breeding and cultivation system.

In this special case, farmers act as breeders (or work closely together with breeders), which results in feedback mechanisms between both social systems. One of

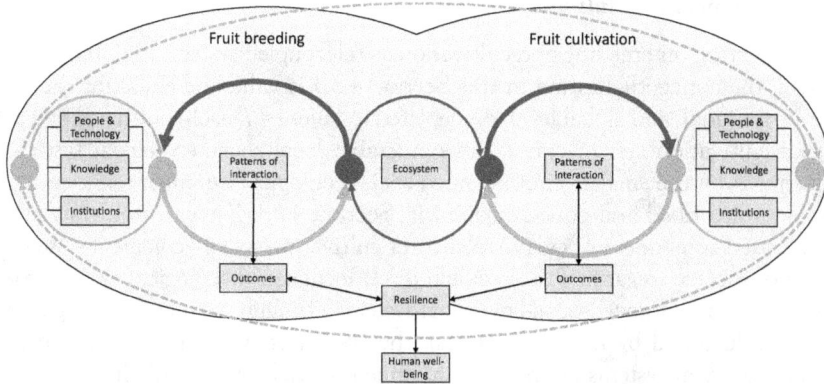

Figure 4.6 Fruit breeding and cultivation as overlapping activities at the local level (own figure).

the consequences of this linkage is that breeding and farming knowledge come together. This is more relevant in the breeding process than in the cultivation process because farmers can directly influence the breeding goals according to their needs and requirements. Institutions are also influencing each other. For example, if the farmer is a member of an organic farmers association, the rules for cultivation also apply to the seedlings in the breeding process. Hence, breeding and cultivation overlap not only in the ecological system, but also in both social systems, in which parts might be the same (farmers as breeders) or shared (farmers cooperate with breeders).

This coupling of breeding and cultivation has consequences for the outcomes of both social-ecological systems. When farmers participate in breeding activities, they usually aim to plant the potentially new cultivars in their orchards. Hence, while the outcomes of the current cultivation system show the farmers connecting factors for desirable changes, they try to induce those changes with the development and finally the planting of new fruit cultivars. For example, if an organic fruit farmer tries to minimize the use of biological plant protection products in cultivation while also getting a minimum level of yield, he/she probably aims for these attributes as outcomes of the breeding process.

On a meta-level, this coupling of both systems is also true for breeding and cultivation in general. Independently of doing breeding on-farm and without a restriction to the local level, both systems share some main ecological resources: fruit cultivars and their budwood. Although both systems can exist as distinct local systems, they are – in any case – linked by telecoupling mechanisms. For example, if a breeder develops a new cultivar, this cultivar can be approved on a national, multinational (EU), or even global level. In the case that this cultivar is approved on a national level, this has effects on all fruit farmers in this nation, because they theoretically have access[11] to this new cultivar – regardless of physical, social, or institutional distances.

4.2.3 Breeding and Cultivation as Telecoupled Social-Ecological Systems in a Multilevel System

By elaborating on breeding and cultivation as telecoupled systems in a multilevel system, the conceptualization in this Section 4.2.3 is valid and equal for the regional, national, and global level. As depicted in Figure 4.7, each multilevel breeding or cultivation system consists of numerable local social-ecological systems (illustrated by the smaller circles named social-ecological systems), representing the ones described and conceptualized in Section 4.2.1. They demonstrate the place-based component of every breeding or cultivation social-ecological system.

All or specific (regarding the regional level) local social-ecological systems together form a multilevel social-ecological system. Through distinct flows of goods or else (illustrated by the dotted lines), the system components of these local social-ecological systems are potentially linked to each other. This linkage can occur in various ways. For example, breeders in the multilevel breeding system can cooperate with each other via the exchange of breeding knowledge or seedlings. This links their social systems. Breeders can also share their ecosystems when different breeding groups conduct fruit breeding in the same location (e.g., at a Fruit-Growing Centre).

In the multilevel cultivation system, farmers can also cooperate with each other through an exchange of knowledge (e.g., on cultivation issues), which links their social systems. Distinct local systems from both domains (breeding and cultivation) can also be coupled as described in Section 4.2.2. In the graph, this special case is shown in the connected dark gray social-ecological systems. However, local social-ecological systems can also totally isolate themselves, in which case no interlinkages between social systems occur. This is illustrated by systems that are not connected with dotted lines to other systems.

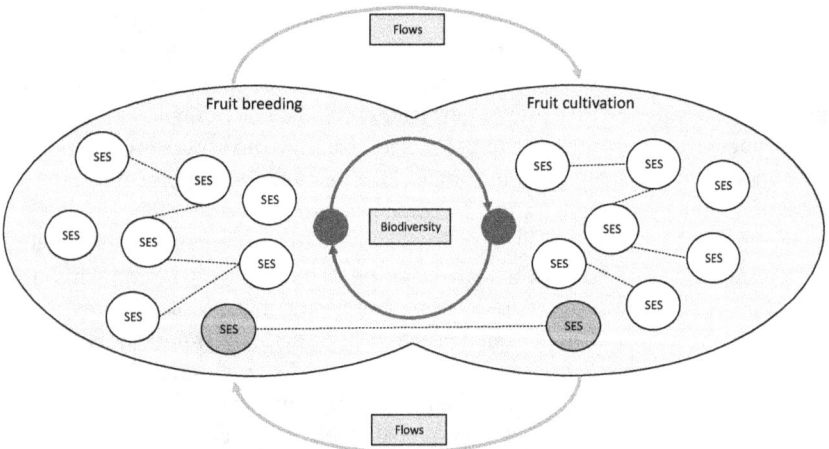

Figure 4.7 Fruit breeding and cultivation as telecoupled social-ecological systems (SES) in a multilevel system (own figure).

Although only a few social or institutional ties between breeding and cultivation exist in this multilevel perspective (See Section 3.2), both are potentially
linked through different flows as depicted in the graph. These could be cultural
(norms of organic farming) or economical (farmers are financing breeders). As
hinted in Section 4.2.2, the biophysical resources fruit cultivars and budwood –
aggregated in the term biodiversity – are in the center of the intersection between
the multilevel systems breeding and cultivation. Biodiversity is thus the main
connecting variable and the common resource base on all levels, emphasizing it
as an important aspect for the functioning of both systems.

These elaborations show that flows in different shapes occur between both multilevel systems as well as inside them. As telecoupling effects are latent and only
emerge or become visible in specific contexts, the systems show certain initial states
(See also Section 4.1). The initial condition of the telecoupled interactions between
multilevel breeding and cultivation systems fits to type 3 in the introduced typology
of Eakin et al. (2017): two overlapping social-ecological systems share the same
resource base (biodiversity) but have (no or) few social or institutional ties. This
initial condition also fits to the distinct multilevel system of breeding, in which few
social and institutional ties exist. For cultivation, the initial condition of type 2 fits
well, whether on a regional, national, or global level: the social-ecological systems
are physically distant (decentral orchards) but share some social (e.g., farming associations) or institutional (e.g., the political-legal framework) ties.

In conclusion, a complex web of interaction can be observed from this multilevel perspective. If one system causes flows, this affects other systems. For example, if breeders develop a new, promising cultivar, this could affect cultivation
systems on all levels if they start planting this new cultivar. History shows these
reciprocal interlinkages very well (Hanke and Flachowsky 2017). With the mainstreaming of chemical plant protection treatments in fruit cultivation in the
1930s, the breeding goal vitality was replaced by other goals like yield and shelf
life. With the usage of chemical fungicides, vitality was not as relevant as before
and breeding goals changed to other, more market-oriented traits. As a result,
low robustness in modern fruit cultivars was perceived as normal. This example
additionally shows telecoupling effects of the chemical sector on fruit cultivation
– a sector that is separate from the multilevel systems of breeding and cultivation.
Thus, not only breeding and cultivation is telecoupled to each other, but both systems are also integrated with larger food systems (because chemical products were
also introduced in the vegetable sector), which are then connected to other systems, such as the chemical industry. Section 4.2.4 tries to capture this complexity.

4.2.4 Breeding and Cultivation as Telecoupled Social-Ecological Systems in a Multilevel System, embedded in a Multilevel Food System

Breeding and cultivation are only single parts of the value chain in fruit production and consumption (See Section 3.2). Other parts of this chain interact with
breeding and cultivation, either through direct interlinkages (e.g., distribution

organizations where farmers are members), or telecoupled interlinkages (e.g., consumer preferences influence breeding goals). The overall value chain of fruit production is in turn only part of a larger multilevel food system. Here, other value chains like carrot or pear production exist and these overall different value chains can also be connected to each other. For example, a fruit farmer could cultivate pears in addition to apples. Further, as it was previously illustrated with the example of the chemical sector, other sectors are constantly influencing the food system. Similar to the local level, as explained in Section 4.2.1, all outcomes influence the resilience of the food system. The integration of multilevel breeding and cultivation in the multilevel food system is illustrated in Figure 4.8.

Overall, Sections 4.2.1–4.2.4 provide an integration of fruit breeding and cultivation into general concepts of social-ecological systems research as well as context-specific research on food systems and telecoupling. While it is always possible to go deeper in describing specific variables of food systems – like Ericksen (2008) did in her food systems framework – the described steps fulfill the purpose of showing the inextricable connection of fruit breeding and cultivation and their conceptual integration into multilevel systems.

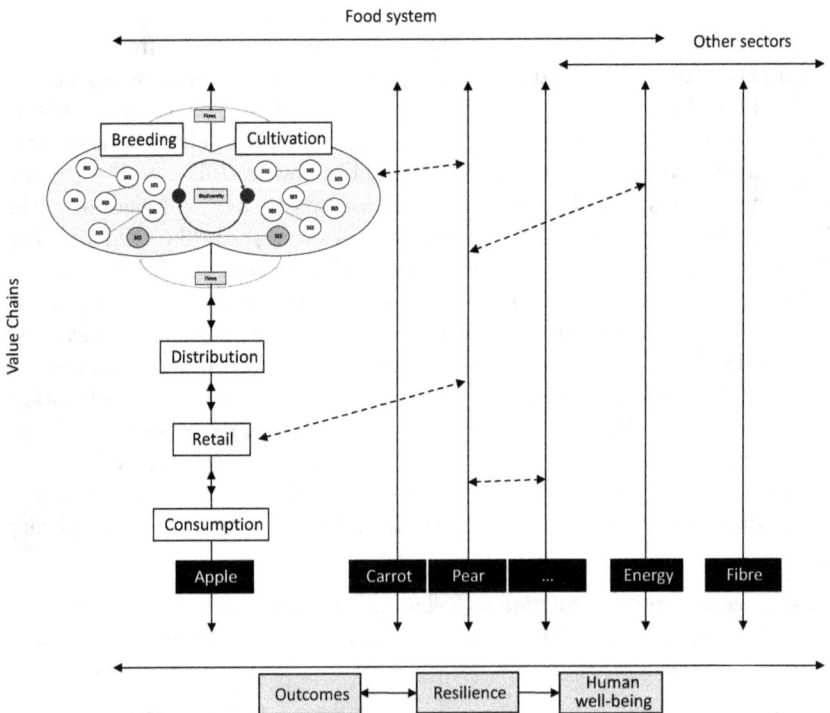

Figure 4.8 Fruit breeding and cultivation embedded in a multilevel food system (based on Tendall et al. 2015).

4.3 Conclusion for this Book

The complex set of outcomes in this conceptualization is the main aspect that decides the resilience of the whole food system. Influencing resilience in this dynamic system is therefore a complex undertaking. Tendall et al. (2015) identify three different levels as entry points for building resilience: (a) national or regional food systems, in which foremost policy interventions can change outcomes; (b) individual value chains such as fruit production from local to global levels, in which individual actors can trigger changes in outcomes; (c) interests and aspirations of individuals such as consumers, whose preferences influence outcomes.

As already explained previously, this book concentrates on the second entry point by conducting empirical research of apple breeding and cultivation in Germany. This reduction of complexity does not result in a disregard of the embeddedness of this national level in a multilevel system, but rather in a clearer focus and operationalization of the research. In the next chapter, the relevant outcomes of breeding and cultivation in the form of ecosystem services and disservices are described in detail.

Notes

1 Herrero-Jáuregui et al. (2018) link the fast growth in publications on social-ecological systems to the foundation of the Stockholm Resilience Centre in 2007 and the incorporation of the term in the Sustainable Development Goals (SDGs) in 2013.
2 In line with that, Cumming (2014) specifically calls for a stronger overarching theoretical framework for the study of social-ecological systems and therefore a more generally applicable framework. Such a framework would enable the consistent generalization of learned lessons from individual case studies and could make sound policy recommendations for managing social-ecological systems. By describing certain criteria of a scientifically well-founded theoretical framework, he proposes his ideal design of such a framework. However, he merely sees attempts for constructing such a framework, none meeting his proposed criteria satisfactorily.
3 Some conceptions perceive social-ecological systems solely as spatially dependent systems (Halliday and Glaser 2011).
4 Research from the perspective of political economy also addresses this complexity with a focus on trade links. Here, food systems are depicted as food regimes (Friedmann 1987; Friedmann and McMichael 1989; McMichael 2009). A recent study by Krausmann and Langthaler (2019) shows the emergence and change of global food regimes over time from a social-ecological perspective, including telecoupling effects.
5 Please note that Gordon et al. (2017) do not literally describe food systems as telecoupled systems. However, their analysis shows the systems, flows, causes, and effects of telecoupling from a global perspective.
6 The usage of the term multilevel in this section includes the regional, national, and global level.
7 At least, this is the case in modern orchards (Robinson 2007).
8 For the empirical context of apples, the relevant political-legal framework is described in Section 3.1.
9 See the elaborations on the legal instrument of variety protection in Section 3.1.
10 See Section 3.2 for a more detailed description of fruit breeding processes at the example of apples. The case studies in Part III also give extensive descriptions of apple breeding processes.

11 In reality, this access can be restricted. For example, the cultivation of the variety *Natyra* is restricted to organic fruit farmers by the license holder. See Section 3.1 for a classification of this procedure in the political-legal framework.

References

Berkes, F.; Folke, C. (1998a): Linking Social and Ecological Systems for Resilience and Sustainability. In Fikret Berkes, Carl Folke (Eds.): *Linking Social and Ecological Systems. Management Practices and Social Mechanisms for Building Resilience.* Cambridge, UK: Cambridge University Press, pp. 1–25.

Berkes, F.; Folke, C. (Eds.) (1998b): *Linking Social and Ecological Systems. Management Practices and Social Mechanisms for Building Resilience.* Cambridge, UK: Cambridge University Press.

Biggs, R.; Schlüter, M.; Schoon, M. L. (Eds.) (2015): *Principles for Building Resilience. Sustaining Ecosystem Services in Social-Ecological Systems.* Cambridge, UK: Cambridge University Press.

Binder, C. R.; Hinkel, J.; Bots, P. W. G.; Pahl-Wostl, C. (2013): Comparison of Frameworks for Analyzing Social-Ecological Systems. In *Ecology and Society* 18 (4). DOI: 10.5751/ES-05551-180426.

Ceccarelli, S. (2012): *Plant Breeding with Farmers – A Technical Manual.* Aleppo: International Center for Agricultural Research in the Dry Areas (ICARDA).

Cumming, G. (2014): Theoretical Frameworks for the Analysis of Social-Ecological Systems. In Shoko Sakai, Chieko Umetsu (Eds.): *Social-Ecological Systems in Transition.* Tokyo: Springer Japan, pp. 3–26.

Eakin, H.; Rueda, X.; Mahanti, A. (2017): Transforming Governance in Telecoupled Food Systems. In *Ecology and Society* 22 (4). DOI: 10.5751/ES-09831–220432.

Ericksen, P. J. (2008): Conceptualizing Food Systems for Global Environmental Change Research. In *Global Environmental Change* 18 (1), pp. 234–245. DOI: 10.1016/j.gloenvcha.2007.09.002.

Fischer, J.; Gardner, T. A.; Bennett, E. M.; Balvanera, P.; Biggs, R.; Carpenter, S. et al. (2015): Advancing Sustainability Through Mainstreaming a Social–ecological Systems Perspective. In *Current Opinion in Environmental Sustainability* 14, pp. 144–149. DOI: 10.1016/j.cosust.2015.06.002.

Friedmann, H. (1987): International Regimes of Food and Agriculture Since 1870. In *Peasants and Peasant Societies* 2, pp. 258–276.

Friedmann, H.; McMichael, P. (1989): Agriculture and the State System. The Rise and Decline of National Agricultures, 1870 to the Present. In *Sociologica Ruralis* 29, pp. 93–117.

Gordon, L. J.; Bignet, V.; Crona, B.; Henriksson, P. J. G.; van Holt, T.; Jonell, M. et al. (2017): Rewiring Food Systems to Enhance Human Health and Biosphere Stewardship. In *Environmental Research Letters* 12 (10), p. 100201. DOI: 10.1088/1748-9326/aa81dc.

Halliday, A.; Glaser, M. (2011): A Management Perspective on Social Ecological Systems: A Generic System Model and its Application to a Case Study from Peru. In *Human Ecology Review* 18 (1), pp. 1–18.

Hanke, M.-V.; Flachowsky, H. (2017): *Obstzüchtung und wissenschaftliche Grundlagen.* Berlin, Heidelberg: Springer Berlin Heidelberg.

Herrero-Jáuregui, C.; Arnaiz-Schmitz, C.; Reyes, M.; Telesnicki, M.; Agramonte, I.; Easdale, M. et al. (2018): What Do We Talk about When We Talk about Social-Ecological Systems? A Literature Review. In *Sustainability* 10 (8), p. 2950. DOI: 10.3390/su10082950.

Hull, V.; Liu, J. (2018): Telecoupling. A New Frontier for Global Sustainability. In *Economy and Society* 23 (4). DOI: 10.5751/ES-10494-230441.

Krausmann, F.; Langthaler, E. (2019): Food Regimes and Their Trade Links. A Socio-ecological Perspective. In *Ecological Economics* 160, pp. 87–95. DOI: 10.1016/j.ecolecon.2019.02.011.

Liu, J.; Hull, V.; Batistella, M.; DeFries, R.; Dietz, T.; Fu, F. et al. (2013): Framing Sustainability in a Telecoupled World. In *Ecology and Society* 18 (2). DOI: 10.5751/ES-05873-180226.

McMichael, P. (2009): A Food Regime Genealogy. In *The Journal of Peasant Studies* 36 (1), pp. 139–169. DOI: 10.1080/03066150902820354.

Ostrom, E. (2009): A General Framework for Analyzing Sustainability of Social-ecological Systems. In *Science (New York, N.Y.)* 325 (5939), pp. 419–422. DOI: 10.1126/science.1172133.

Robinson, T. L. (2007): Recent Advances and Future Directions in Orchard Planting Systems. In *Acta Hortic.* (732), pp. 367–381. DOI: 10.17660/ActaHortic.2007.732.57.

Schlüter, M.; McAllister, R.; Arlinghaus, R.; Bunnefeld, N.; Eisenack, K.; Hölkner, F.Milner-Gulland, E. J. et al. (2012): New Horizons for Managing the Environment: A Review of Coupled Social-Ecological Systems Modeling. In *Natural Resource Modeling* 25 (1), pp. 219–272. DOI: 10.1111/j.1939-7445.2011.00108.x.

Tendall, D. M.; Joerin, J.; Kopainsky, B.; Edwards, P.; Shreck, A.; Le, Q. B. et al. (2015): Food System Resilience. Defining the Concept. In *Global Food Security* 6, pp. 17–23. DOI: 10.1016/j.gfs.2015.08.001.

Turner, B. L.; Kasperson, R. E.; Matson, P. A.; McCarthy, J. J.; Corell, R. W.; Christensen, L. et al. (2003): A Framework for Vulnerability Analysis in Sustainability Science. In *Proceedings of the National Academy of Sciences of the United States of America* 100 (14), pp. 8074–8079. DOI: 10.1073/pnas.1231335100.

Urruty, N.; Tailliez-Lefebvre, D.; Huyghe, C. (2016): Stability, robustness, vulnerability and resilience of agricultural systems. A review. In *Agronomy for Sustainable Development* 36 (1), p. 2. DOI: 10.1007/s13593-015-0347-5.

Walker, B.; Salt, D. (2012): *Resilience Thinking. Sustaining Ecosystems and People in a Changing World*. Washington, DC: Island Press.

Young, O. R.; Berkhout, F.; Gallopin, G. C.; Janssen, M. A.; Ostrom, E.; van der Leeuw, S. (2006): The Globalization of Socio-Ecological Systems. An Agenda for Scientific Research. In *Global Environmental Change* 16 (3), pp. 304–316. DOI: 10.1016/j.gloenvcha.2006.03.004.

5 Ecosystem Services in Fruit Breeding and Cultivation

This chapter gives a comprehensive overview of conducted research on ecosystem services in the context of agriculture. Agriculture as such is one of the main reasons for the loss and decrease of important ecosystem services on a global scale, especially due to the intensive use of pesticides and chemical-synthetic fertilizers (Bennett et al. 2014; MEA 2005). Kremen and Miles (2012) state that "the collective simplification of agroecosystems has led to a loss of biodiversity and to reductions in the supply of key ecosystem services to and from agriculture" (p. 3). The global challenge is to provide food security for the global population while simultaneously maintaining key ecosystem services. Paraphrasing Palomo-Campesino et al. (2018), there are two major positions in the discussions on how to solve this situation: (a) enhancing and innovating agricultural technologies, or (b) adopting agroecological and biodiversity-based approaches.[1] This section sheds light on this discourse in the context of ecosystem services. As a starting point, general ecosystem services and disservices in agriculture are investigated (See Section 5.1). Based on this general model, the impacts of different farming approaches and practices on ecosystem services are explained (See Section 5.2). In addition to the literature on ecosystem services in orchards, this provides the basis to investigate ecosystem services in fruit breeding and cultivation as part of their social-ecological interlinkages (See Section 5.3). Consequently, the relevant set of ecosystem services is defined to further elaborate on the social-ecological resilience of fruit breeding and cultivation.

5.1 Modeling Ecosystem Services to and from Agriculture

During the integration of the concept of ecosystem services into science and policy discourses, Swinton et al. (2007) and Zhang et al. (2007)[2] were the first to specifically integrate ecosystem services into the context of agricultural ecosystems. Agricultural systems are the foremost example of managed ecosystems that are directly shaped to fulfill human needs. The main function of agricultural systems is to provide provisioning services through the production of food, fuel, and fiber. Yet, to accomplish this provision, agriculture depends on a range of ecosystem services and thereby influences other ecosystem services.

DOI: 10.4324/9781003355724-9

Zhang et al. (2007) developed a model to illustrate the flows of ecosystem services and disservices to and from agriculture[3] (See Figure 5.1). They identify several supporting and regulating services, provided by different biotic and abiotic elements. These services flow to agriculture and are beneficial or even necessary for the functioning of agricultural ecosystems. Agriculture in turn provides mainly provisioning services, but also non-marketed services. However, the conduction of agriculture potentially leads to major disservices. *Disservices* describe processes or outputs of actions that reduce the productivity and economic performance of agricultural systems. These disservices can further influence existing ecosystem disservices that independently flow to agricultural ecosystems. For example, habitat loss of beneficial insects can further increase pest damage. All of these (dis)services flow differently across scales – from field to farm, to landscape, or even to a global level (ibid.) – and depend on the site-specific ecological context (Kremen 2005).

It follows that agricultural systems are simultaneously consumers and producers of ecosystem (dis)services (Sandhu et al. 2010). Agricultural systems can receive flows from other ecosystems and also impact other ecosystems with their generated ecosystem (dis)services (Dale and Polasky 2007). For example, water pollution in one farming system may impact other farms if the polluting farm is located upstream. Several specific services and disservices that flow to or from agriculture have been identified and described in the literature (Swinton et al. 2007; Power 2010). A short overview is given in the following.[4]

Major *supporting services* to agriculture are soil provision and nutrient cycling that determine soil fertility. The provision of water in sufficient purity and quantity is also an important supporting service, which in turn depends on soil quality and specifically soil moisture. Maintenance of biodiversity proves an important

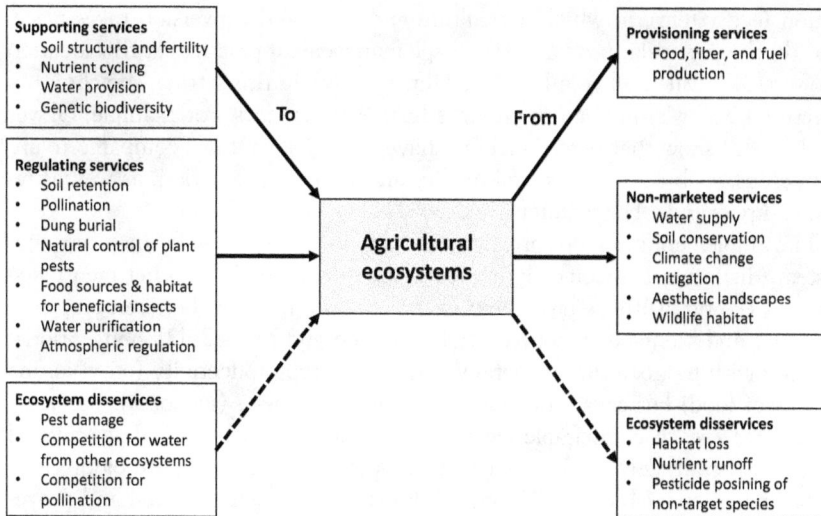

Figure 5.1 Ecosystem services and disservices to and from agriculture (based on Zhang et al. 2007, p. 254).

supporting service for agriculture because it provides "the raw material for natural selection to produce evolutionary adaptations [...], improving or maintaining agricultural productivity" (Zhang et al. 2007, p. 256).

The most prominently discussed *regulating services* to agriculture are pollination and pest control. A large amount (75%) of important fruit and vegetable crop species for global food production, which represent 35% of global food production in total, rely on (natural) pollination services primarily by insects (Klein et al. 2007). Natural pest control by insects, parasitoids, birds, bats, and other beneficial biotic actors suppresses pest damage of cultivated crops and thus improves and stabilizes yield. In the long term, this regulating service can contribute to the development of an "ecological equilibrium" (Zhang et al. 2007, p. 255) by preventing insects or herbs from developing into pests.

Hence, the most crucial *disservices to agriculture* are pests and diseases, involving herbivores, frugivores, seed-eaters, and pathogenic fungal, bacterial, or viral diseases. These hamper the productivity of agricultural systems and, in the worst case, lead to a complete loss of yields (Zhang et al. 2007). *Disservices from agricultural systems* themselves include greenhouse gas emissions, loss of biodiversity and habitats for species, as well as the pollution of soils, water, and non-target species. Pollution largely takes place through the increase in global agricultural production and yields that became possible with the introduction of chemical-synthetic fertilizers and pesticides (Tilman et al. 2002). These pollution effects directly affect the natural disservices of pests and diseases, who then develop resistance to certain pesticides (Zhang et al. 2007; Dale and Polasky 2007).

This shows that some basic dependencies and connections of ecosystem services and disservices exist. Besides, the literature highlights that these (diss-) services highly depend on (a) the specific composition and function of the location (ecosystem), in which agriculture and thus land conversion takes place, and (b) the particular farm and landscape management practices at this location (Power 2010; Dale and Polasky 2007; Tilman 1999). In this setting, the choice of perennial versus annual crops provides further differences. For example, Glover et al. (2012) show that perennial crops have several advantages compared to annual crops, such as improved soil quality and water conservation, as well as reduced inputs and labor requirements.

The implementation of agriculture always involves *trade-offs* between ecosystem (dis)services, specifically between provisioning and the other mentioned services (Power 2010). A prominent example is the trade-off between crop productivity and biodiversity conservation (Kremen and Miles 2012): conventional farming with monocultures generally maximizes crop productivity (provisioning service of food) but causes habitat and biodiversity losses (supporting services). However, to reach comparable levels of yield, alternative diversified agricultural systems generally rely on more land, which potentially causes more biodiversity and habitat losses. Industrial monocultural farming on the other hand proves more harmful to biodiversity than diversified farming systems, which aim to conserve and enhance biodiversity. Nevertheless, the gap in yields and productivity between conventional and organic agriculture is much smaller when performance under different environmental conditions is the key reference (Seufert et al. 2012; Reganold and Wachter 2016).

This example shows that trade-offs between ecosystem (dis)services are an object of debate in the study of ecosystem services. However, as a phenomenon, they are not inevitable and highly depend on the type of farm and landscape management. In the next section, a closer look is taken at the impacts of different farm management approaches on ecosystem services.

5.2 Impacts of Different Farming Approaches on Ecosystem Services

Literature on ecosystem services provides several studies that analyze and compare different farming approaches and practices. In the following, a short review on the impacts of conventional/industrial farming, diversified farming, agroecology, and organic farming on ecosystem services is conducted.[5] Table 5.1 shows a summary of the relevant understandings and definitions used for this review and comparison.

Characteristics and impacts of conventional farming practices have already been mentioned in the foregoing sections and chapters. *Conventional farming* aims at maximizing crop productivity and yield through the farming of monocultures combined with the intensive use of pesticides and chemical-synthetic fertilizers. This results in homogenization and industrialization effects on the field and landscape level that benefit large-scale farming systems and create aggravating circumstances for small-scale farming systems (Kremen and Miles 2012; Tilman et al. 2002). In a meta-study that compares the impacts of conventional and diversified farming systems on ecosystem services, Kremen and Miles (2012) conclude that conventional systems are inferior to diversified systems regarding

Table 5.1 Definitions of farming approaches (based on Kremen and Miles 2012; Migliorini and Wezel 2017; Duru et al. 2015).

Farming approach	Definition
Conventional/ industrial farming	Maximization of crop productivity and yield through the large-scale farming of monocultures combined with the intensive use of pesticides and chemical-synthetic fertilizers.
Diversified farming	Design of agricultural systems is focused on the promotion of agrobiodiversity across ecological, spatial, and temporal scales.
Agroecology	Improving agricultural systems by (a) imitating natural processes, (b) promoting genetic, functional, and species diversity, (c) fostering synergies between biotic and abiotic elements, and (d) implementing knowledge-based participatory and co-creation approaches.
Organic farming	Market-driven approach regulated by EU norms. It demands the use of internal natural resources for the management of agricultural systems, a restriction of external inputs, a strict limit on the use of chemical-synthetic inputs, and aims at an adaptation to regional and local conditions.

a significantly wide range of ecosystem services. The authors define *diversified systems* as production systems that incorporate agrobiodiversity across ecological, temporal, and spatial scales.[6] Overall, they analyzed 12 ecosystem services. Only for crop productivity (provisioning service), the authors found evidence for the superiority of conventional systems against diversified systems.[7] However, they argue that the lower productivity of diversified systems can be compensated with enhanced environmental benefits and the reduction of external effects. Foley et al. (2011; 2005) confirm these results and draw the conclusion that conventional food production clearly trades off against other ecosystem services in agriculture.

Agroecology is jointly defined as a scientific discipline, a set of practices, and a movement. It is a systemic approach and addresses a transformative process of the entire food system (Migliorini and Wezel 2017; Altieri 1995). The core elements of this approach are recycling of biomass; promotion of species and genetic diversity; provision of appropriate natural soil conditions; enhancement of synergies and interactions between biotic and abiotic elements; co-creation and sharing of diverse knowledge; and independence of markets (Migliorini and Wezel 2017). In contrast, *organic agriculture* is strictly regulated by EU norms and a labeling and certification scheme exists. The market-driven approach defines specific techniques for farming. It follows four overall principles (EU Council Regulation 2007):

1. Usage of internal natural resources for the management of agricultural systems
2. Restriction of external inputs
3. Strict limitations on the use of chemical-synthetic inputs
4. Adaptation to regional and local conditions.

Because most literature on the impacts of farming approaches on ecosystem services does not differentiate between agroecological and organic practices, the results of these studies concern both approaches. The main reason for this indifference is the broad overlap of their agricultural practices (Migliorini and Wezel 2017; Kremen and Miles 2012), although they differ in their conceptual and regulative dimensions.

Diversified farming systems as defined by Kremen and Miles (2012) largely overlap with the approaches of agroecology and organic farming. In their meta-study, both approaches were used as synonyms because both incorporate elements of diversified farming. Diversified farming systems prove beneficial for a range of ecosystem services. As already mentioned above, the meta-study identifies a superiority of diversified systems compared to conventional systems because they support several ecosystem services. Strong evidence was found for the support of biodiversity, soil quality, water-holding capacity in soils, carbon sequestration, energy-use efficiency, and resilience to climate change. In relation to the farming of monocultures, control of weeds, diseases, and arthropod pests as well as pollination are supported more strongly. However, the mere support is not equal to a total control of these factors and diversified practices remain insufficient to control pests and diseases and provide sufficient pollination.

In their literature review on agroecological practices and ecosystem services, Palomo-Campesino et al. (2018) conclude that most research concentrates on the environmental and biophysical benefits of these practices. The most frequently analyzed services are regulating services; foremost pest control, soil fertility, and pollination. They show that the use of cover crops, the diversification of crop species, and crop rotations are beneficial practices for climate regulation, the prevention of soil erosion, carbon sequestration, pest control, and pollination. Flower strips or field margins as well as reduced tillage also affect the listed ecosystem services and, additionally, water flow management. On a landscape level, the design and management of complex landscapes benefit pollination, soil fertility, and water flow management. A combination of livestock and crops also enhances soil fertility and pest control (ibid.).

Duru et al. (2015) emphasize the importance of biodiversity in agroecological and organic farming systems. Strengthening and implementing practices that promote biodiversity is a key challenge because although it significantly benefits a range of ecosystem services, it demands re-designing farming systems. Quijas et al. (2010) show the beneficial effects of biodiversity-based practices on the provisioning of plant products, soil fertility, resistance against invasive species, as well as pest and disease control.

Reganold and Wachter (2016) conducted a literature review and specifically compare organic agriculture with conventional agriculture. Similar to Kremen and Miles (2012) for general diversified systems, they conclude that organic agricultural practices outperform conventional practices in all environmental indicators, thus ecosystem services. In detail, they observe higher minimization of pesticide residues, water pollution and energy use, higher nutritional quality of products, higher soil quality, and enhanced biodiversity.

Overall, the literature provides a wide range of case studies and comparisons on the impact of farming approaches and practices on ecosystem services. They show that ecocentric farming approaches, such as agroecology, diversified farming, and organic agriculture, and their respective farming practices are more beneficial for a whole range of ecosystem services than conventional farming. Based on this pool of knowledge, the next section describes and analyzes ecosystem services in the relevant research topic of fruit breeding and cultivation.

5.3 Ecosystem Services in Fruit Cultivation and Breeding

In general, research on ecosystem services in orchards is limited and largely focuses on the effects of agricultural management practices or other specific aspects.[8] While Clothier et al. (2013) and Montanaro et al. (2017) focus on specific ecosystem services, areas, or fruits, Demestihas et al. (2017) were the first ones to give a comprehensive overview of this fragmented literature.

They categorize relevant agricultural management and pedoclimatic conditions, ecosystem functions, and ecosystem services according to the CICES classification (See Section 1.2).[9] Besides a review of existing literature, the authors also analyze connections and trade-offs between the variables to give a coherent conception of ecosystem services in orchards. Figure 5.2 shows the model that evolved out of this categorization in a simplified version.

Agricultural management & pedoclimatic conditions   **Ecosystem functions**   **Ecosystem services**

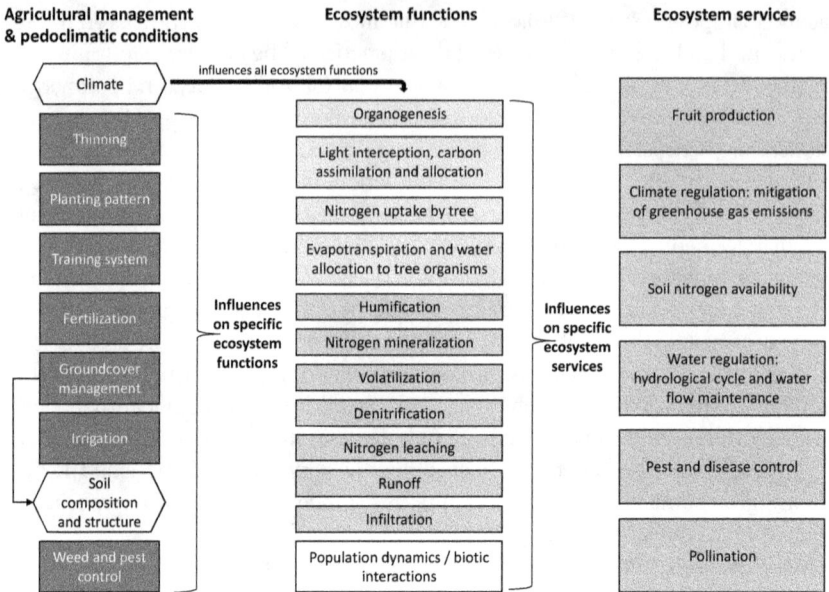

Figure 5.2 Management practices, ecosystem functions, and ecosystem services in orchards (based on Demestihas et al. 2017, p. 4).

In this model, ecosystem functions (middle column) describe basic ecological processes that provide ecosystem services in orchards (right column) and are in turn influenced by agricultural management practices and local environmental conditions (left column). The colored boxes categorize agricultural management practices (dark gray), plant-related ecosystem functions (light gray), soil-related ecosystem functions (medium-light gray), pest-related ecosystem functions (white), and ecosystem services (medium-dark gray). The depicted ecosystem services thus mark the outputs of local ecosystem functions that are influenced by orchard management activities and lead to benefits for humans, for example, fruit production. As described in Section 5.1, these flows of ecosystem services are influenced by fruit cultivation but also influence cultivation and vice versa.

In the following, a short description of these services based on Demestihas et al. (2017) is given because they mark the object of analysis in this book. Relevant literature on the impacts of apple cultivation practices or fruit farming approaches on respective ecosystem services is added when available. This enables an in-depth discussion of all relevant ecosystem services in orchards, including their trade-offs, and scientific insights on the impact of specific farming approaches.

5.3.1 *Provisioning Service: Fruit Production*

The main reason for the existence of men-made orchards is the provisioning service of *fruit production*. The quantity and quality of produced fruits depend on the level of organogenesis,[10] light interception, as well as the assimilation and allocation of carbon, water, and nitrogen. Thus, fruit production is influenced by a number of regulating services (e.g., water cycle) and supporting services (e.g., nutrient cycle). Similar to agricultural systems in general (See Section 5.2), conventional systems of fruit production outperform organic or diversified systems in yields, due to their use of agrochemicals. In a comparison between organic apple cultivation systems and apples cultivated with an integrated pest management (IPM) approach,[11] Samnegård et al. (2019) observe significantly lower yields in organic apple cultivation systems. However, they also observe high variations, with some organic systems having higher yields than average IPM orchards.

5.3.2 *Supporting Service: Soil Nitrogen Availability*

Nitrogen serves as an essential input for plants to grow and thrive. *Soil nitrogen availability* is a supporting service that is influenced by a multitude of ecosystem functions (See Figure 5.2), which makes it a complex service. Healthy soils need to achieve a balance of human or natural nitrogen inputs and nitrogen losses, for example, caused by leaching or denitrification. Thus, neither overprovision nor a shortage of nitrogen should exist for healthy soils. The natural (local) nitrogen cycle on the field is shaped by the local climate and soil conditions. These conditions define the nitrogen need of local soils. Some soils need external nitrogen inputs to be farmed, others do not need any or only a minimum amount. One of the most relevant trade-offs to achieve this nitrogen balance is the use of nitrogen fertilizers that increase fruit yields but have negative impacts on leaching and denitrification (Demestihas et al. 2017), and thus on the natural cycle in the long run. If, for example, an oversupply of nitrogen fertilizers happens, plants cannot uptake the excess nitrogen. It stays in the soil, potentially leaches out with water runoff, and thus pollutes water resources. It follows that management practices have a large influence on nitrogen availability. As further shown (ibid.), several studies confirm the beneficial effect of ecocentric farming practices, such as cover crops, on the soil quality of apple orchards (See also Section 5.2) in comparison to conventional management approaches.

5.3.3 *Regulating Services: Climate Regulation, Water Cycle* *Regulation and Maintenance, Pest and Disease Regulation*

The most numerous ecosystem services in the model of Demestihas et al. (2017) are regulating services. Because of carbon sequestration in fruit trees, *climate regulation* through the mitigation of greenhouse gases is an ecosystem service influenced by fruit cultivation. In their simulation study for apple orchards, Demestihas et al. (2019) found a clear synergy between yield, fruit mass, and carbon sequestration: "[t]he contribution of carbon allocation to fruit in sequestrated carbon at the annual scale was considerable, and this is clearly a specificity of orchards compared to other ecosystems" (p. 10). However, albeit this categorization is correct, the absolute effect of orchards on mitigation actions is negligible (Demestihas et al. 2017). Granatstein and Peck (2017) admit that many orchards need a range of non-renewable resources like fossil fuels or, for example, rain covers made from plastic. These resource demands affect the overall carbon footprint of orchards. It is thus important to analyze this ecosystem service in light of potential trade-offs with orchard management practices.

Water cycle regulation and maintenance are important for water supply and discharge in orchards. Thus, Demestihas et al. (2017) identify drainage and soil water content as relevant indicators for successful water regulation. The water cycle is closely linked to the nitrogen cycle as both can influence each other in positive and negative ways. For example, an oversupply of nitrate can lead to the pollution of baseline water. Together, both ecosystem services are significantly influenced by irrigation practices and by cover crops, which potentially compete with fruit trees (Demestihas et al. 2017, 2019). Irrigation systems for orchards need available water resources: if one orchard relies on non-rechargeable water resources, irrigation systems have major effects on local water systems. Additionally, water quality can be negatively influenced by pesticide use. Thus, for irrigation, it is crucial to "maximize efficiency of water use while also protecting water quality" (Granatstein and Peck 2017, p. 22).

Due to the perennial character of orchards, the regulating service *pest and disease regulation* embodies a "great pest management challenge" (Demestihas et al. 2017, p. 11). Fruit trees provide a habitat for both beneficial arthropods as well as bacterial and fungal diseases or pest insects. Literature provides coherent overviews of these aspects, for example, on beneficial arthropods (Cross et al. 2015) or diseases (Grove et al. 2003) in apple orchards and their interactions. One of the most aggressive and resistant pests is the brown marmorated stinkbug (*halyomorpha halys*), which is highly polyphagous and continuously spreads throughout Europe (Haye et al. 2015). It damages both fruits and trees. Without sufficient control of pests and diseases, orchards potentially experience tremendous losses in yields. A wide range of management practices in organic, diversified, and agroecological farming approaches such as flower strips, cover crops, or hedgerows encourage habitat building for beneficial arthropods in orchards (Demestihas et al. 2017).[12] The conventional strategy for pest control in both organic and conventional orchards is the use of plant protection products because market actors demand high product quality and homogeneity. Using pesticides may contradict habitat building for beneficial arthropods. While many studies show the negative effects of pesticides on habitat supply (Demestihas et al. 2017), Samnegård et al. (2019) observe no significant effects in their investigated apple orchards other than organic orchards having higher plant and insect diversity. General promotion of plant and insect diversity combined with effective management practices that do not rely on pesticides can lead to a high level of purely "ecological pest control" (Demestihas et al. 2017, p. 12). However, reaching this level of ecological control through the use and shape of biodiversity is highly difficult and involves many challenges, such as identifying key pests and selecting synergetic plant assemblages (Simon et al. 2010; Penvern et al. 2019).

In sum, regulating and controlling pests and diseases with natural management practices always involves trade-offs, which highlights the importance of evaluating the net benefits when beneficial arthropods or other animals are introduced.

Peisley et al. (2016) show the importance of this evaluation for bird activity in apple orchards: while birds in apple orchards have positive impacts on pest control because they eat predatory insects (e.g., codling moth larvae), they also damage crops. However, overall the authors argue that bird activity in apple orchards has net benefits for fruit farmers.

Connected with the above-explained ecosystem service of pest and disease control is the regulating service *pollination*. Most fruit species are not self-pollinating and thus rely on pollination services by insects like bees, butterflies, moths, or beetles (See Section 3.2). Higher species richness and diversity can thus increase pollination rates (Demestihas et al. 2017). However, this could also collide with pest control because the population of pest insects potentially increases. The greatest trade-off in the sense of an "obvious antagonism" (ibid., p. 13) exists in the use of pesticides as they can control pests but also decrease pollination rates. This emphasizes the benefits of ecological pest control for balancing pollination rates with pest and disease regulation. Regarding specific management practices, several studies confirm the positive effects of flower strips on both pollination rates and pest control in apple orchards (Campbell et al. 2017; Samnegård et al. 2019)

5.3.4 Trade-offs and Synergies between Ecosystem Services

The descriptions above show that – similar to agricultural systems in general (See Section 5.1) – the management of ecosystem services in orchards always involves trade-offs but also gives opportunities for building synergies. This is especially true for ecocentric farming systems (See Section 5.2), in which the building of natural synergies is the main objective. The evaluation of ecosystem services literature on orchards leads to the conclusion that natural synergies often trade off with yield. However, the choice of agricultural management practices clearly has the greatest impact on trade-offs between ecosystem services, as shown for using pesticides, irrigation systems, or fertilizers. Other practices such as flower strips, hedgerows, or composting seem to promote ecosystem services in a more balanced way.

In their detailed modeling of ecosystem services connections in apple orchards, Demestihas et al. (2019) uncover a range of relationships between ecosystem services. Overall, they conclude that an optimized cultivation system uses organo-mineral fertilizers, comfort irrigation, exclusion nets against codling moths, and scab-resistant cultivars. This kind of system would provide minimum trade-offs between provisioning, supporting, and regulating services, and thus has the highest net benefit. With this result, the authors stress an important aspect: the choice of cultivars, which is mostly overlooked in the literature about ecosystem services in orchards.

5.3.5 Fruit Breeding and Ecosystem Services

In general, few studies exist that connect the concept of ecosystem services to plant breeding. Brummer et al. (2011) argue that besides breeding for general goals such as site-adaptation, maintaining biodiversity, or traits of consumer importance, plant breeders can also breed for the enhancement of distinct eco-system services. These include, for example, breeding for the following services:

- Stable or high yields (provisioning service), often combined with breeding for nutrient use efficiency to improve water quality (regulating service) and minimize nitrogen losses (supporting service).
- Improving soil quality (supporting service), specifically through larger and improved root systems as well as carbon sequestration, thus, climate regula-tion (regulating service).
- Improving pest and disease regulation (regulating service) as well as climate adaptation through tolerance or resistance against biotic and abiotic stresses.

Other studies analyze the impacts of cultivating traditional varieties or landraces on ecosystem services (Waldman and Richardson 2018). Ficiciyan et al. (2018) pro-vide a coherent and extensive review of this body of literature. Their main findings for ecosystem services include the advantageousness of modern varieties over tradi-tional varieties[13] in crop yields and higher performance of traditional varieties un-der harsh environmental conditions. Traditional varieties are also more beneficial for promoting regulating services, especially pest and disease regulation.

However, no studies at all examine fruit breeding as a distinct subsection of plant breeding under the conceptual lens of ecosystem services. The following classification and identification of causal relationships thus aim to close this gap in the literature. It is based on the descriptions given throughout this Section 5.3 and general literature about fruit breeding, hereby mainly apple breeding.

To begin with, it is important to differentiate between (a) the rootstock and (b) the cultivar – at least in commercial farming systems that work with dwarfed fruit trees (See Section 3.2). For meadow orchards or other alternative systems that do not work with rootstocks, only the choice of cultivar is relevant. However, in most commercial orchards the combinational choice of both rootstock and cultivar influences impacts on ecosystem services.

Rootstocks influence the uptake of water and nitrogen into the plant, as indi-cated in the previous section: improved root systems have benefits on soil quality and carbon sequestration. As an example, apple farmers in Germany typically choose the rootstock variety M9, although some alternatives exist (Pfeiffer 2020; Spornberger et al. 2020; Walch et al. 2017). Wang et al. (2019) conducted a review on the progress of rootstock breeding for apples and describes the main breeding goals for specific worldwide territories. The following list gives an impression on breeding goals for rootstocks, trying to connect their findings (if possible) with ecosystem services:

- horticultural traits, for example, grafting compatibility or reproductive capacity
- performance in production, fruit size, and fruit quality (provisioning service)
- resistances to mainly soil-originating pests and diseases, for example, fire blight, collar rot, or woolly apple aphid (regulating service)

- improved uptake of water and nitrogen (regulating and supporting services)
- tolerance to environmental conditions, especially cold and drought

This list connects to the identified aspects by Brummer et al. (2011) as shown above. According to Brown (2012), the most important breeding goals for root-stocks are resistance to pests, diseases, and harsh environmental conditions. Rootstocks thus literally build the base for cultivars to develop and provide ecosystem services.

These most general categorizations of plant and rootstock breeding impacts on ecosystem services are also true for the choice of fruit cultivars. However, this choice of fruit cultivars needs to consider some specific aspects, completing this categorization of ecosystem services in fruit breeding:

- Provisioning services: Referring to the quantity and quality of *fruit production*, the amount and stability of yield, the taste of the product, and visual traits are largely determined by the choice of the cultivar. For example, today's broadly commercialized apple varieties mostly have concurrent beneficial traits but lack robustness and resistance.
- Supporting services: As the rootstocks connect the soil with the fruit trees, their traits largely influence *soil nitrogen availability* or soil quality. Nevertheless, as they are only the 'channel,' nitrogen demands of fruit trees also have relevant impacts on this ecosystem service. Breeding for nutrient use efficiency influences the level of these demands. Improving the efficiency of cultivars regarding their uptake of nitrogen could lead to a reduction in the application of fertilizers, which is one of the main reasons for trade-offs with other ecosystem services (see the previous section).
- Regulating services: Besides the choice of rootstocks, *water cycle regulation and maintenance* as well as *pest and disease regulation* is additionally influenced by the choice of cultivars. Resistances of cultivars against certain pests or diseases (or their general robustness) determine the number of necessary applications of pesticides and fungicides, which in turn determines their potentially negative effects on the water cycle. Breeding for resistance is one of the major breeding goals for fruit cultivars. Recently, robustness also marks a goal that is increasingly pursued by mostly organic fruit breeders (See Chapter 9). No direct or significant impacts on the ecosystem services *climate regulation* and *pollination* can be identified from the results and conclusions above.

Recent discussions in fruit breeding emphasize the role of underutilized and heirloom cultivars for improving beneficial impacts on ecosystem services, especially for pest and disease regulation – similar to traditional varieties and landraces in general plant breeding (Ficiciyan et al. 2018). Kellerhals et al. (2018, 2020) show the importance of these cultivars for breeding activities in the case of apples. They conclude that although heirloom cultivars have limited fruit-eating quality, their resistance and robustness traits could prove vital for further breeding. However, the choice of cultivars and perennial agricultural systems as such also have other functions besides the already explained ecosystem services by affecting people in a cultural way. For a coherent categorization of ecosystem services in fruit breeding and cultivation, the following final section elaborates on cultural ecosystem services.

5.3.6 Cultural Ecosystem Services in Fruit Cultivation and Breeding

Literature on cultural ecosystem services in fruit cultivation and breeding is scarce. Only some of the publications mentioned above add the examination of cultural ecosystem services in fruit cultivation to their elaborations on supporting, regulating, and provisioning ecosystem services. Ficiciyan et al. (2018) argue that traditional crops and landraces have beneficial effects on cultural ecosystem services such as traditional values (cultivars are passed over generations) or cooking characteristics (recipes for these cultivars are passed over generations). This holds also true for underutilized and heirloom fruit cultivars which usually have specific tastes, characteristics, and local or regional values. In their study on ecosystem services in New Zealand's orchards (especially vineyards), Clothier et al. (2013) identify that vineyards provide *aesthetic, spiritual, and recreational services.* As an example, they mention the Rippon Vineyard on the shores of Lake Wanaka where locals and tourists can participate in winery tours or irregular concerts and festivals inside orchards. Other studies also elaborate on the aesthetic value of perennial landscapes, again mostly vineyards, in different countries (Biasi et al. 2012; Farina 2000; Winkler and Nicholas 2016). However, the provisioning of these services only resonates with the corresponding design of orchards or their specific surrounding landscapes, and rather not with the commercial fruit farming of monocultures. Meadow orchards often have this corresponding design and provide cultural ecosystem services, additionally serving as places for *regional identity* (Bieling and Plieninger 2013). This leads to the conclusion that cultural services in fruit cultivation highly depend on the design of orchards, their agricultural management, as well as concurrent tourism or event practices.

Literature on cultural ecosystem services in fruit breeding does not exist at all. Breeding mainly provides cultural ecosystem services for the people that conduct fruit breeding or work in close connection with breeders. This may not be true for breeding that is being done on-farm where trial areas are embedded in orchards (See Section 4.2). In this case, breeding may also provide cultural services for other people such as farm workers, farmers, visitors, or tourists. It is important to note that the perception of cultural ecosystem services for the involved actors is always experience- and context-dependent (Winkler and Nicholas 2016).

In a transdisciplinary workshop with members of *apfel:gut e.V.* (See Chapter 9), sustainability effects of their apple breeding approach were discussed based on the concept of ecosystem services.[14] The results of the conducted workshop with apple breeders and farmers reveal potential cultural services of apple breeding. Similar to fruit cultivation, apple breeding provides *aesthetic values*, for example, by breeders feeling pleasure in growing and tending the seedlings. Breeding can also provide *spiritual values* as it is a highly creative and inspirational performance. It also serves for *identification* with the desirable farming approach the apple cultivars are bred for and thus the *realization of ethical values and norms*, for example, organic agriculture. This is connected to the spiritual values above. Lastly, apple breeding gives *opportunities to learn* about natural interactions in plants and their evolution by using and manipulating these interactions.

Some of the identified cultural services of breeding may also be valid for fruit cultivation. Especially in organic, diversified, or agroecological farming approaches, practicing fruit cultivation may serve as a vehicle for identification, realizes ethical values, and gives opportunities to learn. These aspects are overlooked in current literature but shed light on important facets of the social systems in breeding and cultivation. It follows that orchards potentially provide a range of cultural ecosystem services for fruit farmers, breeders, and other involved actors.

5.4 Conclusion: Complex Web of Ecosystem Services in Fruit Breeding and Cultivation

Literature provides a range of findings for ecosystem services in fruit cultivation, whereby cultural ecosystem services are a rather neglected issue. Elaborations on the topic of fruit breeding are even more scarce or only indirectly connected to the concept of ecosystem services. Especially cultural ecosystem services of fruit breeding are a blind spot in literature. This section aimed to close the research gap on the integration of fruit breeding into the concept of ecosystem services by providing new insights and interpretations to open further research opportunities. Connected with literature results on fruit cultivation, a detailed picture evolves.

The foregoing elaborations have shown a complex web of interactions and trade-offs between fruit breeding and cultivation, as well as their respective impacts on ecosystem services. Figure 5.3 provides an overview of those relationships. It takes the conceptualization of fruit breeding and cultivation into account (See Section 4.2), in which ecosystem services mark the outcomes of interactions in the respective social-ecological system.

The summarizing categorization shows that fruit breeding and cultivation potentially provide very similar cultural, regulating, and supporting services. This emphasizes the connection between both social-ecological systems. However, only fruit cultivation directly produces fruits, thus providing provisioning services, and has an influence on the regulating services of climate regulation and pollination.

The results aggregated in Figure 5.3 serve as an analytical framework for investigating ecosystem services in fruit breeding and cultivation. Because of its direct conceptional link to resilience, the examination of ecosystem services in fruit breeding and cultivation proves the first step to understanding social-ecological resilience in both systems. The next chapter will complete this conception by describing (ideal) characteristics of a resilient fruit breeding and cultivation system.

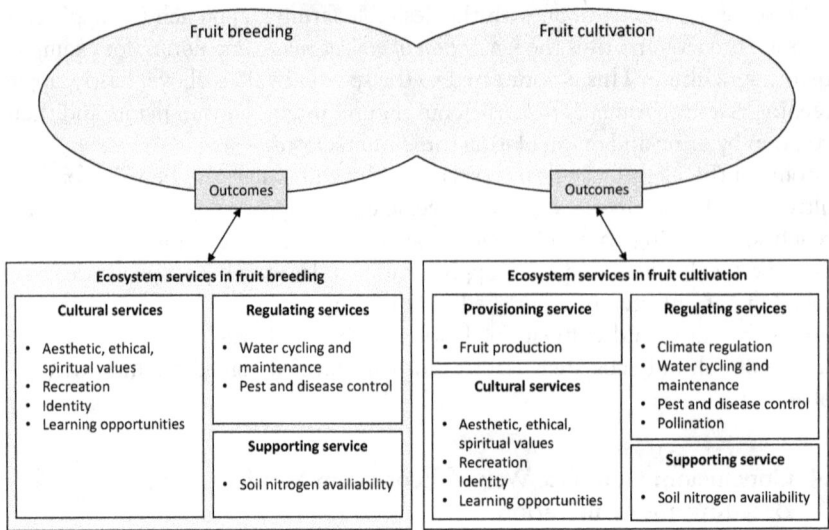

Figure 5.3 Categorization of ecosystem services in fruit breeding and cultivation (own figure).

Notes

1 In more detail, Duru et al. (2015) summarize this discourse as the dualism of weak ecological modernization vs. strong ecological modernization. Weak ecological modernization follows a *technocentric* approach and seeks for general solutions. Strong ecological modernization follows an *ecocentric* approach and seeks for site-specific solutions and the intensification of ecological interactions. Both directions have their strengths and weaknesses. With the "conventionalization of organic farming" (Darnhofer et al. 2010), possible interrelations prove also challenging for the ecocentric approach as such.

2 Both articles were published in the special issue *Ecosystem Services and Agriculture* in the journal *Ecological Economics*.

3 Other authors like Fisher et al. (2009) refer to these flows as *input services* and *output services* which essentially describe the same phenomena.

4 The categorization of the specific services differs in parts from the categorization of TEEB (2019) as discussed in Section 1.2.

5 This typology is rather simplified as it focuses on the main farming approaches. The author recognizes the existence and importance of other and special approaches such as conservation agriculture, agroforestry, climate smart farming, or biodynamic approaches.

6 Agrobiodiversity across ecological scales includes genetic diversity, varietal diversity, multiple intercropped species, and non-crop plantings. Spatial scales include the implementation of agrobiodiversity within, across, and around field level, and at the landscape-to-regional level (Kremen and Miles 2012).

7 There is a large debate on yield-variability in organic and conventional systems. Lesur-Dumoulin et al. (2017) conducted a meta-study and conclude that average yield in organic agriculture is 10–32% lower than in conventional agriculture. However, a large variability exists across the reviewed experiments.

8 Besides, the majority of literature on the analysis of fruit cultivation approaches is not embedded in the ecosystem services concept. Most studies carry out case studies on distinct farming practices on field level that are difficult to scale up. Probably the most popular study, as indicated by the citation count, is the one by Reganold et al. (2001), in which the authors assess and compare the sustainability of organic, conventional, and integrated apple production. Other comparative studies in this style where carried out by Bertschinger et al. (2004), Peck et al. (2006), Simon et al. (2008), or Peck et al. (2010). However, although those studies provide interesting insights, results are very context-sensitive and depend on the respective geographic location, use of cultivars, management practices, and the point in time the study was carried out.

9 Demestihas et al. (2017) were the first to investigate orchards with the particular use of the concept of ecosystem services and thus make an important contribution to include orchards and fruits in the discourse on ecosystem services. In further studies that build on their developed model, this group of researchers analyzed ecosystem services in apple orchards, their trade-offs, and synergies (Demestihas et al. 2018, 2019).

10 Organogenesis describes the process of flowering and fruit initiation.

11 Samnegård et al. (2019) define the IPM approach according to the IPM guidelines, where agroecological structures are implemented, mineral fertilizers are applied during the growing season, and chemical-synthetic pesticides are used for crop protection. The description of integrated apple farming further illustrates this approach (See Section 3.2).

12 However, the effectiveness of specific designs of flower strips remains in a state of debate (Cahenzli et al. 2019; Rodríguez-Gasol et al. 2019).

13 In their review, traditional varieties are defined as "landraces [that] aim to provide genetic resources and plant traits that are well adapted to local environmental and cultural conditions. Landraces have been maintained and selected over time by farmers to meet their personal economic, ecological and cultural needs and cultivated in small-scale farming systems with low input of external factors and high surrounding diversity" (Ficiciyan et al. 2018, p. 2).

14 The transdisciplinary workshop took place on April 13, 2018, in Kassel-Witzenhausen as part of the research project EGON (See Preface). Six members of the association participated in the workshop. It was carried out in an interactive style by discussing and finding consensus on certain categorizations of ecosystem services.

References

Altieri, Miguel A. (1995): *Agroecology: The Scientific Basis of Alternative Agriculture*. Boulder, CO: West View Press.

Bennett, E.; Carpenter, S. R.; Gordon, L. J.; Ramankutty, N.; Balvanera, P.; Campbell, B. et al. (2014): Toward a More Resilient Agriculture. In *The Solutions Journal* 5 (5), pp. 65–75.

Bertschinger, L.; Mouron, P.; Dolega, E.; Höhn, H.; Holliger, E.; Husistein, A. et al. (2004): Ecological Apple Production: A Comparison of Organic and Integrated Apple-Growing. In *Acta Horticulturae* (638), pp. 321–332. DOI: 10.17660/ActaHortic.2004.638.43.

Biasi, R.; Botti, F.; Barbera, G.; Cullotta, S. (2012): The Role of Mediterranean Fruit Tree Orchards and Vineyards in Maintaining the Traditional Agricultural Landscape. In *Acta Horticulturae* (940), pp. 79–88. DOI: 10.17660/ActaHortic.2012.940.9.

Bieling, C.; Plieninger, T. (2013): Recording Manifestations of Cultural Ecosystem Services in the Landscape. In *Landscape Research* 38 (5), pp. 649–667. DOI: 10.1080/01426397.2012.691469.

Brown, S. (2012): Apple. In Marisa Luisa Badenes, David H. Byrne (Eds.): *Fruit Breeding*. Boston, MA: Springer US, pp. 329–367.

Brummer, E. C.; Barber, W. T.; Collier, S. M.; Cox, T. S.; Johnson, R.; Murray, S. C. et al. (2011): Plant Breeding for Harmony Between Agriculture and the Environment. In *Frontiers in Ecology and the Environment* 9 (10), pp. 561–568. DOI: 10.1890/100225.

Cahenzli, F.; Sigsgaard, L.; Daniel, C.; Herz, A.; Jamar, L.; Kelderer, M. et al. (2019): Perennial Flower Strips for Pest Control in Organic Apple Orchards – A Pan-European Study. In *Agriculture, Ecosystems & Environment* 278, pp. 43–53. DOI: 10.1016/j. agee.2019.03.011.

Campbell, A. J.; Wilby, A.; Sutton, P.; Wäckers, F. L. (2017): Do Sown Flower Strips Boost Wild Pollinator Abundance and Pollination Services in a Spring-flowering Crop? A Case Study from UK Cider Apple Orchards. In *Agriculture, Ecosystems & Environment* 239, pp. 20–29. DOI: 10.1016/j.agee.2017.01.005.

Clothier, B. E.; Green, S. R.; Müller, K.; Gentile, R.; Herath, I. K.; Mason, K. M.; Holmes, A. (2013): Orchard Ecosystem Services: Bounty from the Fruit Bowl. In John Dymond (Ed.): *Ecosystem Services in New Zealand. Conditions and Trends*. Lincoln, NE: Manaaki Whenua Press, pp. 94–101.

Cross, J.; Fountain, M.; Marko, V.; Nagy, C. (2015): Arthropod Ecosystem Services in Apple Orchards and Their Economic Benefits. In *Ecological Entomology* 40, pp. 82–96. DOI: 10.1111/een.12234.

Dale, V. H.; Polasky, S. (2007): Measures of the Effects of Agricultural Practices on Ecosystem Services. In *Ecological Economics* 64 (2), pp. 286–296. DOI: 10.1016/j. ecolecon.2007.05.009.

Darnhofer, I.; Lindenthal, T.; Bartel-Kratochvil, R.; Zollitsch, W. (2010): Conventionalisation of Organic Farming Practices. From Structural Criteria Towards an Assessment Based on Organic Principles. A Review. In *Agronomy for Sustainable Development* 30 (1), pp. 67–81. DOI: 10.1051/agro/2009011.

Demestihas, C.; Plénet, D.; Génard, M.; Raynal, C.; Lescourret, F. (2017): Ecosystem Services in Orchards. A Review. In *Agronomy for Sustainable Development* 37 (2), p. 581. DOI: 10.1007/s13593-017-0422-1.

Demestihas, C.; Plénet, D.; Génard, M.; Garcia de Cortazar-Atauri, I.; Launay, M.; Ripoche, D. et al. (2018): Analyzing Ecosystem Services in Apple Orchards Using the STICS Model. In *European Journal of Agronomy* 94, pp. 108–119. DOI: 10.1016/j.eja.2018.01.009.

Demestihas, C.; Plénet, D.; Génard, M.; Raynal, C.; Lescourret, F. (2019): A Simulation Study of Synergies and Tradeoffs Between Multiple Ecosystem Services in Apple Orchards. In *Journal of Environmental Management* 236, pp. 1–16. DOI: 10.1016/j. jenvman.2019.01.073.

Duru, M.; Therond, O.; Martin, G.; Martin-Clouaire, R.; Magne, M.-A.; Justes, E. et al. (2015): How to Implement Biodiversity-based Agriculture to Enhance Ecosystem Services. A Review. In *Agronomy for Sustainable Development* 35 (4), pp. 1259–1281. DOI: 10.1007/s13593-015-0306-1.

EU Council Regulation (2007): EC Regulation 834/2007 of the Council, 28th June 2007. Official Journal of European Communities, L 189/1 of 20th July 2007.

Farina, A. (2000): The Cultural Landscape as a Model for the Integration of Ecology and Economics. In *BioScience* 50 (4), p. 313. DOI: 10.1641/0006-3568(2000)050 [0313:TCLAAM]2.3.CO;2.

Ficiciyan, A.; Loos, J.; Sievers-Glotzbach, S.; Tscharntke, T. (2018): More than Yield. Ecosystem Services of Traditional versus Modern Crop Varieties Revisited. In *Sustainability* 10 (8), p. 2834. DOI: 10.3390/su10082834.

Fisher, B.; Turner, R. K.; Morling, P. (2009): Defining and Classifying Ecosystem Services for Decision Making. In *Ecological Economics* 68 (3), pp. 643–653. DOI: 10.1016/j. ecolecon.2008.09.014.

Foley, J. A.; DeFries, R.; Asner, G. P.; Barford, C.; Bonan, G.; Carpenter, S. R. et al. (2005): Global Consequences of Land Use. In *Science* 309 (5734), pp. 570–574. DOI: 10.1126/science.1111772.

Foley, J. A.; Ramankutty, N.; Brauman, K. A.; Cassidy, E. S.; Gerber, J. S.; Johnston, M. et al. (2011): Solutions for a Cultivated Planet. In *Nature* 478 (7369), pp. 337–342. DOI: 10.1038/nature10452.

Glover, J. D.; Reganold, J. P.; Cox, C. M. (2012): Agriculture: Plant Perennials to Save Africa's Soils. In *Nature* 489 (7416), pp. 359–361. DOI: 10.1038/489359a.

Granatstein, D.; Peck, G. (2017): Assessing the Environmental Impact and Sustainability of Apple Cultivation. In Gayle M. Volk, Amit Dhingra, Sally A. Bound, Dugald C. Close, Peter M. Hirst, M. C. Goffinet et al. (Eds.): *Achieving Sustainable Cultivation of Apples*. 1st ed. Cambridge: Burleigh Dodds Science Publishing (Burleigh Dodds Series in Agricultural Science), pp. 523–549.

Grove, G. G.; Eastwell, K. C.; Jones, A. L.; Sutton T. B. (2003): Diseases of Apple. In David Curtis Ferree, Ian J. Warrington (Eds.): *Apples. Botany, Production, and Uses*. New York: CABI Pub, pp. 459–488.

Haye, T.; Gariepy, T.; Hoelmer, K.; Rossi, J.-P.; Streito, J.-C.; Tassus, X.; Desneux, N. (2015): Range Expansion of the Invasive Brown Marmorated Stinkbug, Halyomorpha Halys: An Increasing Threat to Field, Fruit and Vegetable Crops Worldwide. In *Journal of Pest Science* 88 (4), pp. 665–673. DOI: 10.1007/s10340-015-0670-2.

Kellerhals, M.; Schütz, S.; Baumgartner, I. O.; Andreoli, R.; Gassmann, J.; Bolliger, N. et al. (2018): Broaden the Genetic Basis in Apple Breeding by Using Genetic Resources. In FOEKO (Ed.): *Ecofruit. Proceedings of the 18th International Congress on Organic Fruit Growing*. Ecofruit. Hohenheim, Germany. Fördergemeinschaft Ökologischer Obstbau e.V. (FOEKO), pp. 12–18.

Kellerhals, M.; Tschopp, D.; Roth, M.; Bühlmann-Schütz, S. (2020): Challenges in Apple Breeding. In FOEKO (Ed.): *Ecofruit. 19th International Conference on Organic Fruit-Growing: Proceedings*. Ecofruit. Hohenheim, 17.-19.02.2020. Fördergemeinschaft Ökologischer Obstbau e.V. (FOEKO), pp. 12–18.

Klein, A.-M.; Vaissière, B. E.; Cane, J. H.; Steffan-Dewenter, I.; Cunningham, S. A.; Kremen, C.; Tscharntke, T. (2007): Importance of Pollinators in Changing Landscapes for World Crops. In *Proceedings. Biological Sciences* 274 (1608), pp. 303–313. DOI: 10.1098/rspb.2006.3721.

Kremen, C. (2005): Managing Ecosystem Services. What Do We Need to Know About Their Ecology? In *Ecology letters* 8 (5), pp. 468–479. DOI: 10.1111/j.1461-0248.2005.00751.x.

Kremen, C.; Miles, A. (2012): Ecosystem Services in Biologically Diversified versus Conventional Farming Systems. Benefits, Externalities, and Trade-Offs. In *Ecology and Society* 17 (4). DOI: 10.5751/ES-05035-170440.

Lesur-Dumoulin, C.; Malézieux, E.; Ben-Ari, T.; Langlais, C.; Makowski, D. (2017): Lower Average Yields but Similar Yield Variability in Organic Versus Conventional Horticulture. A Meta-analysis. In *Agronomy for Sustainable Development* 37 (5). DOI: 10.1007/s13593-017-0455-5.

MEA (2005): Ecosystems and human well-being. Synthesis; a report of the Millennium Ecosystem Assessment. Washington, DC: Island Press.

Migliorini, P.; Wezel, A. (2017): Converging and Diverging Principles and Practices of Organic Agriculture Regulations and Agroecology. A Review. In *Agronomy for Sustainable Development* 37 (6), p. 143. DOI: 10.1007/s13593-017-0472-4.

Montanaro, G.; Xiloyannis, C.; Nuzzo, V.; Dichio, B. (2017): Orchard Management, Soil Organic Carbon and Ecosystem Services in Mediterranean Fruit Tree Crops. In *Scientia Horticulturae* 217, pp. 92–101. DOI: 10.1016/j.scienta.2017.01.012.

Palomo-Campesino, S.; González, J.; García-Llorente, M. (2018): Exploring the Connections Between Agroecological Practices and Ecosystem Services. A Systematic Literature Review. In *Sustainability* 10 (12), p. 4339. DOI: 10.3390/su10124339.

Peck, G. M.; Andrews, P. K.; Reganold, J. P.; Fellman, J. K. (2006): Apple Orchard Productivity and Fruit Quality Under Organic, Conventional, and Integrated Management. In *HortScience* 41 (1), pp. 99–107. DOI: 10.21273/HORTSCI.41.1.99.

Peck, G.; Merwin, I. A.; Brown, M. G. (2010): Integrated and Organic Fruit Production Systems for 'Liberty' Apple in the Northeast United States: A Systems-based Evaluation. In *HortScience* 45 (7), pp. 1038–1048, checked on 8/9/2018.

Peisley, R. K.; Saunders, M. E.; Luck, G. W. (2016): Cost-benefit trade-offs of Bird Activity in Apple Orchards. In *PeerJ* 4, e2179. DOI: 10.7717/peerj.2179.

Penvern, S.; Fernique, S.; Cardona, A.; Herz, A.; Ahrenfeldt, E.; Dufils, A. et al. (2019): Farmers' Management of Functional Biodiversity goes Beyond Pest Management in Organic European Apple Orchards. In *Agriculture, Ecosystems & Environment* 284, p. 106555. DOI: 10.1016/j.agee.2019.05.014.

Pfeiffer, B. (2020): Comparison of Rootstocks Geneva 16, M9 and CG11 under organic cultivation at the LVWO Weinsberg – actualized results 2009–2019. In FOEKO (Ed.): *Ecofruit. 19th International Conference on Organic Fruit-Growing: Proceedings*. Ecofruit. Hohenheim, 17.-19.02.2020. Fördergemeinschaft Ökologischer Obstbau e.V. (FOEKO), pp. 27–33.

Power, A. G. (2010): Ecosystem Services and Agriculture. Tradeoffs and Synergies. In *Philosophical Transactions of the Royal Society of London. Series B, Biological Sciences* 365 (1554), pp. 2959–2971. DOI: 10.1098/rstb.2010.0143.

Quijas, S.; Schmid, B.; Balvanera, P. (2010): Plant Diversity Enhances Provision of Ecosystem Services. A New Synthesis. In *Basic and Applied Ecology* 11 (7), pp. 582–593. DOI: 10.1016/j.baae.2010.06.009.

Reganold, J. P.; Glover, J. D.; Andrews, P. K.; Hinman, H. R. (2001): Sustainability of three apple production systems. In *Nature* 410 (6831), pp. 926–930. DOI: 10.1038/35073574.

Reganold, J. P.; Wachter, J. M. (2016): Organic Agriculture in the Twenty-first Century. In *Nature Plants* 2, p. 15221. DOI: 10.1038/NPLANTS.2015.221.

Rodríguez-Gasol, N.; Avilla, J.; Aparicio, Y.; Arnó, J.; Gabarra, R.; Riudavets, J. et al. (2019): The Contribution of Surrounding Margins in the Promotion of Natural Enemies in Mediterranean Apple Orchards. In *Insects* 10 (5). DOI: 10.3390/insects10050148.

Samnegård, U.; Alins, G.; Boreux, V.; Bosch, J.; García, D.; Happe, A.-K. et al. (2019): Management Trade-offs on Ecosystem Services in Apple Orchards Across Europe: Direct and Indirect Effects of Organic Production. In *Journal of Applied Ecology* 56 (4), pp. 802–811. DOI: 10.1111/1365-2664.13292.

Sandhu, H. S.; Wratten, S. D.; Cullen, R. (2010): Organic Agriculture and Ecosystem Services. In *Environmental Science & Policy* 13 (1), pp. 1–7. DOI: 10.1016/j.envsci.2009.11.002.

Seufert, V.; Ramankutty, N.; Foley, J. A. (2012): Comparing the Yields of Organic and Conventional Agriculture. In *Nature* 485 (7397), pp. 229–232. DOI: 10.1038/nature11069.

Simon, S.; Bouvier, J.-C.; Debras, J.-F.; Sauphanor, B. (2010): Biodiversity and Pest Management in Orchard Systems. A Review. In *Agronomy for Sustainable Development* 30 (1), pp. 139–152. DOI: 10.1051/agro/2009013.

Simon, S.; Sauphanor, B.; Buléon, S.; Guinaudeau, J.; Brun, L. (2008): Building up, Management and Evaluation of Orchard Systems: A Four-year Experience in Apple Production. In Jerry Cross, M. G. Brown, J. Fitzgerald, M. Fountain, D. Yohalem (Eds.): Proceedings of the 7th International Conference on Integrated Fruit Production. 7th

International Conference on Integrated Fruit Production. Avignon, October, 27–30, pp. 431–434.

Spornberger, A.; Schüller, E.; Noll, D. (2020): Influence of Geneva Rootstocks on Vegetative and Generative Characteristics of the Apple Cultivar 'Topaz' in a Replanted Orchard. In FOEKO (Ed.): *Ecofruit. 19th International Conference on Organic Fruit-Growing: Proceedings*. Ecofruit. Hohenheim, 17.-19.02.2020. Fördergemeinschaft Ökologischer Obstbau e.V. (FOEKO), pp. 174–175.

Swinton, S. M.; Lupi, F.; Robertson, G. P.; Hamilton, S. K. (2007): Ecosystem Services and Agriculture. Cultivating Agricultural Ecosystems for Diverse Benefits. In *Ecological Economics* 64 (2), pp. 245–252. DOI: 10.1016/j.ecolecon.2007.09.020.

Tilman, D. (1999): Global Environmental Impacts of Agricultural Expansion. The Need for Sustainable and Efficient Practices. In *Proceedings of the National Academy of Sciences of the United States of America* 96 (11), pp. 5995–6000. DOI: 10.1073/pnas.96.11.5995.

Tilman, D.; Cassman, K.; Matson, P. A.; Naylor, R.; Polasky, S. (2002): Agricultural Sustainability and Intensive Production Practices. In *Nature* 418, pp. 671–677.

Walch, B.; Schöneberg, A.; Perren, S. (2017): Apfelunterlagen im Test – Alternativen zu M9. In *Schweizer Zeitschrift für Obst- und Weinbau* 153 (6), pp. 8–12.

Waldman, K. B.; Richardson, R. B. (2018): Confronting Tradeoffs between Agricultural Ecosystem Services and Adaptation to Climate Change in Mali. In *Ecological Economics* 150, pp. 184–193. DOI: 10.1016/j.ecolecon.2018.04.003.

Wang, Y.; Li, W.; Xu, X.; Qiu, C.; Wu, T.; Wei, Q. et al. (2019): Progress of Apple Rootstock Breeding and its Use. In *Horticultural Plant Journal* 5 (5), pp. 183–191. DOI: 10.1016/j.hpj.2019.06.001.

Winkler, K. J.; Nicholas, K. A. (2016): More than Wine: Cultural Ecosystem Services in Vineyard Landscapes in England and California. In *Ecological Economics* 124, pp. 86–98. DOI: 10.1016/j.ecolecon.2016.01.013.

Zhang, W.; Ricketts, T. H.; Kremen, C.; Carney, K.; Swinton, S. M. (2007): Ecosystem Services and Dis-services to Agriculture. In *Ecological Economics* 64 (2), pp. 253–260. DOI: 10.1016/j.ecolecon.2007.02.024.

6 Characteristics of a Resilient Fruit Breeding and Cultivation System

In this chapter, the general resilience characteristics of fruit breeding and cultivation are developed and discussed. In order to get a broad understanding of the conditions and variables of a generally resilient social-ecological system, a detailed description of the *Principles for Building Resilience* by Biggs et al., which serve as the conceptual basis, is given. Besides general research on resilience, insights from resilience research with specific relations to agriculture and food systems are integrated into this description. Beyond a mere descriptive literature review and an explanation of resilience principles, this chapter aims to link existing resilience research with each other and interpret these links in detail. Thereby, the resilience principles are put in a direct context to other relevant resilience literature. Afterward, they are applied to fruit breeding and cultivation to formulate coherent and distinct characteristics of a resilient fruit breeding and cultivation system. These elaborations enable the conception of an analytical framework for evaluating the resilience of fruit breeding and cultivation approaches. The operationalization, assessment, and measurement of resilience have limitations and are challenging due to the abstractness and multidimensionality of the concept (Quinlan et al. 2016; Cumming et al. 2005). As Cabell and Oelofse (2012) put it: "[m]easuring resilience in social-ecological systems has proven to be like aiming at a moving target" (p. 3). In this context, literature on resilience distinguishes between specified and general resilience (Folke 2016).

Specified resilience refers to a particular set of state and external variables in a particular social-ecological system to assess the (specified) resilience of the system or parts of the system against particular shocks or disturbances (Carpenter et al. 2001, 2012). The concept operates under the leading question 'resilience of what to what and for whom?'. For example, the resilience of a park in a certain city district that provides a defined set of ecosystem services could be analyzed in light of heat stress as a particular disturbance. Several indicator-based approaches and models have been developed to assess and measure specified resilience. Quinlan et al. (2016) give an overview of approaches that include both broader ones and approaches that focus on single parts of systems or specific disturbances, for example, climate change. The most widely adopted approach is the workbook for practitioners by the Resilience Alliance[1] (RA) that uses distinct attributes to assess resilience (Resilience Alliance 2010).

DOI: 10.4324/9781003355724-10

However, as it was shown in Section 4.2, fruit breeding and cultivation systems are too complex and different from each other to apply specific indicator sets or measurement points. Specified resilience is therefore no appropriate perspective to develop an understanding of a resilient fruit breeding and cultivation system. Rather, a focus on the general resilience of these systems proves more beneficial. *General resilience* describes the building of the capacity to bounce back, adapt, or transform in response to known and unknown changes (Folke 2016; Walker and Salt 2012; Holling 2004) and is therefore consistent with the understanding of social-ecological resilience (See Section 1.1). This approach is more in line with the general character of agroecological systems:

> [...] building resilience gives agroecosystems the capacity to maintain the ability to feed [...] people in the face of shocks while building the natural capital base upon which they depend and providing a livelihood for the people who make it function
>
> (Cabell and Oelofse 2012, p. 3).

Although the literature on specified resilience is included in the following sections as a central part of resilience research,[2] the focus will be on ideas of the general resilience approach.

In order to get a broad understanding of the conditions and variables of a generally resilient social-ecological system, a detailed description of the *Principles for Building Resilience* (Biggs et al. 2015c) is given (See Section 6.1). Besides general research on resilience, insights from resilience research with specific relations to agriculture and food systems are added to this description. Beyond a mere descriptive literature review and explanation of resilience principles, this section aims to link existing resilience research with each other and interpret these links in detail. Thereby, the resilience principles are put in a direct context to other resilience literature relevant to this book. Afterward, they are applied to fruit breeding and cultivation to formulate coherent and distinct characteristics of a resilient fruit breeding and cultivation system (See Section 6.2). These elaborations enable the construction of an analytical framework for evaluating the resilience of fruit breeding and cultivation approaches Section 6.3).

6.1 Principles for Building Resilience

With the synthesis and generalization of different results and findings from theoretical as well as empirical studies related to resilience, Biggs et al. (2015c) formulated seven principles for building resilience. These principles aim to give a coherent and empirically sound overview about the possibilities for enhancing general resilience in social-ecological systems (Biggs et al. 2015b). The identification of these principles took place with a defined set of methodologies. First, an extensive literature review was carried out by members of the Resilience Alliance Young Scholars (RAYS). This review resulted in the identification of ten resilience principles and the formulation of a background paper for each principle as a base for discussion. The background papers were discussed at a two-day mock-court workshop[3] where primarily members of the RA participated. After the workshop, resilience experts (again mainly from the RA) were interviewed in a modified Delphi process, which was carried out independently from the process before. The experts were asked how the resilience of social-ecological systems can be enhanced and on the basis of the answers, an independent list of principles was created in an iterative process. The comparison of this list with the results of the mock-court workshop showed high concurrence.

As a result of this methodological process, seven generic principles for building resilience were formulated (Biggs et al. 2015b):

1. Maintaining diversity and redundancy
2. Managing slow variables and feedbacks
3. Managing connectivity
4. Fostering Complex Adaptive Systems (CAS) thinking
5. Encouraging learning and experimentation
6. Broadening participation
7. Promoting polycentric governance systems

Principles 1 to 3 refer to the *system-to-be-governed*, and therefore adopt an analytical lens on the state and controlling variables of the system. Principles 4 to 7 refer to the *management and governance system* of social-ecological systems. Hereby, a look at structural as well as dynamic aspects of social-ecological systems and their governance takes place (Biggs et al. 2015c). These distinctions are important because past research only focused on the system-to-be-governed and thereby foremost on static and structural aspects (Quinlan et al. 2016).

The majority of international scholars that carry out research on social-ecological resilience are members of the RA, the main organization for the scientific promotion of this concept. Consequently, these seven principles are a joint outcome of the relevant scientific community and serve as a summary of the current discussion on social-ecological resilience. It is the first in-depth publication that links insights of relevant case studies across scales and condenses broad existing literature into generic principles. By August 2022, Biggs et al. (2015c) were cited about 690 times according to GoogleScholar, exemplifying its relevance in the scientific community. However, some researchers also have a critical view of the RA as such.

Olsson et al. (2015) state that this organization dominates the conceptualization of social-ecological resilience and fails to bridge social and environmental sciences (See Section 1.1). In more detail, two journals particularly dominate the debate on social-ecological resilience: *Ecology & Society* and *Global Environmental Change*. The first journal is published by the RA and the editorial board of the second journal has personnel ties to the RA. Olsson et al. (2015) conclude that albeit social-ecological resilience has a strong but narrow community, outside these two main journals and other purely ecological and environmental journals, resilience theory and thinking are not further included in the main social science journals and contemporary social sciences. They promote the need for more pluralism in the application and development of resilience theory, leaving the path of inter-disciplinarity and rather seeing resilience as a "middle-range theory" (ibid., p. 9). However, although these criticisms are legitimate, the suggestions of Olsson et al. (2015) remain vague. For the integration of different scientific disciplines and on-tological perspectives, it seems more important which scientists or disciplines are working together rather than where exactly they publish their research. However, RA members also include scientists with a background in the social sciences and not only in environmental sciences. Moreover, several researchers across different disciplines already applied the principles to certain research objects and systems (e.g., Choi et al. 2021; Colding et al. 2020; Penny and Goddard 2018), as it is the aim of this book for fruit breeding and cultivation. The resilience principles can thus be classified as generally valid in the scientific community on social-ecological resilience and serve as a fitting basis for further elaborations.

In the following, each principle is described and discussed. Because all prin-ciples are comprehensively explained and analyzed in Biggs et al. (2015c), only a brief overview and explanation of important terms, effects, and implications for every principle are given, using their publication as the main source for this undertaking. If possible, literature on food and agricultural systems that refers to aspects of these principles is included and discussed. Of central importance for this transfer are the works of Cabell and Oelofse (2012), Tendall et al. (2015), Ben-nett et al. (2014), and Darnhofer (2014). Moreover, the case study of MASIPAG,[4] conducted by Gonçalves et al. (2018), will serve as an illustrative example for each principle but principle 3.

6.1.1 Principle 1: Maintaining Diversity and Redundancy

Diversity is an important aspect of social-ecological systems. As Walker and Salt (2012) state: "[r]esilient social-ecological systems would celebrate and encourage diversity" (p. 145). Referring to Sterling (2007), diversity can be defined as (a) the variety, (b) balance, and (c) disparity of elements with regards to a particular variable or function in a system (Kotschy et al. 2015). *Variety* (a) describes the number of different elements that contribute to particular functions or rather eco-system services, for example, the number of different species. *Balance* (b) means the number of each element and therefore the relative quantity of the respective element(s). *Disparity* (c) assesses the differences and similarities between the var-ious elements. Figure 6.1 illustrates these three aspects.

Within this definition of diversity, resilience literature distinguishes between functional and response diversity (Hodbod and Eakin 2015; Bennett et al. 2014; Cabell and Oelofse 2012). *Functional diversity* describes the variety, balance, and disparity of ecosystem services provided by the system. It defines the basic setting inside the system. *Response diversity* is also called redundancy and describes the availability of different possible reactions from the various elements in a system, always regarding a particular disturbance (e.g., changing climate conditions). Dis-parity is therefore the major character of redundancy. Many similarities between elements with respect to a specific function result in high redundancy.

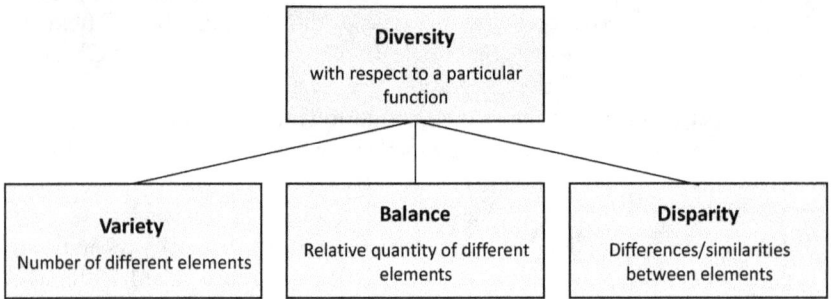

Figure 6.1 Three aspects of diversity (based on Kotschy et al. 2015).

A most illustrating and simple example is the farming of crops: whereas the farming of monocultures has little functional diversity and redundancy, the simultaneous farming of many different crops has high functional diversity and redundancy. By taking a look at food production, all crops fulfill the same function but respond differently to disturbances. Abson et al. (2013) show evidence for the validness of this example for agriculture in lowland regions in the United Kingdom.

Kotschy et al. (2015) provide evidence that low levels of diversity and redundancy in social-ecological systems compromise resilience whereas the increase of diversity and redundancy simultaneously increases resilience. Several authors confirm the importance of high diversity and redundancy in agricultural systems and argue that high rates are beneficial for the adaptive capacity of these systems (Urruty et al. 2016; Cabell and Oelofse 2012). Yet the increase in resilience is only true up to a certain level. After this certain level, more diversity and redundancy of the various elements lead to stagnation of resilience or even compromises it.[5] This is especially the case for social elements (Kotschy et al. 2015). However, trade-offs exist between efficiency as well as coupled diversity and redundancy (Ulanowicz et al. 2009; Bennett et al. 2014). For example, agricultural systems shaped for low diversity and redundancy provide higher (short-term) efficiency than systems shaped reversely. The realization of efficiency in a narrow sense is always coupled with the reduction of redundancy and therefore inevitably leads to a loss of resilience: "optimization (in the sense of maximizing efficiency through tight control) is a large part of the problem, not the solution" (Walker and Salt 2012, p. 140).

This observation emphasizes the importance of formulating the specific goal a social-ecological system aims to fulfill. For food systems, this goal is argued to be food security (Tendall et al. 2015). Hodbod and Eakin (2015) state that

> [t]he biological requirement for life defines the core goal of food systems – i.e. maintaining adequate food security for all humans, at all times – and the only permissible regime state of the food system

(p. 4).

Hodbod and Eakin (2015) developed an analytical framework for the general assessment and effects of diversity and redundancy in food systems. In this assessment, only high response diversity and functional diversity lead to universal food security. However, high redundancy and low diversity enable stability with respect to a particular function, whereas low redundancy and high diversity create multifunctionality (ibid.). Complementary to the aspect mentioned above, the goal of optimizing short-term economic benefits in food systems leads to low diversity and redundancy and therefore low resilience. The authors prove this with the analysis of the Californian food system in times of drought (ibid.). Tendall et al. (2015) also describe high redundancy and diversity (here described as resourcefulness) as vital elements for a resilient food system. Bennett et al. (2014) and Darnhofer (2014) stress the need for diversity of social and ecological variables in agriculture to promote innovation and diverse solutions to agricultural challenges. The example of MASIPAG shows the effects of high diversity and redundancy in social as well as ecological variables very clearly (Gonçalves et al. 2018). A large diversity of farmers, breeders, scientists, and other actors from different regions of the Philippines are joined in this network to foster the sharing and development of diverse rice varieties.

These collected insights lead to some practical implications and recommendations for the implementation of diversity and redundancy into social-ecological systems. Kotschy et al. (2015) make five suggestions:

- monitoring diversity and redundancy
- conserving and valuing redundancy
- maintaining ecological diversity
- building diversity and redundancy into governance systems
- changing the predominant management goals of (short-term) maximum efficiency to (long-term) resilience of ecosystem services

The above-mentioned case study of the Californian food system as a representative case (Hodbod and Eakin 2015) shows the severe consequences of not taking these suggestions into account when designing or managing food systems. As a result, all suggestions seem to be highly relevant for food systems. Gonçalves et al. (2018) strengthen this notion with the results of several case studies in agricultural systems. Both the suggestions and the case studies seem to indicate that the management paradigm (the last suggestion, see above) can be the key factor that eases the implementation of the other four suggestions. However, most food systems are still constructed and managed after the paradigm of maximum efficiency albeit the need for a paradigm shift is slowly recognized (Urruty et al. 2016).

6.1.2 Principle 2: Managing Slow Variables and Feedbacks

In the foregoing principle, system variables have been described according to their quantitative and qualitative characteristics. However, variables have different meanings in different systems and can be differentiated into slow and fast variables. Walker et al. (2012) made the following distinction:

> 'Fast' variables are typically those that are of primary concern to ecosystem users [...]. The dynamics of these fast variables are strongly shaped by other system variables that generally change much more slowly, and hence have been referred to as 'slow', or (because they are not always slow) 'controlling' variables.
>
> (p. 2).

Examples of *fast variables* are provisioning or regulating ecosystem services like crop production or pest regulation. These variables can be changed relatively fast, for example, by replacing one crop variety with another or switching pest regulation products and practices. *Slow variables* are the "underlying structure of social-ecological systems" (Biggs et al. 2015a, p. 109) that change or develop much more slowly than fast variables. Examples are supporting ecosystem services, for example, soil composition, or other variables such as rainfall or social norms. Effects of changes in slow variables are often only seen in the long-term whereas changes in fast variables have direct short-term effects on system dynamics. In turn, slow variables are affected by external variables outside the system. Slow variables are usually the key variables in social-ecological systems. Their current state determines the shape of the current regime and the possible crossing of thresholds into undesirable regimes as well as irreversible regime shifts (Walker and Salt 2012). A number of studies proved this (Biggs et al. 2015a), for example, the case study on the Amudarya River Basin in Central Asia where water usage and water logging have been identified as the key slow variables (Schlüter and Herrfahrdt-Pähle 2011). The case is further explained below. To sum up, slow variables play a vital role in building resilience in social-ecological systems. However, slow and fast variables are relative constructs and their declination is context-dependent. A variable that is slow in one system can be fast in another system (Walker et al. 2012).

Dependent on the connections between slow and fast variables, changes in a particular variable affect other variables, which leads to feedback mechanisms. Biggs et al. (2015a) distinguish between reinforcing/positive feedbacks and dampening/negative feedbacks. *Positive feedbacks* increase the change that took place in the initial variable whereas *negative feedbacks* decrease change. In the example of the Amudarya River Basin (Schlüter and Herrfahrdt-Pähle 2011), increased water usage in agriculture located by the river had positive feedbacks on the desertification of the river area and water logging, which in turn had negative feedbacks on agricultural productivity. In this case, changes in slow variables like water use and resulting water logging in agricultural areas were followed by feedbacks that deteriorated a number of ecosystem services. It is important to understand that feedback mechanisms can generate changes in slow variables that trigger (possibly undesired) regime shifts. In the new regime, other feedback mechanisms may exist.

Different ecological and social slow variables are of importance for agricultural and food systems. On a global level, Tendall et al. (2015) describe the following aspects as major external slow shifts that are negatively affecting the global food system resilience: climate change; soil degradation; pest outbreaks; economic and political crises; and population growth. Additionally, physical infrastructure such as production and storage systems are important social slow variables that determine the resilience of food systems (Worstell and Green 2017). On a local or regional level, one of the most critical ecological slow variables is soil fertility, constituted by different functional plants and animals as well as climate conditions (Chapin et al. 2009). As depicted in the Amudarya River Basin case (see above), water availability in general or water-holding capacity (Duru et al. 2015) is also a relevant ecological slow variable for agricultural systems. Overall, Cabell and Oelofse (2012) see soil fertility, the hydrological cycle, and biodiversity as major ecosystem services that, when strengthened and maintained, enhance the ecological self-regulation of agricultural systems. This self-regulation provides important feedback mechanisms that strengthen general resilience.

On a local level, Gonçalves et al. (2018) show a number of case studies in which soil fertility and water-holding capacity mark important slow variables that need a sustainable management. One example is the Kenyan branch of the association Participatory Ecological Land Use Management (PELUM).[6] In this case, an important social slow variable seems to be the underlying social norms of doing agriculture. This also reflects in the specific (normative) goal, food systems should have: food security (Hodbod and Eakin 2015).

Biggs et al. (2015a) make the following suggestions for managing slow variables and feedbacks to achieve social-ecological resilience.

- Identifying and getting a better understanding of key slow variables and feedbacks in a specific system is a necessary step for successful management.
- Strengthening and building up the feedbacks that keep the system in the desired regime by simultaneously weakening or cutting the feedbacks that keep the system in an undesired regime is important.
- Monitoring key slow variables and feedbacks and establishing governance structures to handle and respond to this monitoring information stabilize social-ecological resilience.

All suggestions are relevant to food and agricultural systems. The key slow ecological variables are the same for all agricultural and food systems (soil fertility, water-holding capacity, biodiversity). These key variables have to be maintained and promoted. In achieving this, feedback mechanisms have to be constructed that generate these maintaining and promoting effects. It appears that a crucial fast variable is the farming management system in agricultural systems that decides on the ecological impact and therefore on the feedbacks of farming management on soils. Farming management is strongly connected with the underlying norms in food systems, which is the key social slow variable.

6.1.3 Principle 3: Managing Connectivity

Connectivity describes the structure and strength by which state variables in a social-ecological system interact (a) with each other according to their different characteristics (See principle 2), and (b) with state variables of other systems (Dakos et al. 2015). In this web of connection, *structure* describes the quantity of connections (also called links) between different state variables (also called nodes). The structure of links occurs in different shapes (ibid.). For example, social variables inside a system or between different systems can be connected to each other by flows of information (See also Section 3.2). Ecological variables inside a system or between different systems can, for example, be connected through the interaction of species. A connection of social and ecological variables typically occurs by managing ecological variables, for example, landscapes.

To further categorize the structure of connectivity, Dakos et al. (2015) differentiate between random, nested, and modular structures. In a *random structure*, the links between the existing nodes are on average the same number for each node and the strength of the links is random, resulting in a largely balanced network with no particular structural characteristics. *Nested structures* are characterized by hierarchies between different groups of nodes. Here, nodes of one particular group (e.g., plants) are linked with a subset of nodes out of another group (e.g., pollinators), but not with each other. In *modular structures*, a diverse set of linked nodes is pooled in specific so-called compartments. These compartments are connected to each other through very few links.

Besides structure as a major characteristic, connectivity *strength* describes the quality or intensity of the links between different nodes. It is crucial to stabilize and intensify important links, for example, the flow of information between social actors. However, which links should be strengthened highly depends on the context and function of the respective links. In short: structure describes high or low connectivity, whereas strength describes strong or weak connectivity (Holling 2001).

High connectivity can both enhance and compromise resilience (Dakos et al. 2015). From an ecological perspective, it can facilitate recovery from disturbances or even block disturbances in all types of social-ecological systems, for example, through strong links to redundant variables (See principle 1). In landscapes, high connectivity also serves to maintain biodiversity by compensation and spillover effects from linked habitats or landscape patches that are activated when species are faced with a possible local extinction. From a social perspective, high connectivity is a necessary state for reaching resilience in other principles as it lays the base for learning (See principle 5) and participation (See principle 6). However, all of these effects only prevail until a certain level of connectivity and highly depend on the quality of the links, their diversity (See principle 1), and the character of disturbances. Therefore, high connectivity can also compromise resilience because confounding external variables like diseases get a better chance to spread through the system. Highly connected social variables can also compromise resilience through maladaptation when diversity of social actors is low, norms get homogenized, and necessary continuous learning effects (See principle 5) are limited. As a result, possibly positive or negative effects of high or low connectivity on resilience highly depend on the diversity and character of state variables in a system (Dakos et al. 2015).

The effects of the level of connectivity hold also true for food and agricultural systems. Depending on the specific context, high as well as low connectivity of different nodes and links can enhance the resilience of food and agricultural systems (Bennett et al. 2014; Tendall et al. 2015; Cabell and Oelofse 2012). For instance, Rotz and Fraser (2015) show the negative effects of high connectivity in their case study of the North American food system. They argue that a rising spatial and economic connectivity across countries (accompanied by declining diversity and decision-making autonomy) between mostly large farms with their oligopolistic structures compromises resilience. Ongoing concentration and integration processes in this food system and resulting commodity concentrations lead to higher vulnerabilities against perturbations like weather, disease, or price volatilities (Rotz and Fraser 2015). However, in another context – the attaining of global food security – Misselhorn et al. (2012) see rising connectivity between (local) actors, scales, and sectors as highly beneficial for the activation of necessary learning processes. These mainly include drawing local conclusions from global experiences and adapting agriculture on local conditions to ensure food security in a sustainable way, for example, with agroecological management approaches of locally adapted crops (See Section 5.2).

Agricultural systems are per se systems that connect biological, technical, and social variables with each other (Urruty et al. 2016). High connectivity can facilitate recovery after disturbing external effects or block these disturbances, as explained above. However, the ecological vulnerability of crop systems against particular diseases possibly increases with high connectivity. From a social perspective, connectivity can also lay the base for access to markets or resources like knowledge. As an example, the organization MASIPAG connects farmers and researchers to share their knowledge and a diversity of rice varieties (Gonçalves et al. 2018). The aim is to (re-)introduce rice varieties that are locally adapted, resilient against pests and climate damages, and require a minimum-use of pesticides. Here, it clearly shows that the establishment of connecting nodes and links is crucial to get the aimed learning and participation processes and outcomes. Participating farmers do not just benefit from introduced structures by simply getting access and using existent rice varieties. They also profit from the quality of the links by learning from other participants' knowledge to adapt their own rice farming system and make their farms resilient. Overall, modular structures seem to be the most appropriate options to build social-ecological resilience in agriculture on a global scale:

> [...] a resilient global agricultural system will be a mosaic of resilient regions, each one unique in some way, interacting through trade, assistance in times of need, and transfer of learning
>
> (Bennett et al. 2014, p. 67).

Cabell and Oelofse (2012) sum this thought up as 'appropriate connectedness.' On a local scale, Darnhofer (2014) additionally evaluates modularity as the key to connecting diverse farms with each other.

While connectivity depends on specific contexts and the characteristics of the system variables, some general implications can be found for managing connectivity (Dakos et al. 2015):

- Getting an overview of the social-ecological system by mapping connectivity and identifying important nodes and links is an important step to understanding its structure and strength.
- Connectivity should be restored where ecosystem services are threatened through not existing or declining nodes or links.
- Strength and structure of connectivity should be optimized to get the desired level of resilience.

These implications are important for designing resilient food and agricultural systems, as for example, indirectly shown by Lamine (2015). Specifically by observing agriculture in the context of global events, Darnhofer (2014) confirms the need to understand existing connectivity.

6.1.4 Principle 4: Fostering CAS Thinking

Whereas principles 1 to 3 focused on the system-to-be-governed, principle 4 and the other following principles are related to the governance system. Resilience literature often describes social-ecological systems as complex adaptive systems (CAS) (See also Section 4.1). This means that they naturally have high connectivity (See principle 3) and shift between different regimes in an adaptive and non-linear way of change because uncertainty and surprises are inherent characteristics of every social-ecological system (Bohensky et al. 2015; Walker and Salt 2012). Viewing and understanding social-ecological systems as CAS is defined as CAS Thinking, which is in turn based on resilience thinking (See Section 1.1). *CAS Thinking* explicitly accepts or even embraces uncertainty and surprise rather than trying to reduce it. As Darnhofer (2014) puts it, this way of thinking understands the dynamics of the earth system as "being fundamentally unpredictable" (p. 462) and hence marks "a radical departure from equilibrium-based approaches" (ibid.).

CAS Thinking is in stark contrast to the mechanistic view of natural resource systems (e.g., by traditional ecology or resource economics) in multiple ways (Schlüter et al. 2012). The most prominent differences are the assumptions of the mechanistic perspectives that system dynamics are linear, that stable and sustainable equilibria exist, and that individual state variables can be optimized by command-and-control approaches. CAS Thinking contradicts all of these assumptions, especially the aspect that components within a social-ecological system can be optimized in isolation: an optimal sustainable state of a system does not exist because system states are always changing (Walker and Salt 2012).

Consequently, "there is no 'normality,' no final equilibrium state that is maintained" (Darnhofer 2014, p. 466).

CAS Thinking can be understood as a mental model. *Mental models* are cognitive frameworks that illustrate how individuals and collectives understand and perceive the world (Bohensky et al. 2015; Biggs et al. 2011). In more detail, mental models "incorporate deeply engrained, often unquestioned, assumptions of the world and how it functions, affect how individuals filter, process and store information, and guide understanding, reasoning, prediction, and ultimately, action" (Biggs et al. 2011, p. 170). However, CAS Thinking is only one possible mental model besides other such as mechanistic thinking. This makes mental models important underlying paradigms that influence actions toward resilience. Ultimately, these models serve to analyze and compare different views on systems and try to answer how different actors within a system think and make sense of system dynamics. Individuals in a particular system often have different (individual) mental models, which can be in parts similar to the mental models of other individuals. Thereby, they can become collective mental models, such as the mechanistic worldview explained above. Mental models as a method are used as planning tools in various fields, for example, water resource management, forestry management, or education (Biggs et al. 2011). Fostering CAS Thinking, therefore, implies integrating this way of thinking, or rather elements out of it, in individual and collective mental models of actors in social-ecological systems.

It follows that CAS Thinking is beneficial for introducing resilient management approaches but has no direct influence on resilience. It rather affects the paradigms (Abson et al. 2017) that have influences on management approaches. However, not much empirical evidence exists for this hypothesis, albeit existing case studies indirectly point to its validity (Bohensky et al. 2015). CAS Thinking fosters holistic approaches that result in a "management of multiple ecosystem services and trade-offs in an integrated way [...] at multiple temporal and spatial scales" (ibid., p. 145). CAS Thinking also eases the implementation of the other resilience principles as it provides their normative and cognitive basis. Management approaches that result from mechanistic or reductionist mental models, for example, the farming of monocultures where only a very small number of ecosystem services is optimized, are rejected. As Bohensky et al. (2015) show, however, CAS Thinking is not easy to implement in larger institutions or groups and may even compromise resilience when attempts to integrate this way of thinking into existing mental models fail. For example, when CAS Thinking is only understood as addressing complexity, concurrent tools could be introduced to better understand this complexity, but the management approach as such is not changed to an adaptive one.

CAS Thinking potentially benefits the systemic management of food systems and farming systems (Urruty et al. 2016; Milestad et al. 2012). First, this way of thinking sensitizes the risks of environmental damages that result from mechanistic agricultural approaches. On the global level, these approaches can push regions or the Earth system itself over critical thresholds into undesired regimes. CAS Thinking makes actors more likely to search for multiple holistic management approaches that build a global agricultural system capable of meeting current and future challenges that are momentarily unknown and uncertain (Bennett et al. 2014; Urruty et al. 2016). Bohensky et al. (2015) show several case studies of agricultural systems in which the non-recognition of these systems as CAS led to agricultural mismanagement and environmental damages. Second, CAS Thinking on the farm level stresses the context-dependency of every (local) management approach. Darnhofer (2014) argues that farms managed based on a mechanistic view fail in these respects, for example, when monocultures with standard-optimizing approaches are cultivated that show no traits of local adaptability and hence lead to losses in biodiversity and farm resilience. In addition, Urruty et al. (2016) describe that bio-geophysical changes (climate change) or socio-economic changes (food price volatilities) continuously create uncertainties. Management approaches need to adapt to these uncertainties to sustain farms over longer time periods. Moreover, they argue that adaptive capacity is currently in general too low and needs to be improved on a farm as well as on a system level. Gonçalves et al. (2018) provide several case studies in which elements of CAS Thinking are implemented into the mental models of farmers. Their example of MASIPAG illustrates this point: the acceptance of risk and uncertainty reflects in the organization's building of adaptive capacity.

Bohensky et al. (2015) make the following recommendations for implementing CAS Thinking into (primarily collective) mental models and governance systems of social-ecological systems.

- It seems most important to develop an uncertainty-tolerant culture by engaging actors in participatory and educational processes (See principles 5 and 6).
- Diverse individual mental models have to be acknowledged and it should be accepted that these models can only partly be influenced.
- Critical thresholds need to be identified that define the operating space of the CAS.
- Institutions should be built up or refined to match CAS processes as closely as possible, which leads to necessary institutional change.

The implementation of CAS Thinking into mental models seems to be foremost relevant for the local level of farming, whereby its implementation into governance systems is important for the systemic level as well.

6.1.5 Principle 5: Encouraging Learning and Experimentation

In the investigation of learning processes, it is important to distinguish between the contents and the processes of how learning takes place. Literature on learning is very broad and involves multiple disciplines. Hence, specific concepts need to be selected that fit into the conception of social-ecological resilience because it only concentrates on particular facets of learning on a systemic and collective level. Cundill et al. (2015) identify two learning concepts that are of foremost relevance for this field of research and application: loop learning and social learning.

   Loop learning categorizes the levels of learning processes in a cyclic model (See Figure 6.2). According to this model, three different learning cycles exist: single-, double-, and triple-loop learning (Cundill et al. 2015; Pahl-Wostl 2009). The basic elements of this model are context, frames, actions, and outcomes (Pahl-Wostl 2009). Actors act in a particular *context* that describes the specific ecological and institutional structure of the system (See principles 1–3) and their views on this system (their mental models, See principle 4). Hereby, all actors apply specific *frames*, which include their basic assumptions derived from the context, that re- sult in specific goals or management priorities. Set in this context and influenced by their frames, actors implement certain *actions* that lead to certain *outcomes*. These outcomes initiate learning loops. *Single-loop learning* affects actions by changing practices, skills, or tools (are we doing things right?), without question- ing the underlying frame. *Double-loop learning* explicitly influences the frames of actors by reframing basic assumptions (are we doing the right things?) and chang- ing management goals or priorities. *Triple-loop learning* refers to the questioning of the context (how do we know what the right thing to do is?), specifically mental models, accompanied values and norms, governance structures, but also learning processes as such (Cundill et al. 2015; Pahl-Wostl 2009).

   These learning cycles – describing what is being learned – either happen in the forms of *social learning*[7] or *individual learning*. Because social-ecological re- silience focuses on systems, social learning – understood as collective learning in groups of actors – seems to be of more relevance than individual forms of learning. Cundill et al. (2015) distinguish three different forms of social learning. First, social learning can deliberately take place through interactive processes in which sharing takes place, for example, of knowledge, perspectives, or values. Second, social learning can deliberately be initiated through shared experimen- tation or reflection processes, such as monitoring programs. Third, it can be an

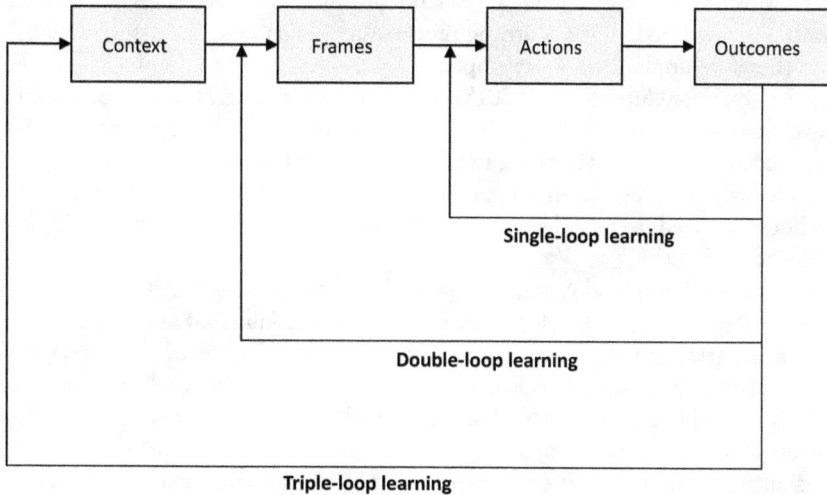

Figure 6.2 Learning cycles in the concept of loop-learning (based on Pahl-Wostl 2009).

unconscious outcome of social interaction as such. All three forms can be mixed in the particularly applied social learning approach.

As knowledge in social-ecological systems is always partial and incomplete due to the system's CAS character (See principle 4), the establishment of continuous learning processes is essential to achieve resilience and necessary to avoid negative regime shifts (Walker and Salt 2012). However, learning only influences resilience when it actually affects management or decision-making processes and therefore closes the knowledge-action gap. This emphasizes the need for fitting management approaches in social-ecological systems, which include learning as an essential element.

According to Cundill et al. (2015), three basic management approaches for social learning in social-ecological systems exist: adaptive management, adaptive co-management, and adaptive governance. *Adaptive management* involves social learning through experimentation and the setting up of monitoring programs to induce learning effects with the continuous testing of hypotheses. This approach emphasizes learning-by-doing as the key to achieve resilience: "[f]lexible social systems that proceed by learning-by-doing are better adapted for long-term survival than are rigid social systems that have set prescriptions for resource use" (Holling et al. 1998, p. 358). *Adaptive co-management* is based on the approach before but also includes elements of continuous knowledge sharing and involves multiple actors. Armitage et al. (2011) define this approach as a "collaborative process of bringing a plurality of knowledge sources and types together to address a defined problem and build an integrated or systems-oriented understanding of that problem" (p. 996). Finally, *adaptive governance* emphasizes social learning on a governance level by sharing knowledge across scales and explicitly aiming for double- and triple-loop learning. This management approach directly aims to shape and change institutions.

However, these management approaches are all planned activities with certain goals. As described above, learning processes can also take place unconsciously and therefore unplanned, for example, in traditional societies that highly rely on local resources (Cundill et al. 2015). Unplanned learning processes potentially have the same learning effects for resilience as planned learning processes. For planned and unplanned learning to be effective, it is important that it takes place on a long-term level and in an appropriate collaborative way. Planned learning additionally needs to be independent of short-term financial funding and political influences (ibid.).

Agricultural and food systems are severely affected by complexities and uncertainties (See principle 4). Moreover, occurring environmental and social breakdowns in agriculture (See Introduction) open up opportunities for loop- and social learning processes. Therefore, learning as such is an important and constant issue to build and shape resilience in food and agricultural systems (Bennett et al. 2014; Urruty et al. 2016). For example, triple-loop learning and social learning processes by sharing knowledge and creating new forms of knowledge can induce necessary changes in actor's mental models to build resilience in agriculture and food systems. Only through learning processes that encourage experimentation and innovation, sustainable and resilient agricultural management approaches may be defined and implemented that lead to a "mosaic of resilient regions" (Bennett et al. 2014, p. 66) and are locally adapted. Gonçalves et al. (2018) give examples of all types of successful loop and social learning in agricultural case studies where learning processes serve as important pieces of the conducted management approaches or governance systems. The authors state that "[learning] ensures that different types and sources of knowledge are valued and considered when developing solutions, and leads to greater willingness to experiment and take risks" (ibid., p. 7). For example, in the organization, MASIPAG social learning takes place by sharing knowledge and also by experimenting and breeding with rice varieties (ibid.). These positive effects of social learning are confirmed by Cabell and Oelofse (2012), who stress the need for shared, diverse, and reflective social learning processes by including these aspects in their set of agroecosystem resilience indicators.

Albeit learning is highly context-dependent, Cundill et al. (2015) give some general recommendations for constructing and managing successful learning processes:

- Long-term monitoring programs are favored tools to detect changes in social-ecological systems, especially with regard to slow variables (See principle 2).
- Institutions should be introduced that give (physical) room for social interaction and learning. For these, long-term funding has to be ensured that is, at best, independent from third parties with particular interests (e.g., politics and single companies).
- Participation in social learning processes should be as broad and diverse as possible to maximize their potential and outcomes (See principle 6).

Especially the second and third suggestions seem highly relevant to the context of food and agricultural systems. Participation puts learners in networks and enhances connectivity (See principle 3).

6.1.6 Principle 6: Broadening Participation

Participation as a concept is in most cases[8] linked with learning processes and elements of it have already been touched in the description of principle 5. For example, social learning as such is also a participatory approach (Stringer et al. 2006) and consequently, participation plays a major role in adaptive co-management or adaptive governance (See principle 5). The description and organization of participation always include (a) who participates (b) to what extent and (c) for what reasons. The respective actors that participate (a) and the reasons for participation (c) depend on the systems' context and can therefore not easily be studied to extract general insights.

The extent of participation (b) is the object of study in much general literature on participation. Probably most prominent is the so-called ladder of citizen participation by Arnstein (1969), in which eight stages of participation are distinguished – from 'manipulation' (citizens only get seats in committees etc. for reasons of teaching them the 'right' opinion) to 'citizen control' (citizens in fact self-manage and control specific processes). Although this model is heavily criticized (Collins and Ison 2006), its basic message has not changed: a whole string of options exists between the lowest possible form of participation, which is the described form of 'fake'-participation, up to the delegation of all decision-making power to participating actors. Rowe and Frewer (2005) condensed this multitude of options into three different forms of engagement, defined by the way information flows between participants: public communication, public consultation, and public participation.

In *public communication*, information only flows one way from decision-makers to participating actors, for example, by just informing actors about changes in the social-ecological system and not requesting feedback. In *public consultation*, information also flows one way, from participating actors to decision-makers. Decision-makers collect information and opinions but the participants have no influence on the afterward decision-making process. Finally, in *public participation*, information flows two-ways and participating actors get decision-making power in resulting forms of dialogue.

As a result, actors can participate in a passive or an active way. *Passive participation* means that actors have no decision-making power in the participatory process (public communication and consultation). *Active participation* refers to the opposite and aims at a transfer of power, as observed in most social learning processes. Therefore, the understanding of participation in the context of social-ecological resilience follows the definition of Leitch et al. (2015) and is defined as the "active engagement of relevant stakeholders in management and governance processes [inside social-ecological systems]" (p. 201).

These forms of participatory processes have essentially three potentials to build resilient social-ecological systems and enhance learning processes (Leitch et al. 2015; Stringer et al. 2006). First, active participation and its direct influence on decision-making bridge information-gathering and action processes. Besides the social learning effect, actors can really participate and are empowered. Second, actions derived from participatory processes can be legitimized through the building of trust and shared understandings, which provide the basis for collective action and successful social learning. This legitimization facilitates democracy because participants can decide on the direction and speed of changes in the system. Third, participation enhances the understanding of the structure and strength (See principle 3) as well as the dynamics of the social-ecological system. It activates scientific and experiential knowledge from different perspectives, whereby the activation of local knowledge is often particularly important for learning effects. Consequently, different social, political, and ethical values, thus, actor's mental models (See principle 4) can be identified. However, the success of participation highly depends on the selected participation process, the participants, and the systems' context (Leitch et al. 2015). Participation may fail responsible actors fail to engage important groups or individuals in the process, or if the process is used by certain groups or individuals to increase their power in the system for their personal benefits.

The implementation of learning processes is necessary to build resilient food and agricultural systems (See principle 5). However, these processes can only be induced in combination with (active) participation processes. This thought reflects in two indicators in the agroecosystem resilience framework of Cabell and Oelofse (2012), although the framework's focus is not on the governance system: social self-organization and the building of human capital. In a self-organized social structure, farmers participate in many local and regional institutions to strengthen their local resilience as well as the resilience of their communities and networks. These can be cooperatives, advisory boards, or farmer's markets. Worstell and Green (2017) confirm the specific importance of self-organization on a local scale for creating resilient food systems. Again, the example of MASIPAG illustrates the effects of participation on resilience by building self-organized structures. At MASIPAG, farmers actively participate in local networks as well as an overall national network. Additionally, other farmers' organizations, NGOs, and scientists also participate in this network (Gonçalves et al. 2018). Institutions of this kind provide the basis to build human capital as a result of participation. Cabell and Oelofse (2012) emphasize the role and relevance of human capital on a systemic level:

It is like a bank account, [...] filled with collective knowledge about how the world works. The more the account is filled with this knowledge, the more return it gets on an investment. In the agroecosystem, humans play a keystone role by influencing their environment to greater and lesser degrees. Both halves of the SES benefit if that influencing role is enacted by a well-informed, well-connected, and well-supported populace.

(p. 9)

With reference to the three facets of participation (who participates to what extent and for which reasons), Leitch et al. (2015) provide several recommendations for integrating participatory processes that enhance resilience:

- It is important to include a diverse range of actors as well as the 'right' people in the participatory setting (dependent on the goal and context of participation) to build an appropriate pool of participants. Hereby, inspiring and locally important leaders as well as competent facilitators, which organize and moderate the participatory process, have to be included.
- (Political) power structures between participants have to be recognized and their effects on the process have to be reduced.
- The particular approach of participation has to be selected according to existing financial and social resources as well as concurrent goals and target groups.
- The goals of the participatory process and expectations of participants should be made transparent and clarified before and throughout the process.

6.1.7 *Principle 7: Promoting Polycentric Governance Systems*

The general concept of polycentricity describes the decentral organization of actors on connected multiple scales (local, regional, national, global). It has been the object of study in a multitude of literature in social sciences (Aligica and Tarko 2012).[9] Polycentricity is also linked to the concept of telecoupling (Oberlack et al. 2018). Many authors conclude that polycentricity can potentially play a major role in shaping governance systems (Schoon et al. 2015). The form and extent of polycentricity in governance systems is defined and described in two respects (ibid.): (a) the breadth of inclusion and (b) the collaborative degree (See Figure 6.3).

The *breadth of inclusion* (a) is best understood when compared with the dichotomous opposite of polycentricity, described as monocentricity (Aligica and Tarko 2012). Monocentric governance systems have only one center of decision-making that encompasses all scales of governance. This center defines and controls the rules as well as the exclusiveness of the governance system top-down. On the contrary, polycentric governance systems have autonomous and decentral centers of decision-making on multiple scales, embedded in an overarching framework of norms and rules. In more detail, the breadth of inclusion in polycentric governance is defined by Schoon et al. (2015):

> [...] multiple interacting governance bodies with autonomy to make and enforce rules within a specific policy arena and geography [...] [that] interact with others at similar scales horizontally and within nested scales vertically.
>
> (p. 226)

However, inside these polycentric systems, the *collaborative degree* (b) can vary. Galaz et al. (2012) describe this continuum as weak and strong polycentricity. The placement of polycentric systems on this continuum depends on the degree of information sharing, the coordination of activities, and the existence of internal mechanisms for problem and conflict solving. Information sharing is the weakest

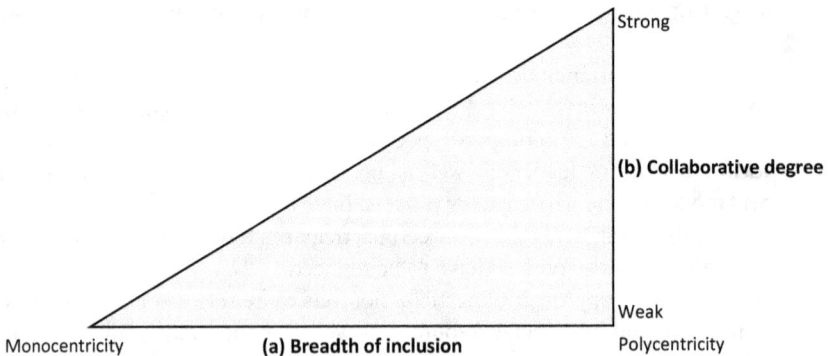

Figure 6.3 Form and extent of polycentricity in governance systems (based on Schoon et al. 2015).

form of polycentricity because it just demands a connectivity structure with little strength that serves for mutual adjustment in an otherwise loose construct of multiple actors. Strong polycentricity needs much higher strength in connectivity to coordinate activities of formal partnerships or joint projects that build new institutions on a large scale. The implementation of internal mechanisms for problem and conflict solving is described as the "strongest, and most demanding processes of polycentric order" (ibid., p. 24). In this case, the polycentric system is able to build shared understandings, carries out a number of joint projects, and incorporates a high degree of institutionalization. According to Schoon et al. (2015), the degree of collaboration can be measured by the level of modularity and connectivity (See principle 3). However, the potential of the collaborative degree is always context-dependent. This makes it difficult to create an exact categorization of weak and strong polycentricity.

Polycentric governance systems contribute to the building of resilience by positively influencing other principles of resilience (Schoon et al. 2015; Walker and Salt 2012):

- Polycentric systems can build or improve response diversity of social variables and thus help to build redundancy in social-ecological systems (See principle 1). For example, if one (decentral) governing body on a specific level in the polycentric system fails due to an external shock, other governing bodies can respond to this shock and provide help to the failing governing body.
- Structure and strength of connectivity can be improved by polycentric systems when such links are built and strengthened that contribute to high polycentricity. Moreover, polycentric systems are beneficial for building modular structures, which have been identified as appropriate to enhance social-ecological resilience (See principle 3).
- The inherent connectivity effects of polycentric systems can potentially provide the basis for the implementation of social learning processes (See principle 5) that affect multiple scales.
- Complementary to the learning processes, polycentricity is also able to broaden participation and activate its potential.

Overall, polycentric governance can serve as a blueprint for the resilient construction of governance in social-ecological systems by integrating a diverse spectrum of actors and institutions and providing the setting for learning and participation. However, both external and internal factors can challenge the stability of polycentric governance systems (Schoon et al. 2015; Galaz et al. 2012; Aligica and Tarko 2012). External environmental changes can lead to disaster situations where a polycentric system may not be able to react as fast as top-down hierarchical systems. Internally, polycentric systems are faced with the problem of high transaction costs if, for example, coordination across scales is too complex or an overlap of authorities is too strong. Moreover, actors may engage in politics and negotiation procedures in which they try to "externalize trade-offs from their area of interest" (Schoon et al. 2015, p. 237). All of these factors may at the worst lead to a breakdown of the polycentric system. However, its vulnerability ultimately depends on its breadth of inclusion as well as its collaborative degree in the specific context of the system.

Polycentric governance also poses an opportunity for the resilient design of food and agricultural systems. Gonçalves et al. (2018) describe these governance systems as "one of the best ways to achieve collective action in the face of disturbance and change [...] [and] well suited for agricultural systems" (p. 7). Some indicators of the framework by Cabell and Oelofse (2012) (See previous sections) include elements of polycentric governance. This encompasses social self-organization and appropriate connectedness. Other indicators, such as optimal redundancy, reflective and shared learning, and the building of human capital, are the potential results of a polycentric governance system. As a general example, elements of polycentric governance are observed at MASIPAG. The network spans local, regional, and national scales under the roof of a distinct construct of certain rules and norms. Joint projects are carried out across these scales. For example, groups of 12 to 60 farmers are responsible for one experimental farm on which local rice varieties are on trial. The best varieties of these experimental farms are then distributed to all members of MASIPAG (Gonçalves et al. 2018).

Overall, polycentric governance may provide a fitting umbrella concept for designing resilient governance systems because it has numerous effects on the other principles of resilience. As a result, this governance mode proves advantageous for agricultural systems. Albeit no direct research has been carried out on the potential of polycentric governance for food systems, this governance mode could possibly solve some challenges for cross-scale and multi-level interactions as posed by Ericksen (2008). However, as Schoon et al. (2015) remark: "polycentricity cannot be designed for but needs to be fostered by creating an accommodating social context" (p. 239).

6.1.8 Interlinkages Between the Principles

Throughout the descriptions of the resilience principles, it already became clear that every principle is linked to other principles in different ways. Table 6.1 provides an overview of the interlinkages between the resilience principles.

These interlinkages are of different shapes (Schlüter et al. 2015): while some principles *facilitate* other principles (e.g., participation facilitates learning), others have potential *antagonistic* effects on each other (e.g., high diversity can make participation also difficult). Another observance is *synergy* effects when principles interact with each other (e.g., diversity and connectivity enhance each other). There can also be facilitating or antagonistic effects between both principles, depending on how they are implemented. For example, diversity can provide the basis for participation, thus facilitating it. However, it can also have antagonistic effects when the diversity of interests and actors is too high, which challenges the effectiveness of participation mechanisms.

Essentially, this illustrates that resilience principles do not work separately. All principles must be applied to build social-ecological resilience. Schlüter et al. (2015)

Table 6.1 Interactions among resilience principles (based on Schlüter et al. 2015)

	P1	P2	P3	P4	P5	P6	P7
P1	▪	Synergistic			Facilitating	Facilitating/ Antagonistic	Synergistic/ Antagonistic
P2	Synergistic/ Antagonistic	▪			Facilitating/ Antagonistic	Synergistic	
P3			▪		Facilitating/ Antagonistic		
P4	Facilitating	Facilitating	Facilitating	▪	Facilitating	Facilitating	Facilitating
P5			Facilitating	Facilitating	▪		
P6		Synergistic			Facilitating	▪	Synergistic
P7	Synergistic	Facilitating			Facilitating	Synergistic	▪

identify three key mechanisms that have to be implemented in shaping resilient systems. First, important system variables, processes, and fitting management options have to be identified and understood. Second, the system has to be prepared for unexpected change by enhancing the awareness of the complexity of systems and providing alternative management options. Third, response capacity in the system-to-be-governed as well as in the governance system is key for dealing with change.

These key mechanisms are also valid for agricultural and food systems: "[a] more resilient agricultural system will need to be persistent, adaptive, and transformative, each at the appropriate moment in time and at the appropriate place" (Bennett et al. 2014, p. 68). For shaping a resilient fruit breeding and cultivation system, it is thus necessary to have a close look on all resilience principles.

6.2 Implications for Fruit Breeding and Cultivation

The aim of this section is to describe the ideal characteristics of resilient fruit breeding and cultivation systems. Consequently, this sheds light on how the relevant ecosystem services of fruit breeding and cultivation (See Chapter 5) can be promoted, maintained, and sustained in the face of disturbance and change. The characterization is carried out by formulating 'resilience attributes' that are derived from the resilience principles (See Section 6.1).

All thoughts and arguments developed in this section are based on the conceptual content of the foregoing sections in this Part II, especially in Section 6.1, and the empirical context (See Chapter 3). Because resilience principles 1–3 explore the system-to-be-governed, direct links to the elaborations in Chapters 4 and 5 can be drawn in this characterization. This proves more difficult for principles 4–7, which refer to the governance system. Here, many back references will be drawn to the system-to-be-governed and its congruent set of ecosystem services. Thus, the resilience attributes discussed in this section are derived from the foregoing elaborations in a logical way and literature references are only included when it is perceived as necessary. Many back references will be given to the previous chapters to make their deduction as comprehensible as possible.

After elaborating on every resilience principle, a short respective summary follows that condenses the thoughts and formulates attributes that describe a resilient fruit breeding and cultivation system. In the end, Figure 6.4 shows an overview of these attributes and their influences on ecosystem services in fruit breeding and cultivation.

6.2.1 Principle 1: Diversity and Redundancy in Fruit Breeding and Cultivation

The uncovering of ecosystem services in fruit breeding and cultivation (See Section 5.3) already showed the important role of diversity in its different facets. 'Planting diversity' with ecological elements such as hedgerows, cover crops, flower strips, etc. promotes several regulating and supporting ecosystem services in fruit cultivation. Using the diversity of traditional and heirloom fruit cultivars in breeding could enhance the robustness and resilience of future fruit varieties. Integrating diverse ecological elements could also provide further cultural ecosystem services by realizing aesthetic, ethical, or spiritual values or creating identity. Examples of those cases were shown in Section 5.3.

In shaping resilient fruit cultivation systems, it is thus necessary to identify and monitor the diversity and redundancy of system variables connected to the provision and sustainability of relevant ecosystem services to give learning opportunities: which ecosystem services are promoted with current agricultural practices and how do they change when these practices are changed? Resilient fruit cultivation systems also conserve and value redundancy, for example, by planting different fruit cultivars with different traits. Robust cultivars that do not provide large yields are redundant in average seasons but prove their value under harsh environmental conditions when high-performing cultivars fail.

Fruit breeding relies on the supply of different varieties that could be used for breeding. Monitoring the diversity and redundancy of fruit varieties and their traits by building gene banks, variety collections, or catalogs is thus an important aspect. However, the conservation of this knowledge and the physical varieties includes giving access to them. Diversity does only prove its value for resilience if this diversity is usable. Overall, this suggests that redundancy should be valued on many levels: any local or special variety could have a gene that potentially proves valuable for a more resilient fruit cultivation system.

Resilience research further suggests the implementation of diversity and redundancy into governance systems to promote and sustain ecosystem services. Thus, resilient fruit cultivation and breeding systems should promote a diversity of actors and organizational models along the whole value chain (See Section 3.2). This additionally involves changing the management paradigm of optimization that is predominant in most fruit cultivation and breeding systems. Rather, eco-centric approaches (See Section 5.1) should be adopted to encourage diversity on a broad level.

Promoting diversity of actors inherently lays the ground for enhanced learning opportunities: breeders, farmers, and marketeers can learn from each other. Any actor could get the opportunity to realize her individual ethical values with the most appropriate cultivation or breeding model. Thus, diversity in governance systems benefits cultural ecosystem services. There is also a direct linkage between the diversity of actors and the diversity of cultivars: many hobby breeders or farmers cultivate and conserve heirloom or 'old' cultivars in their gardens. Activating this diversity and letting those actors participate in the breeding and cultivation system would benefit the systems' overall resilience.

This leads to the following attributes that describe the ideal characteristics of a resilient fruit breeding and cultivation system:

- integrating diverse ecological elements in breeding (genetic diversity) and cultivation ('planting diversity')
- integrating diverse social elements (actors, institutions)
- monitoring diversity and redundancy of fruit varieties

6.2.2 Principle 2: Slow Variables and Feedbacks in Fruit Breeding and Cultivation

Similar to agricultural systems in general, the key ecological slow variables in fruit cultivation systems are the supporting and regulating services of soil nitrogen availability and water cycling and maintenance. Both are slowly changing variables and are influenced by a range of other ecosystem services as well as each other (See Section 5.3). The key social slow variable is the management paradigm. As shown in Section 5.2, the underlying norms that determine the farming approaches and practices directly influence the status and development of ecosystem services. In general, the key ecological slow variables and their interactions with other variables are well understood, as shown by Demestihas et al. (2017) in their study on ecosystem services in fruit cultivation (See Section 5.3). However, when shaping resilient local cultivation systems, it is important to identify the distinct local characteristics (e.g., pedoclimatic conditions) and their specific influence on the local nitrogen and water cycle. Thus, for designing a resilient local fruit cultivation system it is necessary to understand the specific conditions of its slow variables.

Resilience research further suggests that those key ecological variables have to be maintained and promoted, and concurrent feedback mechanisms need to be strengthened (See Section 6.1). This means that the 'right' fast variables (e.g., choice of varieties for cultivation, instruments of pest regulation) have to be activated and linked with the slow variables. Equal to the resilience principles of diversity and redundancy (ibid.), paradigms are key elements in constructing these feedback mechanisms. For example, it was shown that practices such as the use of pesticides/fertilizers or cover crops respectively influence the water cycle – the choice of either agricultural management practice relates to the adopted farming approach or paradigm.

Another important feedback mechanism is the general connection of fruit cultivation to fruit breeding. Strengthening breeding approaches that promote the breeding of cultivars with traits that minimize negative impacts on the nitrogen and water cycle and enable resource-preserving cultivation could prove beneficial for resilience. It becomes clear in the overlap of both social-ecological systems that biodiversity is the major resource base for both systems (See Section 4.2). Strengthening feedbacks with this resource base by using the whole range of its diversity in fruit breeding and cultivation improves the resilience of fruit breeding and cultivation.

Monitoring the identified key ecological slow variables and their feedback mechanisms with other variables by implementing fitting governance structures helps to develop resilience or maintain it. With regards to fruit breeding, some examples of those structures are shown in the case studies on apple breeding approaches (See Part III).

This leads to the following attributes that describe the ideal characteristics of a resilient fruit breeding and cultivation system:

- understanding specific conditions of local water and nutrient cycles
- maintaining, promoting, and monitoring positive feedbacks on slow variables
- promoting ecocentric breeding approaches that maintain slow variables

6.2.3 Principle 3: Managing Connectivity in Fruit Cultivation and Breeding

The connectivity of a social-ecological system is described by the structure and strength of its specific links and nodes. For shaping resilient systems, getting an overview of its connectivity enables the identification of those nodes and links that are important for sustainability. The multi-level conceptualization of fruit cultivation and breeding gives an idea about the structure of connectivity between the state variables ecosystem, people and technology, knowledge, and institutions (See Figure 4.8 in Section 4.2).

Following insights from resilience research, the strength and structure of connectivity in this multi-level system should be optimized to meet resilience demands. Especially in the social system, strong and high connectivity is crucial for building resilience. Here, institutions can serve as nodes for connecting (a) fruit farmers with each other, (b) fruit breeders with each other, and (c) fruit farmers and breeders. Establishing those links for exchanging information, knowledge, and technology promotes learning opportunities. This cultural ecosystem service is important to continuously adapt (and possibly transform) fruit breeding as well as cultivation in changing environments. Finding the appropriate connectedness (Cabell and Oelofse 2012) in this net of institutions provides effective and efficient learning opportunities to better cope with change.

Option (c) mentioned above was already illustrated with the example of MASIPAG in Section 6.1, where researchers and farmers share knowledge and cultivars. (Re-)establishing a strong connectivity between fruit cultivation and breeding by the coupling of both systems is thus a beneficial option to achieve this appropriate connectivity. At the moment, this form of connectivity only exists on small scales. For example, most fruit farmers and breeders in Germany are disconnected from each other (See Part IV). Implementing participatory models of breeding would strengthen connectivity because it creates direct links between farmers and breeders (See Section 4.2).

Besides connectivity in the social system, another option for building resilience in fruit cultivation is the connection of agriculture with other ecosystems to promote diversity and robustness. An example is agroforestry, where agroecological systems are connected with forestry (silvo-arable systems) or additionally with animal husbandry (agrosilvo-pastorale systems). Many researchers highlight the benefits of this connection for promoting mainly provisioning and regulating ecosystem services such as climate regulation, soil nitrogen availability, water regulation and maintenance, or pollination (Böhm et al. 2019; Brown et al. 2018). A modern example of agroforestry is alley-cropping systems, in which fields alternate with (fruit) tree rows. Traditional agroforestry systems in the fruit context are meadow orchards, although most of those systems solely function as nature conservation areas without an economic function (Nerlich et al. 2013).[10]

However, transferring this resilience principle to fruit breeding and cultivation remains, in a way, highly abstract. As shown in Section 4.2, breeding and cultivation are also telecoupled with food systems on a multi-level scale. Mapping connectivity on this global scale and finding its optimal strength and structure is difficult and rather pointless because it is ever-changing. Thus, shaping resilience by connectivity should concentrate on building modular structures on regional and national scales to build the "mosaic of resilient [fruit] regions" Bennett et al. (2014, p. 66) were imagining.

This leads to the following attributes that describe the ideal characteristics of a resilient fruit breeding and cultivation system:

- building and strengthening links and nodes to exchange knowledge and technology
- connecting ecosystems to build synergies

6.2.4 Principle 4: Fostering CAS Thinking in Fruit Breeding and Cultivation

The principle of CAS Thinking refers to the governance of fruit breeding and cultivation. It poses the question of which norms, rules, and instruments influence those domains and how they can be altered to build adaptive capacity and apply adaptive management approaches. Fruit breeding as such is predestined to adopt CAS Thinking because breeding always tries to improve current cultivation in different ways – thus, it is about constantly changing cultivation systems and adaptiveness. However, beyond this basic categorization, it emphasizes the mental models of actors practicing conventional forms of breeding that doing breeding does not necessarily foster resilience. Rather, it could also induce tremendous environmental and social problems, such as the close genetic basis of important apple cultivars or economization trends.

Resilience scholars would argue that CAS Thinking can be implemented into mental models of fruit farmers and breeders by developing an uncertainty-tolerant culture. Part of this culture is to accept the context-dependency of every local fruit farm. Resilient fruit cultivation, therefore, needs a diversity of cultivars and agricultural management practices. Without diversity in fruit breeding, fruit farmers will not be able to establish this context dependency. Similarly, without translating CAS Thinking into distinct adaptive management practices there will be no effect at all. Learning opportunities as a cultural ecosystem service are important to deal with uncertainties and to help establish those management practices. Enabling learning opportunities by connecting breeders, farmers, and other actors with each other is necessary to build adaptive capacity. Without the exchange of knowledge or physical resources (e.g., breeders sharing seedlings), fruit farmers and breeders are not able to build appropriate mental models.

All of these opportunities can emerge by implementing CAS Thinking into norms, or rather collective mental models. Norms can translate into management approaches that consider the whole set of relevant ecosystem services in shaping fruit breeding and cultivation systems. The foregoing elaborations on the impacts of different farming approaches (See Section 5.2) indicate that collective mental models such as diversified farming, agroecology, or organic farming are suitable approaches to sustain and develop relevant ecosystem services. Individual mental models that adopt these norms are beneficial steps toward resilient fruit breeding or cultivation. This goes in line with the necessary institutional change to promote CAS Thinking.

This leads to the following attributes that describe the ideal characteristics of a resilient fruit breeding and cultivation system:

- implementing CAS Thinking into mental models of relevant actors
- translating CAS Thinking into institutions

6.2.5 Principle 5: Encouraging Learning and Experimentation in Fruit Breeding and Cultivation

Learning and experimentation processes provide the basis for a continuous adaptation of fruit cultivation and breeding systems to environmental changes. Scientific research and observations of practitioners are important for learning about variables of the system-to-be-governed. Resilience scholars recommend long-term monitoring programs to detect changes in variables and consequently to better cope with them (See Section 6.1). In fruit breeding, increasing research is carried out on pedigree relations of apple cultivars. Many scientists are connected to each other around the globe and share their knowledge to certain degrees (See Section 3.2). Identifying and monitoring pedigree relations in fruit breeding has the potential to optimize fruit breeding activities because desired traits can be better located in genotypes and uncertainties of crossings between cultivars can be reduced. Fruit cultivation as an agricultural practice is part of several agricultural monitoring programs[11] and much knowledge exists on the impacts of agricultural practices on ecosystem services (See Section 5.2). However, to shape resilient fruit cultivation systems, the interaction of ecosystem services needs monitoring.

Monitoring programs are important institutions to learn about the dynamics of fruit breeding and cultivation. Further promotion of social learning institutions is necessary to collect data for those programs, explore this data, and disseminate concurrent results and insights. Options for shaping resilient learning institutions (See Section 6.1) are (a) learning-by-doing with education programs and by networking between practitioners (adaptive management), (b) continuous knowledge sharing between multiple actors (adaptive co-management), and (c) knowledge sharing across scales that aims at double- and triple-loop learning (adaptive governance). Management and governance institutions already exist in fruit breeding and cultivation (See Chapter 3). However, most of them do not integrate CAS-Thinking in their interaction processes. In this way, many learning processes will not reach beyond single-loop learning (are we doing things right?). For example, breeding institutions that adopt a mechanistic perspective and perceive systems as linear and equilibrium-based can hardly promote resilience. It is thus necessary to tip existing institutions in fruit breeding and cultivation into double- and triple-loop learning processes or build new ones to foster resilience.

Without involving all relevant actors in those learning processes, developing resilient systems may fail. Besides the qualitative dimension of learning as explained above, the quantitative dimension is equally important. Resilience research suggests designing the participation structures in learning processes as broad and diverse as possible (See Section 6.1). Building networks between scientists, farmers, breeders, and other actors along the value chain enhances social connectivity. Fruit farmers should participate in breeding experiments to foster social learning and avoid disconnects between breeding and cultivation. Encouraging resilient learning and experimentation processes between both systems is key to shaping robust and adaptive systems.

This leads to the following attributes that describe the ideal characteristics of a resilient fruit breeding and cultivation system:

- implementing long-term monitoring programs for key ecosystem services (similar to principle 2) and pedigree relations of fruit varieties
- Promoting social learning institutions

6.2.6 Principle 6: Broadening Participation in Fruit Cultivation and Breeding

Combining resilient learning and participation processes in fruit breeding and cultivation means involving the 'right' actors. Who should actively participate in learning processes for what reasons? At the core of participation, fruit breeders and farmers should be located by establishing information flows two-way and transferring power to them. This is crucial for promoting resilience because both actor groups depend on and cannot further develop without each other. Other actor groups along the value chain have also interests to participate in breeding or cultivation decisions: distributors have reasons to participate in breeding and cultivation by sharing knowledge, for example, about the most appropriate fruit sizes or distribution channels. Marketeers and consumers could give valuable insights into consumer demands and trends. However, all of these additional actors participate at the periphery of the breeder–farmer nexus. Without a strong collaboration between both actor groups, the participation of other actors is irrelevant. It is also necessary to reduce the power of participating actors that adopt mechanistic perspectives on fruit breeding and cultivation. Otherwise, participation could have a negative influence on resilience.

The participative methodology needs a narrative to achieve clarity of goals and expectations. It is valuable to adopt PPB as an appropriate approach to be clear about goals, roles, and expectations throughout the whole participatory process. In PPB, fruit farmers actively participate in fruit breeding by being involved in every step of breeding – from the formulation of breeding goals throughout the selection steps to the choice of cultivars for legal assessment. This inclusive concept can be extended to further actors by using the same procedure, hereby fostering their identification with goals and expectations.

Research on agricultural systems and resilience further suggests that social self-organization is beneficial to activate human capital and build robust agro-ecosystems. This is also true for fruit breeding and cultivation: cooperatives, networks, associations, etc. can provide valuable institutions for local resilience. However, those institutions demand financial resources, time, and effort of participating actors. It is thus necessary to develop concurrent business models that compensate for participatory activities. Designing resilient participation thus needs to integrate appropriate business models into current market structures or transform market structures in a way that they may be able to integrate alternative and innovative business models.

This leads to the following attributes that describe the ideal characteristics of a resilient fruit breeding and cultivation system:

- promoting the participation of relevant actors
- implementing a narrative for participation such as PPB
- promoting social self-organization and congruent institutions

6.2.7 Principle 7: Promote Polycentric Governance in Fruit Breeding and Cultivation

Polycentric governance as an umbrella concept for organizing, coordinating, and developing fruit breeding and cultivation could prove advantageous, especially with regard to their telecoupling characteristics (See Section 4.2). Polycentricity fosters the diversity and connectivity of social variables, social learning processes, and participation. It can support the activation and promotion of many other resilience principles. Creating this "accommodating social context" (Schoon et al. 2015, p. 241) by acknowledging and fostering multi-level and cross-scale interactions could provide the basis for designing institutions that promote ecosystem services in a resilient way. This principle in a way bundles many insights of the previous elaborations and is an umbrella principle for principles 4 to 6.

Polycentricity well reflects the governance demands of conceptualizing fruit breeding and cultivation as telecoupled social-ecological systems in a multi-level system (See Section 4.2). It simultaneously emphasizes the local character of fruit orchards or breeding grounds and their impact on regional, national, or global scales. PPB includes elements of polycentricity into its concept by positioning farms that participate in breeding activities as decentral decision-making bodies in an overarching governance framework. Formal partnerships and joint projects between farmers and breeders (and potentially other actors) give shape to a polycentric system and are necessary elements to build nodes for connectivity. Building polycentric governance in fruit breeding and cultivation can translate the mental model of CAS Thinking into an appropriate governance design.

6.2.8 Conclusion: Characteristics of a Resilient Fruit Breeding and Cultivation System

The elaborations in this section shed light on how to understand and design resilient fruit breeding and cultivation systems. Again, it became clear that the resilience principles and their respective attributes are interwoven in a complex way. This needs to be highlighted in connection to ecosystem services.

Figure 6.4 summarizes the influences of the identified resilience attributes on ecosystem services of fruit breeding and cultivation. The left side of the figure

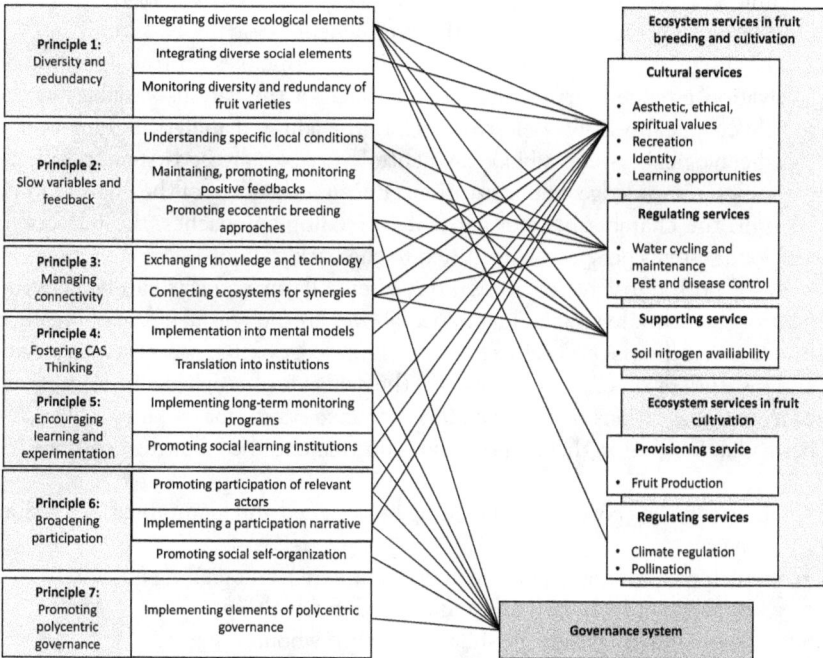

Principle 1: Diversity and redundancy
- Integrating diverse ecological elements
- Integrating diverse social elements
- Monitoring diversity and redundancy of fruit varieties

Principle 2: Slow variables and feedback
- Understanding specific local conditions
- Maintaining, promoting, monitoring positive feedbacks
- Promoting ecocentric breeding approaches

Principle 3: Managing connectivity
- Exchanging knowledge and technology
- Connecting ecosystems for synergies

Principle 4: Fostering CAS Thinking
- Implementation into mental models
- Translation into institutions

Principle 5: Encouraging learning and experimentation
- Implementing long-term monitoring programs
- Promoting social learning institutions

Principle 6: Broadening participation
- Promoting participation of relevant actors
- Implementing a participation narrative
- Promoting social self-organization

Principle 7: Promoting polycentric governance
- Implementing elements of polycentric governance

Ecosystem services in fruit breeding and cultivation

Cultural services
- Aesthetic, ethical, spiritual values
- Recreation
- Identity
- Learning opportunities

Regulating services
- Water cycling and maintenance
- Pest and disease control

Supporting service
- Soil nitrogen availiability

Ecosystem services in fruit cultivation

Provisioning service
- Fruit Production

Regulating services
- Climate regulation
- Pollination

Governance system

Figure 6.4 Influences of resilience attributes on ecosystem services in fruit breeding and cultivation (own figure).

shows the attributes of the respective resilience principles, derived from the elaborations in this section. Lines drawn between these attributes and the ecosystem services on the right side show their promoting and/or sustaining effects on relevant provisioning, regulating, supporting, and cultural ecosystem services that have been identified in Section 5.3. Additionally, some lines show the demands of resilience attributes on the design of the overall governance system.

6.3 Building an Analytical Framework

The foregoing results (See Sections 6.1 and 6.2) show how to build resilient fruit breeding and cultivation systems in order to maintain, promote, and sustain the desired set of ecosystem services that was defined in Section 5.3. Therewith, the results also provide insights into different farming and breeding approaches and their effects on social-ecological resilience.

In order to evaluate specific apple breeding approaches, which is the objective of the following Part III, an analytical framework is necessary. In line with the methodological approach of this book (See Chapter 2), this framework will adopt an explorative and qualitative perspective. Because every apple breeding approach represents a different social-ecological system, this framework is conceptually embedded in the adjusted analytical framework for social-ecological systems (See Section 4.1 and Figure 4.4). The variables of this framework work as guiding posts to identify necessary questions and perspectives for the evaluation of breeding approaches in light of their resilience.

The discussed implications of the resilience principles on fruit breeding and cultivation provide entry points for the design of the various variables in their social-ecological systems. Referring back to the adjusted analytical framework for social-ecological systems (ibid.), the central variables are ecosystem, people and technology, knowledge, and institutions. Because the focus of the analysis in Part III is on the characteristics of particular breeding approaches, the telecoupling and food systems perspective is shifted to the background.

Based on the insights of Section 6.2, essentially two aspects can be derived in light of the central variables mentioned above: (a) the *general effects* of organizing people and technology as well as modeling knowledge exchange, ecosystem services, and institutions on ecosystems; and (b) the *demands* of people, technology, knowledge, and institutions on the overall governance system for building resilience. In particular, insights on these aspects may contribute to the following questions:

- how can the ecosystem of breeding be designed and maintained in a resilient way?
- which people should participate to what extent in breeding processes?
- what kind of technology should be used in which way?
- how should knowledge be shared and with whom?
- which kinds of institutions are necessary for a resilient system?
- how have the interactions between social and ecological systems been organized?

Additionally, the previous section showed which flows between different social-ecological systems (especially between breeding and cultivation) should be implemented, maintained, or strengthened. For example, it seems necessary to establish institutionalized flows of information between fruit breeding and cultivation.

In this context, it is necessary to operationalize the resilience attributes (See Section 6.2) with an analytical framework to be able to explore the described aspects and evaluate breeding approaches. This analytical framework serves as a vehicle for collecting, exploring, and (most importantly) interpreting data and information in light of the specific problem context, here apple breeding in Germany (See Chapter 3). In conclusion, the framework will be able to evaluate if and in which way particular fruit breeding approaches meet the demands of the identified resilience attributes.

The resilience attributes (See Figure 6.4) hence serve as elements of analysis in this framework, similar to a list of indicators. Hereby, two attributes of Principle 2 ("Understanding specific local conditions"; "Maintaining, promoting, monitoring positive feedbacks") and one attribute of Principle 3 ("Connecting ecosystems for synergies") are left out. These attributes particularly refer to fruit cultivation. As apple breeding is the focus of this book, the framework is only relevant for the evaluation of fruit breeding approaches and includes 13 resilience attributes. For gathering information on the effects of the respective approaches on the attributes, a catalog of questions was developed that enables the collection of relevant information and the interpretation of this information in light of the research aim (See Table 6.2).

Table 6.2 Analytical framework for the evaluation of fruit breeding approaches (own table)

	Resilience attributes	Questions for data collection
Principle 1	Integrating diverse ecological elements	How is diversity reflected in agricultural management and breeding practices? How many and which fruit varieties are used in breeding?
	Integrating diverse social elements	Which actors participate in breeding activities?
	Monitoring diversity and redundancy of fruit varieties	How does the breeding approach contribute to monitoring activities? Does the breeding approach monitor the diversity and redundancy of local ecosystem services?
Principle 2	Promoting ecocentric breeding approaches	What is the underlying management paradigm of the breeding approach?
Principle 3	Exchanging knowledge and technology	Do breeders exchange knowledge and/or technology with other breeders/farmers? To what extent? Which kind of institutions exists in this context?

(Continued)

Table 6.2 Continued

	Resilience attributes	Questions for data collection
Principle 4	Implementation into mental models	Are elements of CAS Thinking implemented into the breeding approach?
	Translation into institutions	Do elements of CAS Thinking reflect in the specific management practices?
		Do institutions exist that promote learning opportunities?
Principle 5	Implementing long-term monitoring programs[a]	Does the breeding approach consider long-term monitoring programs of ecosystem services?
		Has the breeding approach access to long-term monitoring programs of pedigree relations?
	Promoting social learning institutions	How does the breeding approach foster social learning?
		Do learning institutions promote single-loop, double-loop, or triple-loop learning?
Principle 6	Promoting the participation of relevant actors	Who participates in learning processes to what extent?
	Implementing a participation narrative	Does the breeding approach promote participatory plant breeding or another narrative?
	Promoting social self-organization	How is participation organized? Do specific institutions exist that promote social self-organization?
Principle 7	Promoting polycentric governance	Is the breeding approach integrated in a polycentric governance system?
		Does it include elements of polycentric governance?

a It should be noted that this resilience attribute is not primarily the concern of fruit breeding organizations. Rather, the responsibility lies on scientific researchers as well as governmental and non-governmental institutions (See Sections 6.1 and 6.2).

It must be emphasized that all evaluations carried out with this framework look at the general character of any breeding approach, not on local or individual specifics. This resonates with the conceptual context of the resilience definition used in this book by observing general resilience and not specified resilience of social-ecological systems.

Notes

1 The Resilience Alliance was established in 1999 and is an international, multidisciplinary research organization that contributes theoretical and empirical research to social-ecological resilience.
2 Specified and general resilience are complementary perspectives and both integrative parts of resilience thinking. Therefore, a complete disregard of one perspective is neither useful nor legitimate.
3 The mock-court workshop as a method for evaluation and discussion originated in Law studies. Participants present their ideas/arguments (in this case, their developed principles) which are then 'put on trial' by the audience members that take different

roles (e.g., defense attorney, devil's advocate). See Biggs et al. (2015b) for a detailed description of the method's application in this context.

4 MASIPAG is a network of farmers, scientists, and NGOs in the Philippines that aims to promote locally adapted sustainable agriculture by sharing knowledge and (mainly) local rice varieties as well as doing plant-breeding on-farm. Gonçalves et al. (2018) conducted several other case studies using the lens of the seven principles of social-ecological resilience.

5 In a systematic review of studies, Dardonville et al. (2020) show that it is difficult to draw specific conclusions about the influence of diversity on the resilience of agricultural systems. However, the review confirms that a diversity of social and ecological variables clearly enhances crop yield dynamics.

6 PELUM is a network that trains African farmers in organic agriculture. Kenyan farmers in the PELUM network specifically use green manure, composting and natural treatments against insects to maintain and increase soil fertility and water-holding capacity of the soils (Gonçalves et al. 2018).

7 Literature provides a wide range of social learning theories, concepts, and applications (See, e.g., Reed et al. 2010; Siebenhüner et al. 2016). Hence, definitions and understandings of social learning are highly diverse. In this book, the understanding by Cundill et al. (2015) is adopted.

8 As Reed et al. (2010) show, participation does not necessarily and automatically lead to social learning processes. However, achieving social learning is usually the goal of participation processes, especially in the context of resilience.

9 For a consistent and historical overview, *See* Aligica and Tarko (2012). Probably most prominent is the work of Ostrom (2008) who argues that polycentric systems are useful to solve collective action problems and included this concept in the analytical frameworks SESF (2009) and IAD (1990) (See also Section 1.3).

10 Besides agroforestry as the most prominent example, other connectivity options are also scientifically discussed. For example, the *Forschungsinstitut Biologischer Landbau* (FiBL) conducts a research project on combining organic laying hen husbandry with apple farming to promote ecosystem services (FiBL 2020).

11 For example, Germany has initiated several monitoring programs for agriculture to generate long-term data on the development of important ecosystem variables such as water cycling and maintenance (*Nitratbericht*, published every four years by the federal government).

References

Abson, D. J.; Fraser, E. D. G.; Benton, T. G. (2013): Landscape Diversity and the Resilience of Agricultural Returns. A Portfolio Analysis of Land-use Patterns and Economic Returns from Lowland Agriculture. In *Agriculture & Food Security* 2 (1), p. 2. DOI: 10.1186/2048-7010-2-2.

Abson, D. J.; Fischer, J.; Leventon, J.; Newig, J.; Schomerus, T.; Vilsmaier, U. et al. (2017): Leverage Points for Sustainability Transformation. In *AMBIO: A Journal of the Human Environment* 46 (1), pp. 30–39. DOI: 10.1007/s13280-016-0800-y.

Aligica, P. D.; Tarko, V. (2012): Polycentricity. From Polanyi to Ostrom, and Beyond. In *Governance* 25 (2), pp. 237–262. DOI: 10.1111/j.1468-0491.2011.01550.x.

Armitage, D.; Berkes, F.; Dale, A.; Kocho-Schellenberg, E.; Patton, E. (2011): Co-management and the Co-production of Knowledge. Learning to Adapt in Canada's Arctic. In *Global Environmental Change* 21 (3), pp. 995–1004. DOI: 10.1016/j.gloenvcha.2011.04.006.

Arnstein, S. R. (1969): A Ladder of Citizen Participation. In *Journal of the American Institute of Planners* 35 (4), pp. 216–224. DOI: 10.1080/01944366908977225.

Bennett, E.; Carpenter, S. R.; Gordon, L. J.; Ramankutty, N.; Balvanera, P.; Campbell, B. et al. (2014): Toward a More Resilient Agriculture. In *The Solutions Journal* 5 (5), pp. 65–75.

Biggs, D.; Abel, N.; Knight, A. T.; Leitch, A.; Langston, A.; Ban, N. C. (2011): The Implementation Crisis in Conservation Planning. Could "Mental Models" Help? In *Conservation Letters* 4 (3), pp. 169–183. DOI: 10.1111/j.1755-263X.2011.00170.x.

Biggs, R.; Gordon, L.; Raudsepp-Hearne, V.; Schlüter, M.; Walker, B. (2015a): Principle 3- Manage Slow Variables and Feedbacks. In Reinette Biggs, Maja Schlüter, Michael L. Schoon (Eds.): *Principles for Building Resilience. Sustaining Ecosystem Services in Social-Ecological Systems.* Cambridge, UK: Cambridge University Press, pp. 105–141.

Biggs, R.; Schlüter, M.; Schoon, M. L. (2015b): An Introduction to the Resilience Approach and Principles to Sustain Ecosystem Services in Social-ecological Systems. In Reinette Biggs, Maja Schlüter, Michael L. Schoon (Eds.): *Principles for Building Resilience. Sustaining Ecosystem Services in Social-Ecological Systems.* Cambridge, UK: Cambridge University Press, pp. 1–31.

Biggs, R.; Schlüter, M.; Schoon, M. L. (Eds.) (2015c): *Principles for Building Resilience. Sustaining Ecosystem Services in Social-Ecological Systems.* Cambridge, UK: Cambridge University Press.

Bohensky, E. L.; Evans, l. S.; Anderies, J. M.; Biggs, D.; Fabricius, C. (2015): Principle 4- Foster Complex Adaptive Systems Thinking. In Reinette Biggs, Maja Schlüter, Michael L. Schoon (Eds.): *Principles for Building Resilience. Sustaining Ecosystem Services in Social-Ecological Systems.* Cambridge, UK: Cambridge University Press, pp. 142–173.

Böhm, C.; Nawroth, G.; Warth, P. (2019): Roadmap Agroforstwirtschaft. Bäume als Bereicherung für landwirtschaftliche Flächen in Deutschland.

Brown, S. E.; Miller, D. C.; Ordonez, P. J.; Baylis, K. (2018): Evidence for the Impacts of Agroforestry on Agricultural Productivity, Ecosystem Services, and Human Well-being in High-income Countries: A Systematic Map Protocol. In *Environmental Evidence* 7 (1). DOI: 10.1186/s13750-018-0136-0.

Cabell, J. F.; Oelofse, M. (2012): An Indicator Framework for Assessing Agroecosystem Resilience. In *Ecology and Society* 17 (1). DOI: 10.5751/ES-04666–170118.

Carpenter, S.; Arrow, K.; Barrett, S.; Biggs, R.; Brock, W.; Crépin, A.-S. et al. (2012): General Resilience to Cope with Extreme Events. In *Sustainability* 4 (12), pp. 3248–3259. DOI: 10.3390/su4123248.

Carpenter, S.; Walker, B.; Anderies, J. M.; Abel, N. (2001): From Metaphor to Measurement. Resilience of What to What? In *Ecosystems* 4 (8), pp. 765–781. DOI: 10.1007/s10021-001-0045-9.

Chapin, F. S.; Kofinas, G. P.; Folke, C. (Eds.) (2009): *Principles of Ecosystem Stewardship. Resilience-based Natural Resource Management in a Changing World.* New York: Springer.

Choi, Y. E.; Oh, C.-O.; Chon, J. (2021): Applying the Resilience Principles for Sustainable Ecotourism Development: A Case Study of the Nakdong Estuary, South Korea. In *Tourism Management* 83, p. 104237. DOI: 10.1016/j.tourman.2020.104237.

Colding, J.; Colding, M.; Barthel, S. (2020): Applying Seven Resilience Principles on the Vision of the Digital City. In *Cities* 103, p. 102761. DOI: 10.1016/j.cities.2020.102761.

Collins, K.; Ison, R. (2006): Dare We Jump Off Arnstein's Ladder? Social Learning as a New Policy Paradigm. *Proceedings of PATH (Participatory Approaches in Science & Technology) Conference,* 4–7 Jun 2006, Edinburgh, checked on 10/9/2018.

Cumming, G. S.; Barnes, G.; Perz, S.; Schmink, M.; Sieving, K. E.; Southworth, J. et al. (2005): An Exploratory Framework for the Empirical Measurement of Resilience. In *Ecosystems* 8 (8), pp. 975–987. DOI: 10.1007/s10021-005-0129-z.

Cundill, G.; Leitch, A. M.; Schultz, L.; Armitage, D.; Peterson, G. (2015): Principle 6- Encourage Learning. In Reinette Biggs, Maja Schlüter, Michael L. Schoon (Eds.): *Principles for Building Resilience. Sustaining Ecosystem Services in Social-Ecological Systems.* Cambridge, UK: Cambridge University Press, pp. 174–200.

Dakos, V.; Quinlan, A.; Baggio, J. A.; Bennett, E.; Bodin, Ö.; BurnSilver, S. (2015): Principle 2- Manage Connectivity. In Reinette Biggs, Maja Schlüter, Michael L. Schoon (Eds.): *Principles for Building Resilience. Sustaining Ecosystem Services in Social-Ecological Systems*. Cambridge, UK: Cambridge University Press, pp. 80–104.

Dardonville, M.; Urruty, N.; Bockstaller, C.; Therond, O. (2020): Influence of Diversity and Intensification Level on Vulnerability, Resilience and Robustness of Agricultural Systems. In *Agricultural Systems* 184, p. 102913. DOI: 10.1016/j.agsy.2020.102913.

Darnhofer, I. (2014): Resilience and Why it Matters for Farm Management. In *European Review of Agricultural Economics* 41 (3), pp. 461–484. DOI: 10.1093/erae/jbu012.

Duru, M.; Therond, O.; Martin, G.; Martin-Clouaire, R.; Magne, M.-A.; Justes, E. et al. (2015): How to Implement Biodiversity-based Agriculture to Enhance Ecosystem Services. A Review. In *Agronomy for Sustainable Development* 35 (4), pp. 1259–1281. DOI: 10.1007/s13593-015-0306-1.

Ericksen, P. J. (2008): Conceptualizing Food Systems for Global Environmental Change Research. In *Global Environmental Change* 18 (1), pp. 234–245. DOI: 10.1016/j.gloenvcha.2007.09.002.

FiBL (2020): Die "Apfelhühner": Biolegehennen in Apfelplantagen. Forschungsinstitut für biologischen Landbau. Available online at https://www.fibl.org/de/infothek/meldung/die-apfelhuehner-biolegehennen-in-apfelplantagen.html, updated on 12/21/2020, checked on 3/4/2021.

Folke, C. (2016): *Resilience - Oxford Research Encyclopedia of Environmental Science*. Oxford: Oxford University Press, p. 1.

Galaz, V.; Crona, B.; Österblom, H.; Olsson, P.; Folke, C. (2012): Polycentric Systems and Interacting Planetary Boundaries – Emerging Governance of Climate Change–ocean Acidification–marine Biodiversity. In *Ecological Economics* 81, pp. 21–32. DOI: 10.1016/j.ecolecon.2011.11.012.

Gonçalves, A.; Höök, K.; Moberg, F. (2018): Applying Resilience in Practice for More Sustainable Agriculture. Lessons Learned from Organic Farming and Other Agroecological Approaches in Brazil, Ethiopia, Kenya, the Philippines, Sweden and Uganda. Stockholm: Swedish Society for Nature Conservation (SSNC).

Hodbod, J.; Eakin, H. (2015): Adapting a Social-ecological Resilience Framework for Food Systems. In *Journal of Environmental Studies and Sciences* 5 (3), pp. 474–484. DOI: 10.1007/s13412-015-0280-6.

Holling, C. S. (2001): Understanding the Complexity of Economic, Ecological, and Social Systems. In *Ecosystems* 4 (5), pp. 390–405. DOI: 10.1007/s10021-001-0101-5.

Holling, C. S. (2004): From Complex Regions to Complex Worlds. In *Ecology and Society* 9 (1). DOI: 10.5751/ES-00612-090111.

Holling, C. S.; Berkes, F.; Folke, C. (1998): Science, Sustainability and Resource Management. In Fikret Berkes, Carl Folke (Eds.): *Linking Social and Ecological Systems. Management Practices and Social Mechanisms for Building Resilience*. Cambridge, UK: Cambridge University Press, pp. 342–362.

Kotschy, K.; Biggs, R.; Daw, T.; Folke, C.; West, P. C. (2015): Principle 1- Maintain Diversity and Redundancy. In Reinette Biggs, Maja Schlüter, Michael L. Schoon (Eds.): *Principles for Building Resilience. Sustaining Ecosystem Services in Social-Ecological Systems*. Cambridge, UK: Cambridge University Press, pp. 50–79.

Lamine, C. (2015): Sustainability and Resilience in Agrifood Systems. Reconnecting Agriculture, Food and the Environment. In *Sociol Ruralis* 55 (1), pp. 41–61. DOI: 10.1111/soru.12061.

Leitch, A. M.; Cundill, G.; Schultz, L.; Meek, C. L. (2015): Principle 6- Broaden Participation. In Reinette Biggs, Maja Schlüter, Michael L. Schoon (Eds.): *Principles for Building*

Resilience. Sustaining Ecosystem Services in Social-Ecological Systems. Cambridge, UK: Cambridge University Press, pp. 201–225.

Milestad, R.; Dedieu, B.; Darnhofer, I.; Bellon, S. (2012): Farms and Farmers Facing Change. The Adaptive Approach. In Ika Darnhofer, David Gibbon, Benoît Dedieu (Eds.): *Farming Systems Research into the 21st Century: The New Dynamic.* Dordrecht: Springer Netherlands, pp. 365–385.

Misselhorn, A.; Aggarwal, P.; Ericksen, P.; Gregory, P.; Horn-Phathanothai, L.; Ingram, J.; Wiebe, K. (2012): A Vision for Attaining Food Security. In *Current Opinion in Environmental Sustainability* 4 (1), pp. 7–17. DOI: 10.1016/j.cosust.2012.01.008.

Nerlich, K.; Graeff-Hönninger, S.; Claupein, W. (2013): Agroforestry in Europe: A Review of the Disappearance of Traditional Systems and Development of Modern Agroforestry Practices, with Emphasis on Experiences in Germany. In *Agroforestry Systems* 87 (2), pp. 475–492. DOI: 10.1007/s10457-012-9560-2.

Oberlack, C.; Boillat, S.; Brönnimann, S.; Gerber, J.-D.; Heinimann, A.; Ifejika Speranza, C. et al. (2018): Polycentric Governance in Telecoupled Resource Systems. In *Ecology and Society* 23 (1). DOI: 10.5751/ES-09902-230116.

Olsson, L.; Jerneck, A.; Thoren, H.; Persson, J.; O'Byrne, D. (2015): Why Resilience Is Unappealing to Social Science. Theoretical and Empirical Investigations of the Scientific Use of Resilience. In *Science Advances* 1 (4), e1400217. DOI: 10.1126/sciadv.1400217.

Ostrom, E. (2008): Polycentric Systems as One Approach for Solving Collective-Action Problems. In *SSRN Journal.* DOI: 10.2139/ssrn.1304697.

Pahl-Wostl, C. (2009): A Conceptual Framework for Analysing Adaptive Capacity and Multi-Level Learning Processes in Resource Governance Regimes. In *Global Environmental Change* 19 (3), pp. 354–365. DOI: 10.1016/j.gloenvcha.2009.06.001.

Penny, G.; Goddard, J. J. (2018): Resilience Principles in Socio-hydrology: A Case-Study Review. In *Water Security* 4–5, pp. 37–43. DOI: 10.1016/j.wasec.2018.11.003.

Quinlan, A. E.; Berbés-Blázquez, M.; Haider, L. J.; Peterson, G. D.; Allen, C. (2016): Measuring and Assessing Resilience. Broadening Understanding through Multiple Disciplinary Perspectives. In *Journal of Applied Ecology* 53 (3), pp. 677–687. DOI: 10.1111/1365-2664.12550.

Reed, M. S.; Evely, A. C.; Cundill, G.; Fazey, I.; Glass, J.; Laing, A. et al. (2010): What Is Social Learning? In *Ecology and Society* 15 (4), p. r1.

Resilience Alliance (2010): Assessing Resilience in Social-Ecological systems: Workbook for Practicioners. Available online at https://www.resalliance.org/files/ResilienceAssessmentV2_2.pdf, checked on 4/11/2018.

Rotz, S.; Fraser, E. D. G. (2015): Resilience and the Industrial Food System. Analyzing the Impacts of Agricultural Industrialization on Food System Vulnerability. In *Journal of Environmental Studies and Sciences* 5 (3), pp. 459–473. DOI: 10.1007/s13412-015-0277-1.

Rowe, G.; Frewer, L. J. (2005): A Typology of Public Engagement Mechanisms. In *Science, Technology, & Human Values* 30 (2), pp. 251–290. DOI: 10.1177/0162243904271724.

Schlüter, M.; Biggs, R.; Schoon, M. L.; Robards, M. D.; Anderies, J. M. (2015): Reflections on Building Resilience - Interactions Among Principles and Implications for Governance. In Reinette Biggs, Maja Schlüter, Michael L. Schoon (Eds.): *Principles for Building Resilience. Sustaining Ecosystem Services in Social-Ecological Systems.* Cambridge, U.K.: Cambridge University Press, pp. 251–282.

Schlüter, M.; Herrfahrdt-Pähle, E. (2011): Exploring Resilience and Transformability of a River Basin in the Face of Socioeconomic and Ecological Crisis: An Example from the Amudarya River Basin, Central Asia. In *Ecology and Society* 16 (1).

Schlüter, M.; McAllister, R.; Arlinghaus, R.; Bunnefeld, N.; Eisenack, K.; Hölkner, F.; Milner-Gulland, E. J. et al. (2012): New Horizons for Managing the Environment: A Review of Coupled Social-Ecological Systems Modeling. In *Natural Resource Modeling* 25 (1), pp. 219–272. DOI: 10.1111/j.1939-7445.2011.00108.x.

Schoon, M. L.; Robards, M. D.; Meek, C. L.; Galaz, V. (2015): Principle 7- Promote Polycentric Governance Systems. In Reinette Biggs, Maja Schlüter, Michael L. Schoon (Eds.): *Principles for Building Resilience. Sustaining Ecosystem Services in Social-Ecological Systems*. Cambridge, UK: Cambridge University Press, pp. 226–250.

Siebenhüner, B.; Rodela, R.; Ecker, F. (2016): Social Learning Research in Ecological Economics: A Survey. In *Environmental Science & Policy* 55, pp. 116–126. DOI: 10.1016/j.envsci.2015.09.010.

Stringer, L. C.; Dougill, A. J.; Fraser, E.; Hubacek, K.; Prell, Christina, Reed, Markus S. (2006): Unpacking "Participation" in the Adaptive Management of Social-ecological Systems: A Critical Review. In *Ecology and Society* 11 (2), p. 39.

Tendall, D. M.; Joerin, J.; Kopainsky, B.; Edwards, P.; Shreck, A.; Le, Q. B. et al. (2015): Food System Resilience. Defining the Concept. In *Global Food Security* 6, pp. 17–23. DOI: 10.1016/j.gfs.2015.08.001.

Ulanowicz, R. E.; Goerner, S. J.; Lietaer, B.; Gomez, R. (2009): Quantifying Sustainability. Resilience, Efficiency and the Return of Information Theory. In *Ecological Complexity* 6 (1), pp. 27–36.

Urruty, N.; Tailliez-Lefebvre, D.; Huyghe, C. (2016): Stability, Robustness, Vulnerability and Resilience of Agricultural Systems. A Review. In *Agronomy for Sustainable Development* 36 (1), p. 2. DOI: 10.1007/s13593-015-0347-5.

Walker, B.; Salt, D. (2012): *Resilience Thinking. Sustaining Ecosystems and People in a Changing World*. Washington: Island Press.

Walker, B. H.; Carpenter, S. R.; Rockstrom, J.; Crépin, A.-S.; Peterson, G. D. (2012): Drivers, "Slow" Variables, "Fast" Variables, Shocks, and Resilience. In *Ecology and Society* 17 (3). DOI: 10.5751/ES-05063-170330.

Worstell, J.; Green, J. (2017): Eight Qualities of Resilient Food Systems: Toward a Sustainability/Resilience Index. In *Journal of Agriculture, Food Systems, and Community Development*, pp. 1–19. DOI: 10.5304/jafscd.2017.073.001.

Interim Conclusion

In Part II, fruit breeding and cultivation were comprehensively defined as social-ecological systems and it was shown how their system variables and governance systems can be designed in a resilient way to maintain and promote their relevant set of ecosystem services.

Conceptualizing fruit breeding and cultivation as social-ecological systems involved four steps (See Chapter 4). (1) Breeding and cultivation take place in social-ecological systems at local levels, in which different actors use certain technological and knowledge resources to shape ecosystems in a specific institutional setting. (2) In the special case of on-farm breeding, both breeding and cultivation overlap at the local level because the ecosystem is the same for both systems. (3) Further, fruit breeding and cultivation are also connected to multilevel scales. Both systems depend on biodiversity as a central component for breeding and cultivating varieties. They are connected by direct interlinkages and telecoupled flows. (4) On this multilevel scales, fruit breeding and cultivation are only single parts of the fruit value chain, which is in turn embedded in larger food systems. Overall, the conceptualization shows that fruit breeding and cultivation are located in a complex web of interaction, in which different outcomes influence their resilience.

Fruit breeding and cultivation provide and influence a broad set of ecosystem services (See Chapter 5). For both systems, these include regulating services (water cycling and maintenance, pest and disease control), a supporting service (soil nitrogen availability), and cultural services (aesthetic, ethical, and spiritual values, recreation, identity, learning opportunities). Fruit cultivation additionally provides a provisioning service (fruit production), and further influences regulating services (climate regulation, pollination). This set of ecosystem services has to be maintained, promoted, and sustained by building suitable buffer, adaptive, and transformative capabilities.

These capabilities are reflected in the identified 16 resilience attributes, derived from an extensive evaluation of relevant resilience literature (See previous sections). The attributes involve elements of the system-to-be-governed and the governance system. Meeting the demands of the identified resilience attributes

DOI: 10.4324/9781003355724-11

enables fruit breeding and cultivation systems to promote or sustain the defined set of ecosystem services.

In the following Part III, 13 resilience attributes will be applied to apple breeding by using the analytical framework developed in Section 6.2. The book thus shifts its lens from the general fruit level to the empirical context (See Chapter 3) of apple breeding and cultivation in Germany.

Part III

Evaluation of Apple Breeding Approaches

How Resilient Are They?

Part III

Introduction

In Part III, an in-depth look is taken at different breeding systems, based on the conceptualization of fruit breeding as a social-ecological system (See Section 4.2). Case study research has been applied to achieve the appropriate analytical depth and to embed this exploration in a specific problem context (See Chapter 2). The object of investigation is apple breeding and cultivation in Germany. While the elaborations in Part II were located on a more general level, this chapter explores more specific empirical cases.

Literature on typologies or categorizations of plant breeding approaches is scarce. Lammerts van Bueren et al. (2018) made a recent attempt to categorize and describe different breeding approaches according to their normative positions and styles of thought, which they name "paradigmatic orientations" (p. 5). They differentiate between (a) community-based breeding, (b) ecosystem-based breeding, (c) trait-based breeding, and (d) corporate-based breeding. These paradigmatic orientations are categorized with the descriptors holism vs. reductionism and subjectivism vs. objectivism as depicted in Figure Part III.

For example, community-based breeding is described as a holistic approach that incorporates the subjective goals of participating actors. Adhering to a different

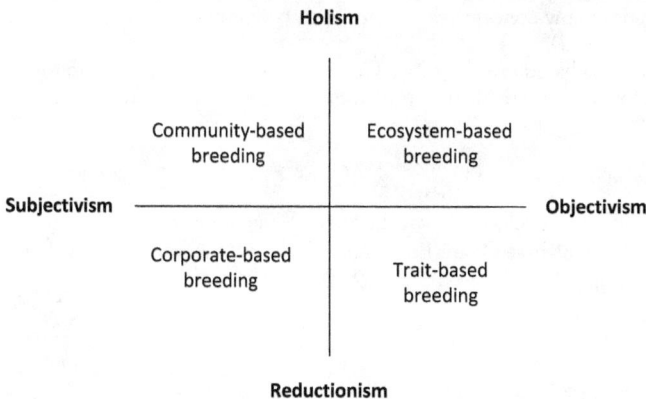

Holism

Community-based breeding | Ecosystem-based breeding

Subjectivism ——————————————— **Objectivism**

Corporate-based breeding | Trait-based breeding

Reductionism

Figure Part III Four paradigmatic orientations of plant breeding (based on Lammerts van Bueren et al. 2018)

DOI: 10.4324/9781003355724-13

logic and moral, trait-based breeding has a reductionist worldview and follows objective goals with the application of technical and genetic sciences. In practice, both of these orientations overlap to certain extents.[1]

For the empirical context of German apple breeding, an alternative categorization is used. It overlaps in many aspects with the one proposed by Lammerts van Bueren et al. (2018) but is more actor-centered to accurately fit the empirical situation. It distinguishes between corporate-based apple breeding (See Chapter 7), public apple breeding (See Chapter 8), and commons-based apple breeding (See Chapter 9). The following chapters each contain a description of the actor-centric approach[2] and a respective case study, followed by the evaluation of the approach with the analytical framework (See Section 6.3). Databases for the evaluations are scientific and gray literature as well as qualitative empirical data on the case studies collected between 2017 and 2019. Empirical data has been collected with three semi-standardized interviews and one focus group (See Chapter 2). The interviews were conducted with the breeders and/or project leaders of the respective case. The focus group with *apfel:gut e.V.* was carried out to specifically understand the institutional dynamics of commons-based fruit breeding. The general goal of data collection was to gather specific insights into the organizational structure, processes, rules, and norms of the breeding organizations.

The practical and conceptual implications of the data analyses are discussed in Chapter 10. First, a comparison of all three evaluations highlights the similarities and differences between the different breeding approaches (See Section 10.1). Thereafter, the final subsection conceptualizes and reflects on commons-based organic apple breeding as an approach that couples breeding and cultivation (See Section 10.2).

Notes

1 The authors propose a new breeding approach (paradigmatic orientation) that combines elements of all four orientations: systems-based plant breeding. Interestingly, this combination includes many conceptual overlaps with commons-based organic apple breeding (See Section 10.2).
2 In the case of commons-based breeding (See Chapter 9), being one of the core topics of this book, a profound theoretical conceptualization will additionally take place.

References

Lammerts van Bueren, E. T.; Struik, P. C.; van Eekeren, N.; Nuijten, E. (2018): Towards Resilience Through Systems-based Plant Breeding. A Review. In *Agronomy for Sustainable Development* 38 (5), p. 42. DOI: 10.1007/s13593-018-0522-6.

7 Corporate-Based Apple Breeding

7.1 Description of the Approach

Corporate-based apple breeding is carried out by organizations that are registered as private companies (e.g., with the legal status GmbH or GmbH & Co. KG) and pursue commercial goals. This approach distinguishes itself from apple breeding by single private actors that also follow commercial goals with a professional style (and possibly the legal status of a GbR), but do not have large financial resources nor a pool of associates. Transferring the findings of Lammerts van Bueren et al. (2018), corporate-based apple breeding follows the long-term aim to generate corporate profit with professional business models and uses instruments like variety protection or club concepts in order to acquire sufficient turnover. This approach often follows a rather centralized, goal-driven approach where specialists and experts make decisions.

Like most countries, Germany does not have a long tradition of corporate-based apple breeding; this is likely due to the long-term nature of apple breeding that contrasts the short-term economic goals of most investors and entrepreneurs. Only two German organizations exist that can be classified as organizations using a corporate-based approach in apple breeding: *Bayerisches Obstzentrum GmbH & Co. KG* (BayOZ),[1] located in southeastern Germany near Munich, and *Züchtungsinitiative Niederelbe GmbH & Co. KG* (ZIN), located in Osnabrück, northwestern Germany. This research focuses on ZIN as a case study to explore the resilience of corporate-based apple breeding.

7.2 Case Study: Züchtungsinitiative Niederelbe GmbH & Co. KG

ZIN was founded in 2002 by fruit farmers, retailers, and the fruit farmer alliance MAL. The company cooperates with a team of researchers from the Osnabrück University of Applied Sciences; Professor Werner Dierend and his colleagues are responsible for the scientific support of the breeding process and conduct the crossbreeding, the cultivation of seedlings, and species selections. ZIN further cooperates with the tree nursery *Carolus* in Belgium, which is responsible for the (later) cultivation and processing of apple trees. The goal of ZIN is to breed apple

DOI: 10.4324/9781003355724-14

varieties that are broadly appealing to consumers and retailers and generate premium prices for producers. Fruit farmers and retailers may acquire membership in the company by acquiring shares in the ZIN KG as part of the ZIN GmbH & Co. KG.[2] Each member is required to pay annual fees to finance the long-term breeding process. As a result, the members of ZIN are both an early target group and profiteers of novel varieties.

ZIN's breeding goals for apples are the following:

- Fruit qualities: appealing visual traits and texture, good taste, crispy and firm flesh, good storability, low allergenic potential.
- Tree qualities: stable yield, resistance against diseases and pests (specifically apple scab), no alternation, robustness, and adaptability.

Additionally, they look out for niche varieties like ornamental apples and cultivars with special traits, such as deep red-black color (Interview ZIN 2019). Breeding goals were developed collectively by the ZIN members at the time of the foundation of ZIN in several discussions (ibid.). All breeding goals are open for discussion and members continuously have the opportunity to criticize and discuss them, especially during the annual general assembly. In reality, however, occasional amendments and discussions of the breeding goals are only carried out by a "close circle" (ibid., own translation) consisting of the chief breeder and two other members. The majority of members do not make use of their right to participate in breeding goal discussions.

The breeding process can roughly be divided into two phases, in which different actors participate, according to their responsibilities and decision-making competences (Interview ZIN). Phase 1 includes crossings, seedling cultivation, and the first selection step. Phase 2 includes the second and third selection steps as well as the following legal steps to introduce the novel variety to the market.

- Phase 1: Fitting parental lines are identified and subsequently crossing plans are created and carried out in the trial areas at the university. Harvested apple seeds out of the crossings are stratified for several weeks. Cultivation of seedlings takes place in foliage greenhouses at the same location. After 1–1.5 years of cultivation, budwood is cut from every seedling and processed by tree nursery *Carolus* on M9 rootstocks. After a one-year cultivation of these clones, all are planted in a selection area next to the river Niederelbe in northern Germany. Here, the first selection step (max. six years) takes place according to the criteria of tree growth, fruit size, look, and taste. Fruits of good-performing clones are further tested, for example, with sensory measures and compositional analyses in laboratories at the university.
- Phase 2: Budwood is cut of clones that successfully completed the first selection step. Afterward, it is again processed by *Carolus* on M9 rootstocks. Those clones are also cultivated for one year and then planted in the selection areas both at the river *Niederelbe* and at the technical college. Thus, the first differences in adaptability can be observed. In this second selection step

(max. eight years), the selection criteria regarding fruit and tree quality are much more extensive. These include breeding sub-goals with further specifications (e.g., the composition of ingredients) and criteria such as consumer acceptance. Clones that fully meet selection criteria continue on to the third and final selection step, where they are planted in other trial areas and fruit farms of ZIN members. There the final step is an assessment considering suitability for variety protection and the final introduction into the market.

In phase 1, the responsibility solely lies with the chief breeder – especially in the first selection step that he carries out alone (or at maximum with another university colleague or ZIN member) according to a defined selection scheme (Interview ZIN).[3] Only in phase 2 do more actors get involved in the decision-making processes. In the second selection step, regular discussions on promising cultivars take place in two panels with 10–25 ZIN members following a specific selection scheme (ibid.). In these panels, apple farmers are more likely to participate than apple retailers (ibid.). Current selection results and the state of discussions on cultivars are presented to ZIN members at the annual general assembly, where they have the chance to give feedback on the breeding process (ibid.). Additionally, scientists and students from the university are involved in any stage of the breeding process if there are opportunities for research projects or theses, for example, in the study areas of agricultural marketing and sensory studies (ibid.). Furthermore, ZIN has networked cooperation with many German actors in the breeding and farming sector, sharing knowledge to certain degrees (ibid.).

ZINs apple breeding has no underlying normative orientations. They breed "free of ideology" (Interview ZIN, own translation) and take apple farming as a whole as a target group which addresses integrated and organic farming alike. Farmers with both integrated and organic approaches are members of ZIN. In 2020, ZIN introduced its first own variety *Deichperle*[4] into the market.

7.3 Analyzing the Resilience of Corporate-Based Apple Breeding

7.3.1 *Resilience Attribute 1a: Integrating Diverse Ecological Elements*

How many and which varieties ZIN exactly uses in the breeding process is confidential. This makes it difficult to assess the ecological diversity in breeding. However, some information on the use of traditional and heirloom varieties is available.

ZIN uses some traditional and heirloom apple varieties in its breeding program[5] but had had no successful results so far. In a presentation on the status quo of breeding, Dierend (2017) reports that out of crossings with traditional and heirloom apple varieties, approx. 3.500 seedlings were planted in the first selection step. The performance of these clones was not satisfying and only seven clones made it to the second selection step. ZIN data and experience show that usually about 35 clones from this quantity of crossings prove to be suitable for the second selection step. It is important to include varieties with traits that fit the breeding goals in the crossings – regardless of them being traditional or modern varieties (Interview ZIN). As a result, traditional and heirloom varieties are periodically included in breeding but do not play a major role.

Different apple cultivation practices are represented in the pool of members, which includes both integrated and organic farmers. This gives ZIN the potential to breed varieties for different cultivation demands, practices, and specificities. However, if this potential can actually be exploited depends on the participation of those actors in the breeding process (See below). Exact results will only be observed in the coming years or even decades if ZIN potentially introduces new varieties with different characteristics.

7.3.2 Resilience Attribute 1b: Integrating Diverse Social Elements

As already mentioned in the previous section, a range of actors participate in the breeding process to different extents. Besides the chief breeder and his close circle, the following actors are involved:

- Fruit farmers: Integrated and organic fruit farmers are the majority group of ZIN members. This ranges from small farmers to "big players" (Interview ZIN, own translation) from the *Altes Land*, the main fruit farming area in northern Germany. They primarily participate in phase 2 in the breeding process (See the previous section).
- Fruit retailers and marketeers: Retailers and marketeers, foremost the alliance MAL, are also members of ZIN. Similar to the fruit farmers, they primarily have the opportunity to participate in phase 2 of the breeding process.
- Scientists: Especially by performing research or exchanging knowledge, scientists from different disciplines participate in the breeding process. Additionally, cooperation via research projects takes place with partners outside of Osnabrück, for example, on the allergenic potential of apple with the technical university in Munich (Interview ZIN). The chief breeder is also in constant exchange with other fruit breeding colleagues, for example, from the JKI[6] (ibid.).

In general, a diversity of actors participates in ZIN's breeding process.

7.3.3 Resilience Attribute 1c: Monitoring Diversity and Redundancy of Fruit Varieties

In general, responsibility for the preservation of apple cultivars is recognized (Interview ZIN). The organization preserves and monitors some cultivars – including traditional and heirloom varieties – and is a partner location of the German National Fruit Gene Bank *Deutsche Genbank Obst* (ibid.). However, the main purpose of ZIN is to use the diversity of varieties and develop new ones: "everyone talks about the importance of gene pools. As a breeder, I see thousands of new clones every year. Experiencing this diversity is amazing [...] and has a calming effect – nature still has a great scope" (ibid., own translation). In conclusion, ZIN contributes to a small extent to both the preservation and monitoring processes of fruit varieties as part of the larger German fruit breeding network.

7.3.4 Resilience Attribute 2: Promoting Ecocentric Breeding Approaches

Promoting the breeding of cultivars with traits that minimize negative impacts on the nitrogen and water cycle aiming at resource-preserving cultivation is not directly a part of ZIN's breeding goals (See the previous section). Only robustness as a breeding goal contributes to resource preservation, but its explicit understanding remains vague. However, the benefits of possibly breeding cultivars with a reduced need for plant protection products are acknowledged, although those cultivars are not seen as "the solution for every problem" (Interview ZIN, own translation).

Traditional varieties have been included in the breeding process (Interview ZIN), but the general gene pool remains rather narrow because most included cultivars are related to each other.[7] It follows that ZIN does not particularly adopt an ecocentric breeding approach. ZIN explicitly breeds free of ideology, adopting a pragmatic approach that does not connect to a certain management paradigm.[8]

7.3.5 Resilience Attribute 3: Exchanging Knowledge and Technology

Inside the organization of ZIN, fruit farmers, breeders, and scientists connect with each other. Formal and informal opportunities exist to share knowledge and start discussions. Technological and human resources as well as trial areas are exchanged or shared with the technical college of Osnabrück via research projects. This exchange is crucial to the overall success of ZIN's breeding efforts (Interview ZIN).

Knowledge of the breeding process is theoretically transparent to all ZIN members at all times. The chief breeder presents the status quo of breeding at every annual general assembly, but not in every detail. Not every crossing combination is presented because a certain risk exists in information leakage with a large number of ZIN members. As most members are more interested in phase 2 of the breeding process, this seems to be no problem for the members (Interview ZIN). Thus, an actual knowledge exchange with ZIN members only takes place in phase 2 of the breeding process (See the previous section), and not every member participates. However, the chief breeder constantly meets ZIN farmers in the trial areas while doing and observing selections, informally exchanging opinions and thoughts, and including those in his breeding plans. In conclusion, knowledge exchange primarily follows a top-down approach but is peppered with elements of bottom-up processes via informal exchanges. The chief breeder mainly informs about the breeding process and organizes the knowledge exchange, making him the central actor in ZIN's connectivity net.

Knowledge exchange also takes place with actors outside of ZIN. This only happens to a certain extent: "Usually we are a bit cautious [...] because there is a private company behind us" (Interview ZIN, own translation). The chief breeder exchanges and discusses knowledge with other breeding colleagues, operators of other trial stations, and former ZIN members (ibid.). He also participates in the working groups *Leistungsprüfung im deutschen Obstbau* (Performance testing in German fruit farming) in the association of German chambers of agriculture and *Arbeitskreis Züchtung*[9] (task force breeding) in the *Bundesfachgruppe Obstbau* (ibid.). Additionally, the chief breeder sometimes presents breeding results at conferences.

7.3.6 Resilience Attribute 4a: Implementation into Mental Models

ZIN does not particularly implement CAS Thinking into its breeding approach. This is due to two reasons. First, the organization does not directly adopt an ecocentric form of breeding. Second, ZIN follows a rather static breeding process that goes in line with past efforts to breed varieties suitable for current market conditions, for example, by largely crossing modern cultivars with each other (low diversity). As explained earlier (See Introduction), this reduced the resilience of fruit cultivation and breeding.

Nevertheless, some elements can be identified that reflect aspects of CAS Thinking. This primarily includes the effort to establish learning processes between farmers, breeders, and retailers. Establishing those processes opens up opportunities to breed varieties adapted to the needs of specific actors and management practices. However, this does not necessarily lead to resilience in breeding as demands could generate from mental models that follow non-resilient paradigms. Additionally, organic farmers practicing ecocentric approaches are members of ZIN, building potential for the transfer of their individual norms into ZINs organization. Robustness, of major importance for organic fruit farmers, is already a breeding goal of ZIN, although not of primary importance. However, as this transfer did not happen on a broad scale since the founding of the organization, it remains questionable if future opportunities open up.

7.3.7 Resilience Attribute 4b: Translation into Institutions

As explained above, CAS Thinking is not implemented in the mental models of relevant actors which are primarily responsible for ZIN's breeding efforts. As a result, a translation of CAS Thinking into institutions cannot be observed.

7.3.8 Resilience Attribute 5a: Implementing Long-Term Monitoring Programs

ZIN has not implemented an ecosystem-based long-term monitoring program. The database used for monitoring breeding efforts and controlling the selection process concentrates on traits of fruits and trees, consistent with the breeding goals (See the previous section). Indirectly, it could be argued that effects on regulating and supporting ecosystem services are monitored because breeding for resistance is of relevance in their efforts (Interview ZIN). Resistances potentially reduce the usage of plant protection products, thus preserving ecosystem services.

Pedigree relations of apple varieties are considered in ZIN's breeding efforts. For example, they avoid to cross varieties where a high vulnerability for apple cancer exists in both parental lines (Interview ZIN). Network connections of the chief breeder with other breeders and scientists potentially give access to information on pedigree relations beyond knowledge from scientific publications. However, although the chief breeder is continuously involved in scientific publications on apple and general fruit breeding, he does not seem connected to an (international) consortium that does research on pedigree relations in apples.

7.3.9 Resilience Attribute 5b: Promoting Social Learning Institutions

ZIN's organizational structure promotes social learning institutions to a certain point. First, adaptive management is fostered through networking between practitioners. Different practitioners are included in phase 2 of the breeding process (See above), where selection results are discussed. Those institutionalized discussions foster learning-by-doing. Second, this networking provides opportunities for continuously sharing knowledge between breeders, practitioners, and scientists. Scientists share their knowledge with the chief breeder who presents it to the ZIN members. Discussing selections involves farmers and the close circle. Knowledge sharing also takes place in other settings besides discussing selection results (See above) whereby the chief breeder seems to be the key actor who links others.

Social learning institutions are not implemented in phase 1 of the breeding process, where the chief breeder (sometimes together with his close circle) takes decisions alone (Interview ZIN) and only individual learning takes place. Literature and interview data also do not reveal forms of adaptive governance where knowledge sharing takes place across scales aiming for double- and triple-loop learning. Rather, social and individual learning processes do not reach beyond single-loop learning (are we doing things right?). Actions already mentioned above like the testing of RFID chips as a tool for the selection process or the implementation of traditional and heirloom cultivars in some crossings show examples of single-loop learning. Those actions, however, do not reframe or question basic assumptions about the breeding goals[10] or the breeding process (double-loop learning), or values and norms (triple-loop learning). This is consistent with the finding, that ZIN does not incorporate CAS Thinking into its breeding approach.

7.3.10 Resilience Attribute 6a: Promoting Participation of Relevant Actors

In general, the following actor groups are included in the breeding process: fruit farmers, fruit retailers, and scientists. Scientists do only participate in a passive way, similar to public communication (See Section 6.1), by informing on their research results. In phase 1 of the breeding process, the ZIN members also participate in the form of public communication as the chief breeder informs them about the status of breeding (See the previous section). In phase 2 of the breeding process, active participation takes place as a dialogue between the chief breeder and the members is established. Fruit farmers and retailers get decision-making power, although according only the farmers really participate (See above).

This transfer of decision-making power to all relevant actor groups reflects in the organizational structure of ZIN. Important decisions regarding the introduction of new varieties, the definition of production limits for varieties, or guidelines for cultivation and marketing, have to be taken via a two-thirds majority in the plenary of ZIN GmbH & Co. KG. In this plenary, the actors are divided into the partner groups ZIN founders (three votes), fruit farmers (one vote), fruit farmer alliance MAL (one and a half votes), and fruit retailers (one and a half votes). The obligation of a two-thirds majority secures the participation of all partner groups as it is necessary to find a consensus between the founders, farmers, and retailers.

Forms of participation are coherent with the forms of learning processes. In phase 1 of the breeding process, the level of social learning is low and participation is passive. In phase 2 of the breeding process, the level of social learning is higher and participation is active.

7.3.11 Resilience Attribute 6b: Implementing a Participation Narrative

ZIN does not implement participatory plant breeding. However, with their promotion of a privately funded form of apple breeding involving the cooperation of farmers and retailers it could be argued that the organization adopts a kind of (corporate-based) participation narrative.

7.3.12 Resilience Attribute 6c: Promoting Social Self-Organization

Empirical data on ZIN does not reveal any forms of social self-organization. The breeding process is executed in a top-down manner and physically takes place in central trial areas. Local networks or institutions that work self-organized are not included in the active part of the participation process.

7.3.13 Resilience Attribute 7: Promoting Polycentric Governance

ZIN's breeding approach is not integrated in a polycentric governance system and does not include elements of polycentric governance. No autonomous and decentral centers of decision-making on multiple scales can be observed in the governance system of ZIN.

7.4 Interim Conclusion

Figure 7.1 shows a summary of analyzing the resilience of ZIN's breeding approach. In shaping the system-to-be-governed, ZIN performs well in integrating diverse social elements and exchanging knowledge and technology. By integrating diverse ecological elements to a certain point and supporting the monitoring of diversity and redundancy of fruit varieties, the organization shows further elements of resilience. However, ZIN does not sufficiently recognize slow variables and ecological feedback processes in breeding by not directly promoting an ecocentric breeding approach. However, some elements of this approach are included. The analysis of ZIN's governance system shows that its breeding approach does not foster CAS Thinking and does not promote social self-organization in its participatory structure. Mainly because of those aspects, polycentric governance plays no role for ZIN. Implemented learning processes show elements of resilience as long-term monitoring programs are implemented to a certain degree and social learning is fostered in phase 2 of the breeding process. ZIN also established a corporate-based participation narrative which results in connecting relevant actors from the fruit value chain and promoting their active participation in parts of the breeding process.

Overall, ZIN shows some strengths in the participatory structure of its breeding approach by recognizing the social diversity of business actors in the apple sector and trying to integrate this diversity into the initiative. The missing promotion of CAS Thinking and ecocentricity translates into some drawbacks regarding the sufficient integration of ecological diversity and active participation in phase 1 of the breeding process. The missing participatory extent in phase 1 probably

Principle 1: Diversity and redundancy	Integrating diverse ecological elements	Principle 4: Fostering CAS Thinking	Implementation into mental models
	Integrating diverse social elements		Translation into institutions
	Monitoring diversity and redundancy of fruit varieties	Principle 5: Encouraging learning and experimentation	Implementing long-term monitoring programs
			Promoting social learning institutions
Principle 2: Slow variables and feedback	Promoting ecocentric breeding approaches	Principle 6: Broadening participation	Promoting participation of relevant actors
			Implementing a participation narrative
			Promoting social self-organization
Principle 3: Managing connectivity	Exchanging knowledge and technology	Principle 7: Promoting polycentric governance	Implementing elements of polycentric governance

| Color coding scheme: | Good performance | Includes some elements of resilience attribute | Includes no elements of resilience attribute |

Figure 7.1 Resilience analysis of ZIN's breeding approach (own figure). Color codings only illustrate tendencies.

reflects the corporate structure of ZIN as being an organization that ultimately aims at generating profit. Thus, participating actors are indeed shareholders with certain cost-benefit demands.

Notes

1 BayOZ was founded in 2002 and, besides apple varieties, also breeds pear, berry, cherry, and other fruit varieties. Selections and trials are done by the organization itself. To secure financing, BayOZ also cultivates and sells fruit trees and bushes (Neumüller and Dittrich 2016).
2 At the moment, no additional memberships will be granted to avoid free-riding. The only way to get access is by purchasing existing memberships traded in internet portals (Interview ZIN). By 2019, ZIN had about 170 members (ibid.).
3 At the start of ZIN, it was the idea to use RFID (Radio Frequency Identification) chips in the selection field to efficiently collect and store information on any seedling in a digital database in the whole breeding process (Dierend and Schacht 2009). Due to practical challenges, this technology is only used in the first selection step to assist the chief breeder in the selection work (Interview ZIN).
4 The variety is an offspring of a crossing of *Topaz* x *Dalinbel*, two modern varieties.
5 For example, *Dülmener Herbstrosenapfel*, *Finkenwerder*, *Horneburger*, or *Rote Sternrenette* (Dierend 2017). All are traditional varieties that were first classified in the 19th century.
6 JKI provides the case study for public apple breeding, See Chapter 8.
7 This is a general problem in fruit breeding, See Introduction.
8 Although this breeding without ideology in general holds true, the interview partner categorically excludes the potential usage of genetic engineering methods now and in the future. It is emphasized that ZIN does breeding in a classical way (Interview ZIN).
9 This committee meets once a year but the chief breeder did not actively participate in the last years (Interview ZIN).
10 One exemption is the inclusion of niche varieties as side-goals beyond the initial breeding goals. For example, ornamental apples have been selected on the demand of ZIN farmers when the chief breeder guided a group of them through the trial areas and got into discussion (Interview ZIN).

References

Dierend, W. (2017): Beurteilung von Nachkommen alter Apfelsorten im Rahmen der Züchtungsarbeit der Züchtungsinitative Niederelbe. Symposium Obstsortenvielfalt. Berlin, 11/14/2017.

Dierend, W.; Schacht, H. (2009): Züchtungsinitiative Niederelbe. In *Erwerbs-Obstbau* 51 (2), pp. 67–71. DOI: 10.1007/s10341-009-0083-6.

Interview ZIN (2019). Qualitative interview with a breeder from ZIN. Transcription from the original language, Germany.

Lammerts van Bueren, E. T.; Struik, P. C.; van Eekeren, N.; Nuijten, E. (2018): Towards Resilience Through Systems-based Plant Breeding. A Review. In *Agronomy for Sustainable Development* 38 (5), p. 42. DOI: 10.1007/s13593-018-0522-6.

Neumüller, M.; Dittrich, F. (2016): Das Bayerische Obstzentrum: Obstzüchtung aus Leidenschaft, 1/26/2016. Available online at https://docplayer.org/27610899-Das-bayerische-obstzentrum-obstzuechtung-aus-leidenschaft.html, checked on 11/3/2020.

8   Public Apple Breeding

8.1 Description of the Approach

Public apple breeding is carried out by organizations that are publicly financed and thus, in theory, follow public interests. They could be financed by governmental actors on a national, federal, or even regional level. In comparison to corporate-based breeding, public apple breeding does not necessarily follow commercial goals.[1] Organizations that carry out this form of breeding are often scientific institutes that simultaneously do research on apple breeding. Because of this proximity to research institutions, which generally focus on incremental, technical innovations, Lammerts van Bueren et al. (2018) argue that public breeding organizations often adopt trait-based breeding approaches. Those approaches incorporate technocentric, highly specialized forms of breeding and are embedded in a centralized organizational structure.

In Germany, a long tradition of public fruit and specifically apple breeding exists. Hanke and Flachowsky (2017) locate the late 1920s as the starting point of systematic fruit breeding and fruit breeding research in Germany. The first public institute for breeding research, the *Kaiser-Wilhelm-Institut für Züchtungsforschung*, was founded in Müncheberg (located in what is today the federal state of Saxony-Anhalt) in 1928. Locations, organizational structures, and numbers of research institutes as well as their research objectives changed throughout the Second World War and the German partition (Hanke and Flachowsky 2017). Since the German reunification in 1990, public fruit breeding is concentrated at the national Julius Kühn-Institute (JKI)[2] in Dresden-Pillnitz, in the east of Germany. Besides JKI, several smaller public breeding institutions exist in southwestern Germany. The most prominent ones are the *Staatliche Lehr- und Versuchsanstalt für Wein- und Obstbau* (LVWO) in Weinsberg[3] and the *Kompetenzzentrum Obstbau Bodensee* (KOB) in Bavendorf.[4] The Institute for Breeding Research on Fruit Crops, *Institut für Züchtungsforschung an Obst* (ZO), at JKI will serve as a case study to explore the resilience of public apple breeding.

8.2 Case Study: *Institut für Züchtungsforschung an Obst*

Following the organizational restructuring of public plant breeding activities and research in Germany, the JKI was founded in 2008. Since then, the institute is

DOI: 10.4324/9781003355724-15

doing research in several fields and incorporates 17 specialized institutes. The ZO is one of these specialized institutes but has its historical origins as an independent organization in the 1920s when fruit breeding started at the location of Dresden-Pillnitz.

According to JKI (2019), the three core competences of the ZO are (a) collecting, conserving, evaluating, and documenting fruit genetic resources in its role as coordination center of the *Deutsche Genbank Obst*; (b) breeding fruit cultivars and rootstocks for organic and integrated fruit cultivation; and (c) doing research on innovative fruit breeding methods. These three competence areas are not separated from each other in practice. For example, the gene bank serves as a resource base for breeding programs, new breeding methods are applied to practical breeding, or breeding demands influence research on breeding methods.

JKI is publicly funded and acquires further financial and human resources for breeding and research through externally funded research projects. At the moment, the ZO employs two permanent fruit breeders and concentrates on breeding new apple, pear, as well as sweet and sour cherry varieties. One breeder is particularly responsible for apple breeding. The overarching goal is to apply resistance breeding to develop healthy and productive fruit varieties that need a low amount of plant protection.

Apple breeding is generally targeted at the fresh market. The ZO aims to breed a "variety for the world" (Interview ZO, own translation) that is suitable for any cultivation site.[5] However, during the selection process, the apple breeder still looks out for niche varieties that are suitable for hobby gardeners although this is not the focus of breeding activities (Interview ZO 2019). Specific breeding goals of ZO for apples are the following:

- Fruit qualities: high juiciness, crispness, and aroma, good storability and shelf-life, appealing visual traits and texture.
- Tree qualities: stable yield, breeding for resistance to apple scab, powdery mildew, fire blight, and apple blotch, stress tolerance, and robustness.

Breeding for resistance is one of the main goals in the ZO's apple breeding program. Whereas in past breeding efforts the focus was on monogenic resistances, the ZO now aims at pyramidizing genes[6] to build strong resistances against diseases and pests (Interview ZO). Overall, this seems to be their interpretation of robustness in tree qualities.

The breeding process follows a regular scheme in seven steps (Interview ZO). In all steps, biotechnological breeding methods are applied via molecular genome analyses (especially the method of marker-assisted selection). This enables the analysis and evaluation of genotype data throughout the whole breeding process. The seven steps are the following (ibid.):

1. At first, the breeding team defines a specific project and decides on the desired combination of characteristics according to the breeding goals. On this basis, fitting parental lines are identified and crossing plans are developed.

2. Apple seeds harvested from the crossings are stratified and planted in foliage greenhouses in the ZO's area.
3. Every seedling is inoculated with a pest or disease (e.g., apple scab, depending on the goal of the respective breeding project). Subsequently, the genotypes of all seedlings that show vulnerabilities are identified and analyzed to generate and chart data on vulnerable genes. Afterward, these seedlings are terminated.
4. The most promising seedlings are further cultivated with fast breeding methods (Flachowsky et al. 2011), resulting in a first flowering after one year of cultivation. Budwood is cut from every seedling, processed on rootstocks, and cultivated for 1–2 years in a tree nursery.
5. All trees are then planted on the trial areas and the selection process starts. Normally, the clones are evaluated and selected after three years, when they bear fruits. After these three selection phases, the best-performing trees are cloned several times. They are evaluated again in subsequent fruit-bearing years, according to the wanted combination of characteristics, and further selected.
6. Thereafter, tastings of the fruits of the surviving clones take place with colleagues and market actors.
7. Lastly, the most promising clones are assessed for variety protection and introduced into the market.

Besides the apple breeder, the head of trial areas and other scientists from JKI are involved in the breeding process (Interview ZO). The head of trial areas is responsible for coordinating the trials and organizing cultivation activities throughout the whole process. Marker-assisted selections are supported by molecular geneticists and a technical assistant supports the crossing stage. The apple breeder is in regular contact with all actors and coordinates the overall breeding activities. Especially in the later stages of the breeding process, the ZO discusses selection results with colleagues in specific external fruit commissions, market actors (especially farmers), pomologists, and also with colleagues from other countries like China or New Zealand (ibid.). It remains unclear if these discussions take place according to regular schedules and schemes or rather randomly.

The apple breeding programs of ZO have no underlying normative orientations. Breeding is generally addressed at fruit farmers and not targeted on specific farming management approaches. In the last decades, the ZO has already introduced several varieties into the market. Variety protection licenses are administered by the *Deutsche Saatgutgesellschaft mbH*.

8.3 Analyzing the Resilience of Public Apple Breeding

8.3.1 Resilience Attribute 1a: Integrating Diverse Ecological Elements

How many and which varieties the ZO exactly uses in its breeding projects is under disclosure. Although being a public organization, the organization is still in competition with other fruit breeding organizations. Similar to ZIN, this makes it generally difficult to assess the ecological diversity in breeding (See Section 7.3).

As the ZO coordinates the *Deutsche Genbank Obst*, the Institute has direct access to a wide variety of genetic resources and cultivars.[7] The interviewee calls this a "reservoir for breeding" (Interview ZO, own translation) where one can search for cultivars with suitable traits for parental lines. The gene bank holds over 1.000 apple cultivars. Historically, the ZO has included traditional and heirloom cultivars in their breeding programs and even developed and registered several varieties that emerged out of crossings with those varieties (Peil 2017). However, it is important to note that the definition of traditional and heirloom varieties differ from those of other organizations. The ZO categorizes cultivars first classified before 1900 as "old varieties" (ibid.) and those first classified between 1900 and 1931 as "middle-old varieties" (ibid.). According to this categorization, the inclusion of *Golden Delicious* as a parent in crossing plans would be interpreted as using traditional and heirloom varieties in breeding. In sum, the ZO potentially uses a diverse range of cultivars for its breeding projects but interview data did not reveal exact information.

8.3.2 Resilience Attribute 1b: Integrating Diverse Social Elements

Besides the responsible breeder for apple breeding, the following actors are included in the breeding process:

- Scientists: Because the ZO also performs research on fruit breeding, a range of scientists from ZO and other external organizations take part in the breeding process (Interview ZO). Internally, this primarily includes colleagues from the fields of molecular genetics, biotechnology, and the head of trial areas. Externally, the apple breeder consults and cooperates with colleagues from other scientific organizations, both in Germany and other countries.
- Practitioners: The apple breeder consults on selection results with other breeding colleagues via the fruit commissions *Sächsische Sortenkommission* and *Fachkommission Kernobst – Arbeitskreis Züchtung*. Moreover, the apple breeder consults pomologists, apple farmers, and other market actors at various stages of the breeding process (Interview ZO).

In conclusion, the general diversity of actors included in the breeding process regarding their variety is high. However, it remains unclear (a) if this inclusiveness is true for every breeding project, and (b) if the inclusion takes place randomly or based on regular schemes (See above).

8.3.3 Resilience Attribute 1c: Monitoring Diversity and Redundancy of Fruit Varieties

Because the ZO is the coordinator of the *Deutsche Genbank Obst*, it significantly contributes to monitoring and preserving the diversity of fruit varieties. The gene bank captures a high quantity of data on varieties including naming and synonyms, first references, origin country, the preserving institutions, and several different fruit and tree characteristics if available (JKI 2020).

8.3.4 Resilience Attribute 2: Promoting Ecocentric Breeding Approaches

With the integration of robustness as a breeding goal, the ZO indirectly aims to contribute to resource preservation. The interviewee sees a growing future demand for robust cultivars due to coming further restrictions on using plant protection products in cultivation (Interview ZO). However, the approach of the ZO does not explicitly aim at minimizing negative environmental impacts.[8]

The ZO does not adopt an ecocentric but rather a trait-based orientation (See Section 8.1). As the description of the breeding process showed, the breeding philosophy is technocentric. Additionally, breeding activities do not connect to a certain management paradigm. The interviewee sees no dichotomy between integrated and organic apple farming because both approaches demand robust varieties (Interview ZO). Thus, the ZO aims to breed varieties for all fruit farmers.

8.3.5 Resilience Attribute 3: Exchanging Knowledge and Technology

The internal structure of the ZO and JKI as such connects scientists very closely to each other (interview ZO). Formal and informal opportunities for knowledge exchange exist through research projects, sharing of technological devices, and meetings of working groups.

Knowledge exchange with German breeding actors outside of the ZO takes place in the fruit commissions *Sächsische Sortenkommission* and *Fachkommission Kernobst – Arbeitskreis Züchtung* as well as with other federal state actors (e.g., so-called *Landesanstalten* that engage in agriculture and horticulture). In these institutions, breeding aspects and selection results are discussed with colleagues or material exchange for testing reasons takes place (Interview ZO). Furthermore, the ZO is well connected with European fruit breeders through joint research projects. Outside Europe, for example, colleagues from New Zealand are important partners for exchanging knowledge on breeding cultivars with resistance to fire blight (ibid.). In cooperation modes with partners from South Africa or China, the ZO sends budwood of promising cultivars to these partners for tests in different locations and climatic settings (ibid.). Overall, the ZO is well connected with (scientific) breeding actors inside and outside of Germany, thus engaging in formal and informal knowledge exchange processes.

Knowledge exchange with farmers takes place in the *Sächsische Sortenkommission*, where this actor group also participates. This seems to be the only formal occasion for knowledge exchange with farmers. Informally, the ZO gets regular visits from fruit farmers and visits fruit orchards. They also organize 'shows' at the ZO where market actors are invited to test and evaluate current selection results.

On these occasions, the ZO aims to acquire knowledge on farmers' and other market actors' demands (Interview ZO).

8.3.6 Resilience Attribute 4a: Implementation into Mental Models

Regarding the implementation of CAS Thinking, the case for ZO is similar to the one of ZIN (See Section 7.3). The ZO does not adopt an ecocentric but rather a trait-based form of breeding. Additionally, the generalist approach to breed a "variety for the world" (Interview ZO, own translation) counteracts the context-dependency that CAS Thinking signifies. Still, some elements can be identified that reflect aspects of CAS Thinking. The ZO established and uses several institutions that promote learning processes between breeders, primarily on a scientific level. This web of connection helps to build adaptive capacity in breeding and acknowledges the need for locally adapted and niche varieties (see the previous section).

8.3.7 Resilience Attribute 4b: Translation into Institutions

As explained above, CAS Thinking is not implemented into the mental models of relevant actors that are primarily responsible for the ZO's breeding efforts. As a result, a translation of CAS Thinking into institutions cannot be observed.

8.3.8 Resilience Attribute 5a: Implementing Long-Term Monitoring Programs

The ZO has not implemented an explicit ecosystem-based long-term monitoring program. Similar to ZIN (See Section 7.3), the database used for monitoring the selection process involves only traits of fruits and trees that are consistent with the breeding goals (Interview ZO). However, as resistances are the focus of their breeding efforts, monitoring indicators for apple scab or powdery mildew indirectly also monitor effects on regulating and supporting ecosystem services.

Besides the fact of choosing parental lines due to the desired combination of traits, it is unclear if pedigree relations play a major role for the respective breeding projects of the ZO. The approach rather focuses on genetic traits, especially by using marker-assisted selection techniques, not specifically on pedigree relations. However, the apple breeder has some expertise in this field as he was involved in several scientific publications on genomic mappings of fruit cultivars and pedigree analyses (JKI 2020). It is thus likely that pedigree relations play a role in the ZO's breeding, albeit not an explicit one.

8.3.9 Resilience Attribute 5b: Promoting Social Learning Institutions

The ZO fosters social learning processes with other scientists and breeders by using several formal and informal institutions. On a scientific and practical breeding level, these networking and cooperation activities promote learning-by-doing, especially through the exchange of breeding material or knowledge on selection results. Knowledge sharing with those actors takes place continuously, foremost via research projects with partners around the world. Thus, elements of adaptive management are included in the ZO's breeding approach. Additionally, knowledge and opinions are shared with fruit farmers, mostly on irregular and informal occasions. In conclusion, the ZO fosters social learning intensively on a scientific level but rather narrowly on a practical level.

Social learning institutions are not implemented until the start of the selection process (the fifth step of the breeding process; See the previous section). The apple breeder primarily discusses and shares selection results and clones with breeding actors (Interview ZO). Before the selection process starts, mostly individual learning processes play a role in the breeding process.[9] Further, literature and interview data do not indicate that individual and social learning processes reach beyond single-loop learning, as basic assumptions as well as values and norms are not questioned or reframed.[10] Similar to ZIN (See Section 7.3), this is consistent with the finding that the ZO does not incorporate CAS Thinking into its breeding approach.

8.3.10 Resilience Attribute 6a: Promoting Participation of Relevant Actors

In general, scientists, breeders, and fruit farmers are included in the breeding process. Fruit farmers and breeders participate in a passive way, equivalent to public communication (See Section 6.1). They give feedback on formal and informal occasions in the later stages of the breeding process, but a transfer of decision-making power does not take place. Scientists also participate in a rather passive way. Although active participation takes place in cooperation modes and projects on fruit breeding research, scientists outside the ZO do not continuously engage in the breeding process in an active way. To conclude, the ZO established no strong collaborations with relevant actors – especially no active participation of fruit farmers that has been identified to significantly promote resilience of breeding efforts (See Section 6.2).

8.3.11 Resilience Attribute 6b: Implementing a Participation Narrative

The ZO does not implement participatory plant breeding. Interview data and gray literature also do not indicate that the organization implements an alternative participation narrative.

8.3.12 Resilience Attribute 6c: Promoting Social Self-Organization

The ZO does not promote social self-organization as the breeding process is organized in a top-down manner. Local networks or institutions that work self-organized are not included in the breeding process.

8.3.13 *Resilience Attribute 7: Promoting Polycentric Governance*

The breeding approach of ZIN is not integrated in a polycentric governance system and does not include elements of polycentric governance. No autonomous and decentral centers of decision-making on multiple scales can be observed in the governance system.

8.4 Interim Conclusion

Figure 8.1 shows a summary of analyzing the resilience of the ZO's breeding approach. In the system-to-be-governed, the ZO integrates diverse social elements and highly contributes to monitoring the diversity and redundancy of fruit varieties.[11] The organization shows further elements of resilience by integrating diverse ecological elements to a certain point and establishing a net of connectivity to exchange knowledge and breeding material. However, the ZO does not sufficiently recognize slow variables and ecological feedback processes in breeding by not promoting an ecocentric but rather a trait-based breeding approach. The analysis of the ZO's governance system revealed that neither CAS Thinking nor participation is fostered. Participation remains passive, a participation narrative or forms of social self-organization are not implemented. Consequently, polycentric governance plays no role either. However, implemented learning processes show elements of resilience as long-term monitoring and institutions for social learning are established and used to a certain point.

Principle 1: Diversity and redundancy	Integrating diverse ecological elements	Principle 4: Fostering CAS Thinking	Implementation into mental models
	Integrating diverse social elements		Translation into institutions
	Monitoring diversity and redundancy of fruit varieties	Principle 5: Encouraging learning and experimentation	Implementing long-term monitoring programs
			Promoting social learning institutions
Principle 2: Slow variables and feedback	Promoting ecocentric breeding approaches	Principle 6: Broadening participation	Promoting participation of relevant actors
			Implementing a participation narrative
			Promoting social self-organization
Principle 3: Managing connectivity	Exchanging knowledge and technology	Principle 7: Promoting polycentric governance	Implementing elements of polycentric governance

Color coding scheme:	Good performance	Includes some elements of resilience attribute	Includes no elements of resilience attribute

Figure 8.1 Resilience analysis of the ZO's breeding approach (own figure). Color codings only illustrate tendencies.

Overall, the ZO shows beneficial structures in its breeding approach regarding principle 1 by consulting diverse actors and monitoring the diversity of fruit varieties. Knowledge exchange and learning play a major role but are not sufficiently integrated into the breeder-farmer nexus. The missing promotion of ecocentricity and CAS Thinking leads to a non-sufficient participatory structure that is not able to address resilience demands. It becomes clear that while the overlap between apple breeding and breeding research in the ZO has certain benefits, it also generates drawbacks because the organizational structure does not sufficiently include practitioners in an active way.

Notes

1 However, public breeding organizations could still generate income by registering variety protection for their bred varieties.
2 This does refer to the organization as such. The naming as Julius Kühn-Institute only took place in 2008.
3 LVWO was historically founded in 1868 and is the oldest German school for fruit and wine cultivation. Since the late 1990s, LVWO conducts apple breeding, although this is not the prior objective of the organization. The focus lies on education and research.
4 KOB Bavendorf was established in 2001 as a foundation by several public and private partners. The goal of the organization is to promote fruit cultivation and the preservation of the cultural landscape of Lake Constance through research, consultation, and networking activities. Thereby, KOB also participates in fruit breeding activities and projects.
5 Still, the interviewee acknowledges that this is a challenge because varieties ripen differently under different and changing environmental conditions. An example is *Golden Delicious*, a variety which ripens perfectly in Southern Tyrol but not in Northern Germany (Interview ZO).
6 Pyramidization describes the building of more than one resistance gene against a particular pest or disease to diversify the genetic sources of resistances (Hanke and Flachowsky 2017).
7 Of course, any breeder has access to this gene bank, as it is a public good. Here, direct access refers to the local proximity of personnel, technological, and genetic resources.
8 It has to be noted that on their website, the ZO states that they conduct "innovative breeding for sustainable resource preservation and healthy diet" (JKI 2020). Interview data does not reveal what is actually meant by that in detail besides their approach of breeding for resistance.
9 It can be argued that publishing scientific papers on breeding methods, etc., which refer to the breeding stages before the start of the selection process, can trigger social learning processes within the scientific community. However, as shown in Chapter 6, social learning in the context of resilience has a practical direction that involves the direct participation of practitioners in learning processes.
10 It has to be noted that, on a large temporal scale, double-loop learning took place. According to the interviewee, in the past the ZO has concentrated on breeding varieties with monogenic resistances. Today, the ZO tries to breed varieties with polygenic resistances (Interview ZO). However, as the ZO implemented no forms of adaptive governance that aim for double-loop learning across scales, it is argued here that learning processes do not reach beyond single-loop learning.
11 The ZO is a special case because it is simultaneously the coordinator of the *Deutsche Genbank Obst*. Public breeding approaches do not necessarily play this role, so this point of analysis should not be generalized.

References

Flachowsky, H.; Le Roux, P.-M.; Peil, A.; Patocchi, A.; Richter, K.; Hanke, M.-V. (2011): Application of a High-speed Breeding Technology to Apple (Malus × Domestica) Based on Transgenic Early Flowering Plants and Marker-assisted Selection. In *The New Phytologist* 192 (2), pp. 364–377. DOI: 10.1111/j.1469-8137.2011.03813.x.

Hanke, M.-V.; Flachowsky, H. (2017): *Obstzüchtung und wissenschaftliche Grundlagen*. Berlin, Heidelberg: Springer Berlin Heidelberg.

Interview ZO (2019): Qualitative telephone interview with a breeder from ZO. Transcription from the original language, Germany.

JKI (2019): Institut für Züchtungforschung an Obst. Julius Kühn-Institut. Available online at https://www.julius-kuehn.de/media/Veroeffentlichungen/InstitutsbroschFlyer/JKI_Institutsbroschuere_ZO.pdf, checked on 7/16/2019.

JKI (2020): Institut für Züchtungsforschung an Obst. Edited by Julius Kühn-Institut. Available online at https://www.julius-kuehn.de/zo/, checked on 11/17/2020.

Lammerts van Bueren, E. T.; Struik, P. C.; van Eekeren, N.; Nuijten, E. (2018): Towards Resilience Through Systems-based Plant Breeding. A Review. In *Agronomy for Sustainable Development* 38 (5), p. 42. DOI: 10.1007/s13593-018-0522-6.

Peil, A. (2017): Nutzung alter Sorten in der Pillnitzer Obstzüchtung. Symposium Obstsortenvielfalt. Berlin, 2017.

9 Commons-Based Apple Breeding

9.1 Description of the Approach

Commons-based apple breeding is carried out by organizations that concentrate on community aspects, collaboration, and an integration of different perspectives. Different actors participate in the breeding process and a strong focus is laid on the breeder-farmer nexus by promoting on-farm breeding structures. Conceptually, this approach is based on institutional-economic commons studies (See Section 1.3). It involves forms of traditional, knowledge, and global commons because resources are managed as common goods in different ways:

> Commons-based fruit breeding […] involves, in its core, three different types of goods: (1) cultivars in form of budwood of selected seedlings and fruits as their physical products; (2) cultivars as the result of an innovation process and hence an intellectual, cultural resource as well as (3) genetic diversity as input to every fruit breeding process.
>
> (Wolter and Sievers-Glotzbach 2019, p. 321)

These three different resource types are connected with the different commons conceptions (Wolter and Sievers-Glotzbach 2019): natural resources such as apple seeds, seedlings, and scions are managed in a breeding community according to common-property regimes within defined community boundaries (traditional commons). Knowledge resources are shared within the breeding community and, to certain degrees, with actors outside the community (knowledge commons). Agrobiodiversity is categorized as a global commons because access to plant genetic resources is demanded on a global scale. Thus, newly bred apple varieties should be open for use and further development by any interested actor (global commons).

A first attempt to characterize commons-based fruit breeding was carried out by Sievers-Glotzbach and Wolter (2018). Recently, Sievers-Glotzbach et al. (2020) summarized their transdisciplinary discussions on the conceptualization of commons-based plant breeding, which further develops the first thoughts on commons-based plant and fruit breeding (Kliem and Tschersich 2017; Sievers-Glotzbach and Wolter 2018). Sievers-Glotzbach et al. (2020) define

DOI: 10.4324/9781003355724-16

the core criteria of so-called *Seed Commons*.[1] Adapted to commons-based apple breeding, these four major characteristics can be derived:

1. **Collective responsibility for agrobiodiversity**: Actors within the breeding community are committed to conserve and enhance agrobiodiversity by protecting, provisioning, and developing diversity on the plant species and genetic level. This implies free access to developed apple varieties and a special focus on developing regionally adapted varieties.
2. **Collective, polycentric management**: Rules and norms for the development, usage, and sharing of resources are developed collectively. Overarching aspects, such as breeding goals and the target groups of breeding results, are agreed upon at the community level. Apple breeding takes place on-farm in decentral structures, based on independent decision-making powers regarding the day-to-day business of cultivating the seedlings.
3. **Knowledge sharing**: Within the breeding community, formal and practical knowledge on apple breeding and cultivation is shared between participating actors throughout the whole breeding process. Knowledge that is important for promoting further breeding activities and innovation (e.g., crossing combinations) is made transparent and shared with a global community.
4. **Collective ownership**: The relevant resources described above are in collective ownership of the breeding community. Intellectual property rights or variety protection on developed varieties is restricted due to the collective responsibility on agrobiodiversity. New varieties and certain knowledge resources should thus be in collective ownership of the global community.

The commons-based approach has large conceptual overlaps with PPB and community-based breeding as defined by Lammerts van Bueren et al. (2018). Especially farmer-led participatory plant breeding approaches resonate with the collective management character of commons-based approaches. Community-based breeding follows a holocentric worldview by respecting cultural and ecological diversity (ibid.). It aims for collaboration between multiple actors, knowledge integration, and bottom-up management. Both concepts are thus incorporated in commons-based apple breeding.

In Germany, only the association *apfel:gut e.V.* can be classified as an organization that carries out commons-based apple breeding. This organization will serve as a case study to explore the resilience of commons-based apple breeding.

9.2 Case Study: *apfel:gut e.V.*

The association *apfel:gut e.V.* was founded in 2019 by a group of fruit farmers, breeders, and other interested actors.[2] Before its foundation, this group of actors already worked as the project *apfel:gut* under the umbrella of the association *Saat:gut e.V.* since 2010. The latter conducts organic plant breeding, foremost of vegetables, in a community of farmers, plant breeders, and distributors, to counteract privatization and commercialization developments in the seed sector.

Finally, the *apfel:gut* group founded their own association because their project grew ever larger over the years and needed more autonomy. However, personal and normative connections to *Saat:gut e.V.* remain close.

apfel:gut focuses on apple breeding but is also interested in breeding other fruit cultivars. For example, they established an extensive pear breeding program. Breeding takes place on-farm on selection areas that are integrated into the cultivation fields of participating fruit farmers all over Germany.[3] All fruit farmers conduct organic fruit cultivation, certified by either *Bioland* or *demeter* as major organic farming associations in Germany. Thus, the seedlings for breeding are decentrally cultivated by organic farming rules. Overall, *apfel:gut* has four goals (Wolter and Sievers-Glotzbach 2019): (1) development of apple and pear varieties specifically suited for organic fruit farming with a minimized use of plant protection products; (2) preservation and enhancement of diversity in fruit breeding through the use of traditional and heirloom cultivars in breeding; (3) promotion of organic fruit breeding and farming through communicating the breeding process and results as well as participating on regional and national events; (4) creating transparency and public benefit.

Especially goals (1) and (2) translate into the following specific breeding goals for apple and pear varieties:

* Fruit qualities: medium size, appealing visual traits and texture, special taste, diversity in taste.
* Tree qualities: vitality, robustness, site adaptation.

Compared to the cases of corporate-based (See Chapter 7) and public apple breeding (See Chapter 8), the breeding goals are more general, albeit diverse specific goals are constantly derived from them (Ristel et al. 2016). Vitality describes the health conditions of the whole plant (tree, leaves, fruits, etc.) whereas robustness is defined as less susceptibility against a wide range of pests and diseases through a high level of diversity inside the fruit. Primarily, *apfel:gut* aims to breed varieties for the organic fresh market. However, they are also open for developing varieties for cider and apple juice as well as niche varieties for hobby gardeners.

apfel:gut's breeding process is described in a nutshell by Wolter and Sievers-Glotzbach (2019). It divides into three phases. Phase 1 includes crossings, pre-raising of seedlings, and the distribution of seedlings to participating farms. Phase 2 comprises the first, second, and third selection steps. Phase 3 describes the registration and assessment of potential new varieties.

* Phase 1: The definition of crossing plans always takes place with the credo of crossing a modern variety with a traditional or heirloom cultivar. Crossings take place at every participating fruit farm with location-specific crossing plans and are carried out by mobile fruit breeders that travel through Germany in the crossing seasons. The harvested apple seeds as result of the crossings are pooled in two locations in northern Germany, stratified, and pre-raised in foliage greenhouses. Afterward, the seedlings are distributed to

selection areas on the different participating farms where the fruit farmers cultivate them.

- Phase 2: All selection steps take place in a decentral way on the specific locations. The first selection step (max. two years) solely focuses on tree characteristics, namely vitality and robustness. Seedlings are evaluated on their susceptibility to pests and diseases. All seedlings that pass the first selection step are then re-planted to give them more room for growing and enter the second selection step (max. four years). In this step, the focus lies on fruit qualities consistent with the breeding goals. Promising seedlings with satisfying tree and fruit characteristics enter the third and last selection step (max. seven years). Budwood is cut from those seedlings to produce clone, process them on M9 rootstocks, and distribute them to different *apfel:gut* locations. The farmers then evaluate their yield and site-specificities under commercial farming conditions.
- Phase 3: Clones with promising results and characteristics are finally registered for legal assessment, which includes an additional official evaluation and multiplication (max. five years) to get permission for commercial farming and distribution.

Fruit farmers, breeders, and other active members of the association (e.g., orchard consultant, biologist) are involved in all phases of the breeding process. The chairwoman of *apfel:gut e.V.*, Inde Sattler, has the particular role of project coordinator and can be described as a farmer-breeder because, as a fruit farmer, she is particularly invested in breeding. Initially, the general breeding goals described above were developed by a small group of eventual members and additional breeding experts from across Europe. Crossing plans are not developed by all members but by an inner circle that has the most professional knowledge on breeding. However, breeding goals, crossing plans, selection criteria, and cultivation procedures are constantly discussed in the breeding community in regular formal and informal meetings.[4] The members deliberately perform participatory plant breeding (Ristel and Sattler 2014).

Because of its legal status and non-profit character, *apfel:gut* has certain responsibilities and obligations. New varieties are bred for the public and should be open to use for anyone interested in the possibilities of the current legal framework. In 2019, *apfel:gut* registered their first variety *Wanja*, Germany's first apple amateur variety.

9.3 Analyzing the Resilience of Commons-Based Apple Breeding

9.3.1 *Resilience Attribute 1a: Integrating Diverse Ecological Elements*

apfel:gut uses a range of traditional and heirloom apple cultivars in their apple breeding program: "[...] robust and high yielding old cultivars (cvs) are mainly crossed with not too susceptible new varieties. Additionally, old cvs are crossed with other old cvs and open-pollinated seeds are sown from time to time" (Ristel et al. 2016, p. 137). It is important to note that *apfel:gut* defines those 'old' cultivars according to their characteristics (e.g., level of inbreeding) and not primarily on a temporal scale. Thus, they would not define *Golden Delicious* as a traditional or heirloom variety like the ZO (See Chapter 8). Ristel and Sattler (2014) give an overview of the specific cultivars integrated in breeding and list about 30 traditional and heirloom cultivars.[5] The actual benefits of implementing these cultivars in their crossing plans is continuously evaluated. Ristel et al. (2018) published observations of positive and negative traits and their heritability regarding some used cultivars. Overall, *apfel:gut* uses a diversity of apple varieties in their breeding program, reflecting their general credo to cross modern with traditional and heirloom cultivars.

Different organic apple cultivation practices are represented in the association as participating farmers are bound by formal guidelines and rules of *Bioland* and *demeter* (See the previous section). Connected to the overall goals, *apfel:gut* thus aims to breed varieties that are suitable for different organic farming practices and consider site-specificities. Because the association is an organic initiative, no integrated fruit farmers participate but are rather excluded. As a result, *apfel:gut* reflects the diversity of farming practices only within their normative scope.

9.3.2 Resilience Attribute 1b: Integrating Diverse Social Elements

Based on the analysis of Wolter and Sievers-Glotzbach (2019), actors that participate in the breeding process can be located in an inner layer (breeding community) and an outer layer (stakeholders). This includes the following actors:

- Fruit breeders: Some active members of *apfel:gut* have a professional background as fruit breeders and work as mobile breeders who organize, document, and carry out crossings. They mainly participate in phase 1 but also consult farmers by discussing ideas and results in phase 2, when the selection process starts.
- Fruit farmers: Organic fruit farmers throughout Germany participate in the breeding process by cultivating and selecting seedlings. Primarily, they participate in phase 2 because the selection areas are part of their fruit farms. In regular meetings and discussions, they also have the opportunity to participate in the other phases of the breeding process.
- Other members of the association: Other *apfel:gut* members (e.g., an orchard consultant or a biologist) also periodically participate in all breeding phases. They take part in discussions during the breeding process or help with seedling cultivation.
- Stakeholders: The German organic fruit breeding and cultivation sector is a dense network of practitioners (Focus Group 2017). *apfel:gut* constantly consults with other organic (and also non-organic) fruit breeders and farmers on their selection results. Additionally, they consult with organic fruit and food retailers. Consultations with stakeholders mainly take place on issues in phase 2 of the breeding process.

In general, a high diversity of actors actively participates in *apfel:gut*'s breeding process.

9.3.3 Resilience Attribute 1c: Monitoring Diversity and Redundancy of Fruit Varieties

Many participating fruit farmers preserve and monitor apple cultivars by maintaining variety collections or even being a partner location of the *Deutsche Genbank Obst*. For example, the fruit farmer and pomologist Hans-Joachim Bannier, being an active member of *apfel:gut*, maintains an arboretum with over 300 fruit varieties. Preservation and enhancement of fruit diversity are one of *apfel:gut's* main goals. For some fruit farmers, their variety collections and a general sensitivity for diversity served as an initial motivation to become a member of *apfel:gut* (Focus Group 2017). Because of the breeding strategy to combine modern with traditional and heirloom cultivars, the association has a comprehensive interest in preserving and monitoring fruit varieties. This strategy depends on the direct access to a diversity of cultivars. Thus, they foster cooperation with conservators or tree nurseries. In sum, *apfel:gut* contributes to preservation and monitoring processes of fruit varieties – both through individual activities of members and through the inherent goals of the organization.

9.3.4 Resilience Attribute 2: Promoting Ecocentric Breeding Approaches

Congruent with the overarching goals of *apfel:gut* and its breeding goals, the organization adopts an ecocentric breeding approach. *apfel:gut* aims at breeding varieties for resource-preserving cultivation that need as few (organic) plant protection measures as possible. The objective is to minimize negative impacts on the nitrogen and water cycle. Robust varieties shall be bred, specifically for organic farming practices: "[v]itality and loss of susceptibility are the keys for an organic breeding strategy" (Ristel et al. 2016, p. 138). Site-specificity is one of the main breeding goals of *apfel:gut*. Only actors that identify with this ecocentric breeding approach can become active members of the association. Hence, their approach directly aims at developing varieties suitable for ecocentric farming approaches such as organic fruit farming.

9.3.5 Resilience Attribute 3: Exchanging Knowledge and Technology

Knowledge and technology are exchanged within the *apfel:gut* community between breeders and farmers (and the other active members) as well as with stakeholders outside the community to certain degrees. Wolter and Sievers-Glotzbach (2019) elaborated on the established rules-in-form and rules-in-use as well as the character of the shared goods in their case study on *apfel:gut*.

Within their breeding community, *apfel:gut* members freely share seeds, seedlings, budwood, and fruits (biophysical resources) according to crossing and selection plans. Breeding and farming knowledge are also freely shared and discussed in regular virtual meetings or on demand. All knowledge about breeding processes and results is theoretically open to anyone inside the community. The breeders maintain breeding books for every participating farm, in which they document crossing and selection data throughout the breeding processes.

Biophysical resources are not shared with stakeholders unless they are cooperation partners of *apfel:gut*, for example, scientific institutions or associations. In this case, a formal submission agreement on the handling of the resources is signed. Sharing of breeding knowledge with stakeholders takes place to certain degrees. General information on crossing plans and selection results is shared with all interested parties, mostly with other fruit breeders. This information is made accessible through personal communication or the publication and presentation of results at conferences or other occasions. However, specific knowledge is shared individually and irregularly:

> Specific knowledge about the breeding process (e.g. the breeding books) is only shared in individual cases and the sharing is considered carefully regarding the norms and personal connections to the respective person, organization or community to avoid misuse. Feedback from actors in the outer layer to the different types of shared knowledge is greatly appreciated for the further breeding process (e.g. ideas for crossings, possible reasons for and learnings out of selection results).
>
> (Wolter and Sievers-Glotzbach 2019, p. 320)

apfel:gut members are well connected with other organic fruit farmers, breeders, representatives from associations, or market actors. Cooperation and a constant exchange with other associations like the *Fördergemeinschaft Ökologischer Obstbau e.V.* (FÖKO), the Öko-Obstbau Norddeutschland Versuchs- und Beratungsring e.V. (ÖON), or *Saat:gut e.V.* exist. Beyond German actors, knowledge exchange also takes place with breeding colleagues from the Netherlands, Belgium, or Switzerland. In some cases, *apfel:gut* members fulfill additional roles in other associations, for example, as working group representatives in the FÖKO. Overall, knowledge exchange on breeding takes place both on formal occasions (publications, conferences) as well as in informal ways (personal communication, interdependences with other institutions in the organic fruit sector).

9.3.6 Resilience Attribute 4a: Implementation into Mental Models

apfel:gut implements elements of CAS Thinking into its breeding approach in several ways. First, the organization adopts an ecocentric breeding approach as explained above. Second, context-dependency as the most important aspect of CAS is incorporated in the organization's breeding goals. Third, on-farm breeding practices reflect those two points by valuing local and regional specificities both from a social and an ecological perspective. Fourth, learning processes are established both within the community as well as with stakeholders. Fifth, the organization shows elements of an uncertainty-tolerant culture: "the members hold great respect for ecological diversity, complexity, and aspects of chance, inherent to nature as an ecological system" (Wolter and Sievers-Glotzbach 2019, p. 317). This results in a culture of error-friendliness, beneficial for accepting breeding as a complex adaptive process. As Ristel et al. (2018) point out: "[...] creativity in parental selection and diverse environmental conditions seem to be key factors for finding resilient cultivars for future needs" (p. 99).

9.3.7 Resilience Attribute 4b: Translation into Institutions

apfel:gut adopts the norms of organic fruit farming and breeding, which have been identified as a beneficial management paradigm for CAS Thinking (See Section 6.2). This management paradigm and accompanying norms shape the whole organizational structure and trickle down into specific processes and institutions. Although the *apfel:gut* members have different individual motivations and preferences for participating in the association (Focus Group 2017), organic farming serves as a collective mental model for all of them.

Adaptive capacity is built through the described knowledge and resource exchange between breeders and farmers inside the breeding community and with stakeholders. Especially on-farm breeding is an important translation of CAS Thinking into institutions. The decentral work with the seedlings is carried out by participating farmers and is characterized by a rather independent form of cultivation and a "willingness to experiment" (Focus Group 2017, own translation). Learning opportunities as key cultural ecosystem services for CAS Thinking have already been described loosely and will be categorized below with the description of social learning institutions.

9.3.8 Resilience Attribute 5a: Implementing Long-Term Monitoring Programs

apfel:gut has not directly implemented an ecosystem-based long-term monitoring program. The breeding books that serve as monitoring databases for every breeding location only capture details on traits of fruits and trees congruent with the breeding goals. As robustness and vitality are major breeding goals, susceptibility against a range of pests and diseases as well as plant health aspects are documented and monitored. The more promising the distinct selection results are, the more details are documented in the breeding books for these seedlings or trees. Overall, similar to the case of ZIN (See Chapter 7), it could be argued that effects on regulating and supporting ecosystem services are monitored.

Pedigree relations of apple varieties are of central importance for *apfel:gut* to accomplish robustness of future varieties by avoiding inbreeding. Crossing plans are constructed in ways that minimize inbreeding tendencies. Active *apfel:gut* members are keen to broaden their knowledge on pedigree relations and include this knowledge into the breeding process (Focus Group 2017). The member Hans-Joachim Bannier published a paper on pedigree relations of modern apple varieties and inbreeding tendencies (Bannier 2011) that serves as a base for the development of crossing plans since. With the start of the research project EGON (See Preface), knowledge on pedigree relations of important varieties for *apfel:gut*'s breeding efforts have been further broadened and published (Howard et al. 2021). Pedigree relations of 28 varieties have been analyzed that play a major role in the crossing plans of the organization and where information on relations was missing. Interestingly, the analysis revealed that many varieties had the same grandparents (ibid.), and thus are more closely related than anticipated by *apfel:-gut* members. Crossing plans were adapted subsequently.

9.3.9 Resilience Attribute 5b: Promoting Social Learning Institutions

As *apfel:gut* conducts breeding in a "community of practitioners" (Focus Group 2017, own translation), social learning is an inherent aspect of its institutional structure. Knowledge is continuously shared between active members inside the association, particularly between farmers and breeders, and with stakeholders beyond. In phase 1 of the breeding process, mainly the inner circle is involved to formulate crossing plans. In phase 2, the included selection phases represent the heart of social learning, where farmers and breeders are in a permanent exchange. Selection results are regularly discussed within the community and with cooperation partners to avoid a disconnect between breeding and cultivation. The whole process of social learning can be defined as adaptive co-management:

> Regarding breeding, the formally agreed-upon breeding goals function as important qualitative objectives for operational decision-making in the whole breeding process. However, they only mark a rough plan. Breeding is merely seen as an evolutionary process, contrary to a strict scientific breeding process as carried out by other breeding organizations. The community members discuss the basic instructions as well as the framework for working and selection procedures (selection criteria). [...] [Ideas] are welcome and every farmer is encouraged to make experiments on his or her own farm upon consultation with a breeder.
>
> (Wolter and Sievers-Glotzbach 2019, p. 319)

Literature and interview data do not reveal any forms of adaptive governance. Thus, social learning processes inside the breeding community do not reach beyond single-loop learning (are we doing things right?) because of the fixed management paradigm the organization has adopted. However, through the interactions with stakeholders and the goal to promote organic fruit breeding and farming, *apfel:gut* aims to induce double-loop and triple-loop learning processes in other organizations and individuals. This includes the reframing of breeding goals or the questioning of organizational approaches, values, and norms.

9.3.10 Resilience Attribute 6a: Promoting Participation of Relevant Actors

In general, the following actors are included in the breeding process: fruit farmers, fruit breeders, other active members of *apfel:gut*, and stakeholders. A strong collaboration between fruit breeders, fruit farmers, and other active members exists. This collaboration takes place across all breeding phases, whereby some actor groups play larger roles in respective phases than others (see above). Because all participants are active members of the association, everyone has the opportunity to actively shape the breeding process. Decision-making power formally lies within the community. In practice, fruit breeders and the inner circle mainly execute decisions in phase 1 of the breeding process. Farmers adopt decision-making power in phase 2 in a decentralized way through the cultivation and selection of seedlings. Across this, permanent and spontaneous discussions take place: "the community emphasizes that communication between farmers and breeders is the crux in the everyday work with the seedlings. Liability and the possibility of spontaneous agreements are necessary obligations [...]" (Wolter and Sievers-Glotzbach 2019, p. 320).

Participation of stakeholders takes place in form of dialogues and consultations, thus in a passive way. Breeding issues are discussed at conferences, meetings, or exhibitions. In most cases, this consultation is not regularly planned but rather emerges on spot.

9.3.11 Resilience Attribute 6b: Implementing a Participation Narrative

apfel:gut deliberately performs PPB and titles their individual narrative as "participatory organic fruit breeding" (Ristel and Sattler 2014). On their website, they state:

> Congruent with organic farming, our breeding activities are process-oriented and transparent. With the non-profit status of our association, everyone shall have access to our varieties. In our community, we work on an equal footing. The participatory approach aims at breeders working together with commercial fruit growers. Thus, they can directly propose wishes and ideas for breeding goals. We want to expand this approach to further include fruit retail, processing companies, and customers.
>
> (apfel:gut 2020, own translation)

This clarifies goals, roles, and expectations throughout the whole breeding process. The participatory narrative strengthens the breeder-farmer nexus and also includes relevant actors beyond this nexus.

9.3.12 Resilience Attribute 6c: Promoting Social Self-Organization

As already explained above, social self-organization takes place on the farm level in phase 2 of the breeding process with the start of selection processes (Wolter and Sievers-Glotzbach 2019):

> The operational level largely consists of the practical decentral work with the seedlings on the farms. Day-to-day-decisions regarding the practical breeding

on-farm are characterized by intuition [...] as well as a constant exchange between the individual members in the community. The farmers, who have the relevant farming knowledge, cultivate the seedlings rather independently [...].

(p. 320)

Human capital is thus activated on the operational level and pooled in the association of *apfel:gut*. The whole association emerged in a socially self-organized way beyond state and market structures out of a "societal need" (Focus Group 2017, own translation). However, because of the non-profit status, especially the allocation of financial resources is challenging for *apfel:gut*:

[...] Organization of funding is a primary concern, limiting our breeding efforts. [...] In our opinion, public sponsorship should also be given to organic breeding work itself and not mainly the breeding research.

(Ristel et al. 2018, p. 99)

At the moment, the association is primarily financed with sponsorships from foundations, donations, and through the participation in research projects such as EGON (See Preface).

9.3.13 Resilience Attribute 7: Promoting Polycentric Governance

The breeding approach of *apfel:gut* shows elements of polycentric governance:

[T]he evolutionary character of the breeding process and the normative attributes characterizing the breeding community are mirrored in the polycentric decentralized decision-making system, nested on different levels.

(Wolter and Sievers-Glotzbach 2019, p. 321)

While the association remains the coordinating center of breeding efforts, individual farms mark diverse units that act autonomously to certain degrees. This governance structure is a result of the organization's norms and participation narrative:

Polycentric decision-making in the development of new fruit varieties appears to be motivated by two aspects. First, by allowing people with diverse motivations but shared normative and ideological convictions to work collectively towards shared goals, and second by the rationale to increase the genepool of cultivars in overall fruit growing through breeding in diverse environments.

(ibid., p. 323)

An important element that holds this governance system together is a "basic trust" (Focus Group 2017, own translation) between the active members of *apfel:-gut*. As one member puts it: "i have no desire to work with control mechanisms. I have an absolute trust in the willingness of the people. That we all make mistakes and learn from them" (Focus Group 2017, own translation).

9.4 Interim Conclusion

Figure 9.1 shows a summary of analyzing the resilience of *apfel:gut*'s breeding approach. In shaping the system-to-be-governed, *apfel:gut* performs well in integrating diverse ecological and social elements, exchanging knowledge and technology, and promoting an ecocentric breeding approach labeled as participatory organic fruit breeding. Both the organization and individual members support the monitoring of diversity and redundancy of fruit varieties to certain degrees. The analysis of *apfel:gut*'s governance system shows that the breeding approach fosters CAS Thinking, encourages learning and experimentation, and broadens participation by strengthening the breeder-farmer nexus with a specific participation narrative and the promotion of social self-organization. Further, the organization incorporates elements of polycentric governance.

Overall, the analysis of *apfel:gut* reveals that commons-based apple breeding is a promising approach to foster resilient fruit breeding based on the identified resilience attributes (See Section 6.2). The core of this approach marks the strong connection between farmers and breeders that is reflected in its different facets in almost all resilience principles. However, this approach is highly challenged by acquiring sufficient financial resources and it remains open if this breeding approach is actually able to deliver resilient apple varieties that will be adopted and used by fruit farmers.

Principle 1: Diversity and redundancy	Integrating diverse ecological elements	**Principle 4:** Fostering CAS Thinking	Implementation into mental models
	Integrating diverse social elements		Translation into institutions
	Monitoring diversity and redundancy of fruit varieties	**Principle 5:** Encouraging learning and experimentation	Implementing long-term monitoring programs
			Promoting social learning institutions
Principle 2: Slow variables and feedback	Promoting ecocentric breeding approaches	**Principle 6:** Broadening participation	Promoting participation of relevant actors
			Implementing a participation narrative
			Promoting social self-organization
Principle 3: Managing connectivity	Exchanging knowledge and technology	**Principle 7:** Promoting polycentric governance	Implementing elements of polycentric governance

| Color coding scheme: | Good performance | Includes some elements of resilience attribute | Includes no elements of resilience attribute |

Figure 9.1 Resilience analysis of apfel:gut's breeding approach (own figure). Color codings only illustrate tendencies.

Notes

1 The paper also provides an overview of literature about commons-based seed systems, connecting it with empirical data on commons-based breeding initiatives and consolidating the core criteria on a coherent theoretical and empirical basis. Hence, the author does not include an extensive literature review on commons-based seed systems here but rather refers to Sievers-Glotzbach et al. (2020) and Kliem and Tschersich (2017).
2 A case study on *apfel:gut* that analyzes the organization in light of commons conceptions was already carried out and published by Wolter and Sievers-Glotzbach (2019). The following elaborations largely base on this publication and the qualitative data that was collected in this context.
3 This is consistent with the theoretical conceptualization of on-farm breeding in the context of social-ecological systems in Section 4.2.
4 Wolter and Sievers-Glotzbach (2019) show a detailed overview of rules-in-form as well as rules-in-use on the constitutional, collective choice, and operational level. The elaborations reveal that only a limited set of rules-in-use and control mechanisms exist.
5 For example, *Alkmene, Seestermüher Zitronenapfel, Karminer*, or *Strauwalds neue Goldparmäne*.

References

apfel:gut (2020): Was wir wollen. Unsere Ziele. Available online at http://www.apfel-gut.org/was_wir_wollen.cfm, updated on 2020, checked on 12/31/2020.

Bannier, H.-J. (2011): Moderne Apfelzüchtung. Genetische Verarmung und Tendenzen zur Inzucht. In *Erwerbs-Obstbau* 52 (3–4), pp. 85–110. DOI: 10.1007/s10341-010-0113-4.

Focus Group (2017): Six members of apfel:gut. Transcription from the original language, Germany.

Howard, N. P.; Luby, J. J.; van de Weg, E.; Durel, C.-E.; Denancé, C.; Muranty, H. et al. (2021): Applications of SNP-based Apple Pedigree Identification to Regionally Specific Germplasm Collections and Breeding Programs. In *Acta Horticulturae* (1307), pp. 231–238. DOI: 10.17660/ActaHortic.2021.1307.36.

Kliem, L.; Tschersich, J. (2017): From Agrobiodiversity to Social-ecological Transformation. Defining Central Concepts for the RightSeeds Project. In *Working Paper Series on Environment and Sustainability Issues*. Available online at https://uol.de/fileadmin/user_upload/coast/Arbeitspapiere_Poster/COAST_Working_Paper_Series_-_01-_From_Agrobiodiversity_to_Social-Ecological_Transformation__2017_.pdf, checked on 2/21/2021.

Lammerts van Bueren, E. T.; Struik, P. C.; van Eekeren, N.; Nuijten, E. (2018): Towards resilience through systems-based plant breeding. A review. In *Agronomy for Sustainable Development* 38 (5), p. 42. DOI: 10.1007/s13593-018-0522-6.

Ristel, M.; Bornemann, M.; Sattler, I. (2018): Apfel:gut – Preliminary Results. In FOEKO (Ed.): *Ecofruit. Proceedings of the 18th International Congress on Organic Fruit Growing*. Ecofruit. Hohenheim, Germany. Fördergemeinschaft Ökologischer Obstbau e.V. (FOEKO), pp. 96–99.

Ristel, M.; Sattler, I. (2014): Apfel: Gut–Participatory Organic Fruit Breeding. In FOEKO (Ed.): *Ecofruit. 16th International Conference on Organic Fruit-Growing: Proceedings*. Ecofruit. Hohenheim. Fördergemeinschaft Ökologischer Obstbau e.V. (FOEKO), pp. 158–161.

Ristel, M.; Sattler, I.; Bannier, H.-J. (2016): Apfel: gut–More Vitality, Genetic Diversity and Less Susceptibility as an Organic Fruit Breeding Strategy. In FOEKO (Ed.): *Ecofruit.*

17th International Conference on Organic Fruit-Growing: Proceedings. 2016. Ecofruit. Hohenheim. Fördergemeinschaft Ökologischer Obstbau e.V. (FOEKO), pp. 136–139.

Sievers-Glotzbach, S.; Tschersich, J.; Gmeiner, N.; Kliem, L.; Ficiciyan, A. (2020): Diverse Seeds – Shared Practices: Conceptualizing Seed Commons. In *International Journal of the Commons* 14 (1), pp. 418–438. DOI: 10.5334/ijc.1043.

Sievers-Glotzbach, S.; Wolter, H. (2018): Bringing Commons Elements into Fruit Breeding. In FOEKO (Ed.): *Ecofruit. Proceedings of the 18th International Congress on Organic Fruit Growing*. Ecofruit. Hohenheim, Germany. Fördergemeinschaft Ökologischer Obstbau e.V. (FOEKO), pp. 19–28.

Wolter, H.; Sievers-Glotzbach, S. (2019): Bridging Traditional and New Commons: The Case of Fruit Breeding. In *International Journal of the Commons* 13 (1), p. 303. DOI: 10.18352/ijc.869.

10 Comparison of Apple Breeding Approaches

The evaluation of the three breeding approaches – corporate-based (See Chapter 7), public (See Chapter 8), and commons-based (See Chapter 9) apple breeding – provides several insights. A comparison of the resilience analyses (See Section 10.1) sheds light on the specific advantages and disadvantages of the different breeding approaches. Because of the huge impact of breeding on cultivation and vice versa, the findings further need a discussion in the context of cultivation, generalizing the empirical findings beyond the German case. This leads to the development and reflection of the concept of commons-based organic apple breeding (See Section 10.2).

10.1 Comparison of Apple Breeding Approaches

The interim conclusions of the three case studies on apple breeding approaches (See Sections 7.3, 8.3, and 9.3) show a respective wrap-up of their evaluation.

Figure 10.1 shows a comparison across the resilience attributes of the system-to-be-governed. The only similarity between all approaches is a good performance in integrating diverse social elements, although the specific elements differ in variety, balance, and disparity. While commons-based apple breeding is the only approach that sufficiently integrates ecological diversity by intensively valuing the (potential) benefits of traditional and heirloom cultivars, public apple breeding is the only approach that performs well in monitoring the diversity and redundancy of fruit varieties. Commons-based apple breeding performs well in promoting an ecocentric breeding approach while corporate-based and public apple breeding approaches only partly include elements of this attribute. Both corporate-based and commons-based apple breeding approaches show a good performance in exchanging knowledge and technology – again to a different extent with different actors and different rules.

Overall, commons-based apple breeding best meets the resilience demands of the system-to-be-governed. Most striking for corporate-based and public apple breeding approaches is the missing direct implementation of an ecocentric breeding approach that also determines the integration of diverse ecological elements.

Figure 10.2 subsequently shows a comparison of the performances in the governance system. Commons-based apple breeding fosters CAS Thinking

DOI: 10.4324/9781003355724-17

	Corporate-based apple breeding	Public apple breeding	Commons-based apple breeding
Principle 1: Diversity and redundancy	Integrating diverse ecological elements	Integrating diverse ecological elements	Integrating diverse ecological elements
	Integrating diverse social elements	Integrating diverse social elements	Integrating diverse social elements
	Monitoring diversity and redundancy of fruit varieties	Monitoring diversity and redundancy of fruit varieties	Monitoring diversity and redundancy of fruit varieties
Principle 2: Slow variables and feedback	Promoting ecocentric breeding approaches	Promoting ecocentric breeding approaches	Promoting ecocentric breeding approaches
Principle 3: Managing connectivity	Exchanging knowledge and technology	Exchanging knowledge and technology	Exchanging knowledge and technology

Color coding scheme: | Good performance | Includes some elements of resilience attribute | Includes no elements of resilience attribute |

Figure 10.1 Comparison of resilience attributes in the system-to-be-governed (own figure). Color codings only illustrate tendencies.

	Corporate-based apple breeding	Public apple breeding	Commons-based apple breeding
Principle 4: Fostering CAS Thinking	Implementation into mental models	Implementation into mental models	Implementation into mental models
	Translation into institutions	Translation into institutions	Translation into institutions
Principle 5: Encouraging learning and experimentation	Implementing long-term monitoring programs	Implementing long-term monitoring programs	Implementing long-term monitoring programs
	Promoting social learning institutions	Promoting social learning institutions	Promoting social learning institutions
Principle 6: Broadening participation	Promoting participation of relevant actors	Promoting participation of relevant actors	Promoting participation of relevant actors
	Implementing a participation narrative	Implementing a participation narrative	Implementing a participation narrative
	Promoting social self-organization	Promoting social self-organization	Promoting social self-organization
Principle 7: Promoting polycentric governance	Implementing elements of polycentric governance	Implementing elements of polycentric governance	Implementing elements of polycentric governance

Color coding scheme: | Good performance | Includes some elements of resilience attribute | Includes no elements of resilience attribute |

Figure 10.2 Comparison of resilience attributes in the governance system (own figure). Color codings only illustrate tendencies.

while corporate-based and public apple breeding approaches fail to promote it. This connects with their missing direct promotion of an ecocentric breeding approach. All approaches encourage learning and experimentation in different ways and to different extents. Hereby, commons-based apple breeding is the only approach to sufficiently implement social learning institutions that meet resilience demands. This is due to the adopted PPB approach, which leads to a good performance in broadening the participation of different actors. Here, public apple breeding fails to meet resilience demands at all while corporate-based apple breeding promotes the participation of relevant actors – albeit in a different way than the commons-based approach, reflecting its organizational specificities. Except for commons-based apple breeding, the governance modes follow a rather monocentric form where breeding is conducted in a top-down manner.

Overall, commons-based apple breeding delivers a wholly resilient governance system. Corporate-based apple breeding shows some promising elements while the public approach does not meet the demands of a resilient governance system. The governance system of the commons-based approach has conceptual advantages to react and adapt to changing environmental conditions compared to the other approaches. The key seems to be the normative orientation of the active actors, inherent to the commons-based organizational character: ecocentricity and elements of CAS Thinking translate into corresponding governance institutions as well as the valuing of diversity on an ecological and social level. However, commons-based apple breeding still has to show results in the form of successfully breeding resilient cultivars.

Interestingly, most of the aspects that result in the good performance of the commons-based approach are not only determined by its organizational characteristics, but rather by the actor's norms. When the resilience attributes of principles 2 and 4 are categorized as key attributes, changing norms of key actors in public or corporate-based breeding organizations could foster resilience in the context of these approaches:

> A shift in mindsets towards approaches that better recognize the features and behaviours of SES [social-ecological systems] as CAS [...] appears to be particularly fundamental [...]
>
> (Schlüter et al. 2015, p. 278).

This finding has value beyond the empirical context of Germany, specifically for breeding organizations in important breeding centers such as the US or the Netherlands (See Section 3.3). However, the influences of actors on institutions and vice versa remain a sociological issue that cannot be resolved in the context of this book.

A clear advantage of corporate-based breeding and specifically public apple breeding is a (relative) financial security of their breeding approaches. While corporate-based breeding aims at generating profits through the introduction of new varieties, public breeding is publicly funded. Both breeding approaches use variety protection as an instrument for generating income. This is also not a German-specific phenomenon but a business and finance model that can be observed globally (Clark et al. 2012). Commons-based breeding, on the other hand, continuously applies for public funding and generates financial resources through donations. Challenges for commons-based breeding in this context will be further elaborated in Part IV.

This comparison of different apple breeding approaches is based on qualitative case studies in a specific empirical context, which results in methodological limitations (See Section 2.3 and Appendix B). The relatively clear and significant result of this evaluation should thus be placed into this context. It remains open if other case studies from Germany (at least for public and corporate-based apple breeding) or other countries would result in similar conclusions. Other cases of apple breeding exist that (could) show characteristics of commons-based breeding or conduct organic breeding (Koutis et al. 2020). For example, the initiative NOVAFRUITS carries out organic apple breeding on-farm across borders in Northern France and Belgium in cooperation with public breeding institutes. Over 25 farmers participate in the initiative to breed new apple varieties (Lateur 2019). It would be valuable to analyze further cases such as NOVAFRUITS with the developed analytical framework (See Section 6.3).

10.2 Commons-Based Organic Apple Breeding

The evaluation of different apple breeding approaches already showed that conceptual linkages to cultivation approaches exist. This is congruent with the theoretical conceptualization of fruit breeding and cultivation as social-ecological systems (See Section 4.2): apple breeding organizations are embedded in a multilevel system and connected to fruit cultivation organizations through cultural, social, or economic links. All case studies indicate that breeding goals aim to address specific cultivation practices or demands. In the case study on commons-based apple breeding, *apfel:gut* explicitly integrates organic fruit farming in their approach.

This finding suggests to re-conceptualize commons-based breeding as *commons-based organic apple breeding*.[1] The re-conceptualization is valuable for breeding initiatives across the world because the basic goals and values of organic farming match with the major characteristics of commons-based apple breeding:

> The international guidelines of organic agriculture (IFOAM 2014) emphasize the need for robust varieties and breeding under organic conditions. Moreover, organic farming aims at ensuring free access to genetic resources and at new organizational structures, based on cooperation between farmers, traders and breeders [...]
>
> (Wolter and Sievers-Glotzbach 2019, pp. 307).

This connects with the literature evaluation in Section 5.2, where the impacts of different farming approaches on ecosystem services have been compared and evaluated. Results showed that ecocentric approaches such as agroecology or organic agriculture generally prove most beneficial for the promotion of ecosystem services. Those guidelines and abbreviated standards have implications for plant breeding and, for example, call for participatory structures and transparency (Lammerts van Bueren 2010). Sievers-Glotzbach and Wolter (2018) already showed that organic agriculture is able to supply a fitting normative frame for commons-based fruit breeding.[2]

It is recognized that it would also be of scientific value to incorporate other cultivation directions in this re-conceptualization. Integrated fruit production is a globally adopted cultivation approach and also aims at fruit cultivation with lesser environmental impacts through the adoption of different pest management regimes (Damos et al. 2015). From an entrepreneurial perspective, meadow orchards experience a renovation as a sustainable alternative to contemporary fruit cultivation practices (Barde et al. 2019). Implementing organic cultivation in the commons concept thus only serves as an example to link breeding and cultivation in a single concept.

In conclusion, commons-based organic apple breeding can be defined as a breeding approach that aims to consequently implement the IFOAM principles by adopting collective responsibility for agrobiodiversity, sharing knowledge between a diversity of actors, breeding resilient cultivars with collective and polycentric management practices, and enabling collective ownership of important resources and varieties. The breeding community breeds cultivars for fruit farmers that adopt the IFOAM principles in their farming practices.

Commons-based organic apple breeding could be beneficial for several "practical ways forward"[3] (Schlüter et al. 2015, p. 269) in designing resilient agricultural systems. The IFOAM norms and commons principles clarify the goals of breeding by specifying which ecosystem services are to be enhanced and what governance structures are to be pursued. Trust and the building of social capital are inherent characteristics of the commons-based approach that foster collective action. Commons-based organic apple breeding poses an integrative approach that emphasizes the coupling of social and ecological systems by focusing on their interactions and feedbacks. Learning and participation structures additionally provide room for spontaneous explorations, uncertainty, and surprise.

However, the exclusive focus on one cultivation perspective could also lead to drawbacks in resilience. Schlüter et al. (2015) argue that exclusively managing for efficiency and control (which they call the conventional paradigm) turns a blind eye to resilient management approaches that foster diversity, learning, or participation. An exclusive focus on organic cultivation and market integration of robust, organic varieties also potentially disregards suitable alternatives both on the organizational and regime levels. For example, the disintegration of integrated fruit production as concept that is widely adopted in Germany and in other countries could be challenging when the aim is to achieve a regime shift toward more resilient cultivation and breeding systems.[4] Commons-based organic apple breeding is thus "confronted with the difficulty of finding the right balance between too much of a given principle and too little" (ibid., p. 271).

This dilemma of exclusivity is addressed by Lammerts van Bueren et al. (2018) with the proposal of a systems-based plant breeding approach that integrates beneficial characteristics of different breeding approaches[5] and creates a balance between different styles of thought and orientations:

> In the envisioned concept of systems-based breeding, 'system' is defined as the space that encompasses the civil society (with its diversity of cultural

norms and values), policy (with various governance institutions), nature (including the diversity of pedo-climatic conditions and habitats), agriculture (including the diversity of agro-ecosystems and farming systems), and value chains and markets as interrelated and mutually dependent components of the entire system [...]

(ibid., pp. 11).

In this sense, resilience is achieved by integrating a diversity of actors across different sectors (fruit, vegetable, etc.) and levels. This is in line with one of the fundamentals of resilience theory: "exposure to the full range of social and environmental variation is necessary for maintaining and building resilience" (Walker 2020, p. 1). However, it remains questionable if such a holistic concept adequately addresses the context-sensitivity of fruit and particularly apple breeding that was identified as an important element for achieving resilience (See Chapter 6).

Beyond specific cultivation and breeding approaches, the reflections on commons-based organic apple breeding show the benefits of a further coupling of breeding and cultivation on a local scale in the context of resilience. In Section 4.2, those strong feedback mechanisms were illustrated with special case (2) "Breeding and cultivation as overlapping activities at local levels." Contemporary studies of resilient agricultural systems still widely neglect this important coupling of both systems and their conceptual relevance. Even innovative studies on resilience in agricultural systems such as Lade et al. (2020), where different conceptual perspectives on resilience are combined, exclude breeding in their observations. Overall, the results of this Part III suggest that a strengthening of the coupling of breeding and farming benefits the resilience of both systems. Moreover, it would be valuable to include the social-ecological system of breeding in the discussion on resilient agricultural systems.

Notes

1 While commons-based ecocentric apple breeding would probably be more conceptually accurate – as this would directly include other farming concepts besides organic apple farming – it would not sufficiently emphasize the market- and practitioner-orientation. Commons-based organic apple breeding better connects to participatory organic fruit breeding, which is already implemented in scientific and practitioner discourses (See Chapter 9).

2 Interestingly, Sievers-Glotzbach et al. (2020) did not include the cultivation perspective into their conception of *Seed Commons* – although both case studies in their paper refer to organic breeding initiatives. However, their developed *Seed Commons* criteria as well as the adaptation of these criteria to apple breeding (See Section 9.1) indirectly reflect ecocentric cultivation practices by taking certain norms and values into account. It remains questionable if commons-based breeding approaches would exist without their coupling with ecocentric values and cultivation practices.

3 These include: (a) clarify goals and develop and monitor relevant metrics for each principle; (b) take an integrative approach that builds on multiple knowledge resources; (c) shift away from exclusively managing for efficiency toward planning for uncertainty and surprise; (d) create spaces for spontaneous exploration; (e) build trust and social capital (Schlüter et al. 2015).

4 Nevertheless, it should be noted that integrated apple production puts the focus on adapting the pest management regime on the conventional management paradigm (cultivating high-yielding varieties on a large scale) and designing it in a way to reduce negative environmental and health impacts. The other way around, commons-based apple breeding rather focuses on the choice of cultivars by breeding robust and vital apple varieties. Integrated apple production does not include any norms or guidelines on a social level that address participation or learning aspects. It is thus questionable how and if an integration of this cultivation approach should be pursued.

5 These are community-based, ecosystem-based, corporate-based, and trait-based breeding. Several characteristics of all approaches were already discussed in Chapters 7–9.

References

Barde, M.; Hochmann, L.; Barde, M. (2019): *Streuobstwirtschaft. Aufbruch zu einem neuen sozialökologischen Unternehmertum.* München: Oekom-Verl. Ges. für Ökologische Kommunikation.

Clark, J. R.; Aust, A. B.; Jondle, R. J. (2012): Intellectual Property Protection and Marketing of New Fruit Cultivars. In Marisa Luisa Badenes, David H. Byrne (Eds.): *Fruit Breeding.* Boston, MA: Springer US, pp. 69–96.

Damos, P.; Colomar, L.-A. E.; Ioriatti, C. (2015): Integrated Fruit Production and Pest Management in Europe: The Apple Case Study and How Far We Are from the Original Concept? In *Insects* 6 (3), pp. 626–657. DOI: 10.3390/insects6030626.

IFOAM (2014): The IFOAM Norms for Organic Production and Processing. Version 2014. Available online at http://www.ifoam.bio/sites/default/files/ifoam_norms_version_july_2014.pdf, checked on 3/23/2021.

Koutis, K.; Warlop, F.; Bolliger, N.; Steinemann, B.; Rodriguez Burruezo, A.; Mendes Moreira, P.; Messmer, M. (2020): Perspectives on European Organic Apple Breeding and Propagation Under the Frame of LIVESEED Project. In FOEKO (Ed.): *Ecofruit. 19th International Conference on Organic Fruit-Growing: Proceedings.* Ecofruit. Hohenheim, 17.-19.02.2020. Fördergemeinschaft Ökologischer Obstbau e.V. (FOEKO), pp. 104–107. Available online at https://www.ecofruit.net/wp-content/uploads/2020/04/7_Koutis_104-107.pdf, checked on 8/16/2022.

Lade, S. J.; Walker, B. H.; Haider, L. J. (2020): Resilience as Pathway Diversity: Linking Systems, Individual, and Temporal Perspectives on Resilience. In *Ecology and Society* 25 (3). DOI: 10.5751/ES-11760–250319.

Lammerts van Bueren, E. (2010): Ethics of Plant Breeding: The IFOAM Basic Principles as a Guide for the Evolution of Organic Plant Breeding. In *Ecology & Farming* (February), pp. 7–10, checked on 3/28/2018.

Lammerts van Bueren, E. T.; Struik, P. C.; van Eekeren, N.; Nuijten, E. (2018): Towards resilience through systems-based plant breeding. A review. In *Agronomy for Sustainable Development* 38 (5), p. 42. DOI: 10.1007/s13593-018-0522-6.

Lateur, M. (2019): Experiences from 'NOVAFRUITS': an apple trans-border organic participatory breeding program based on robust and disease tolerant old local cultivars. EGON-Abschlussveranstaltung. Jork, Germany, 2019. Available online at https://uol.de/f/2/dept/wire/fachgebiete/oekogueter/EGON_Abschlussveranstaltung/EGON_Abschlussveranstaltung_Vortrag_Lateur.pdf, checked on 3/6/2021.

Schlüter, M.; Biggs, R.; Schoon, M. L.; Robards, M. D.; Anderies, J. M. (2015): Reflections on Building Resilience – Interactions among Principles and Implications for Governance. In Reinette Biggs, Maja Schlüter, Michael L. Schoon (Eds.): *Principles for Building*

Resilience. Sustaining Ecosystem Services in Social-Ecological Systems. Cambridge, UK: Cambridge University Press, pp. 251–282.

Sievers-Glotzbach, S.; Tschersich, J.; Gmeiner, N.; Kliem, L.; Ficiciyan, A. (2020): Diverse Seeds – Shared Practices: Conceptualizing Seed Commons. In *International Journal of the Commons* 14 (1), pp. 418–438. DOI: 10.5334/ijc.1043.

Sievers-Glotzbach, S.; Wolter, H. (2018): Bringing commons elements into fruit breeding. In FOEKO (Ed.): *Ecofruit. Proceedings of the 18th International Congress on Organic Fruit Growing.* Ecofruit. Hohenheim, Germany. Fördergemeinschaft Ökologischer Obstbau e.V. (FOEKO), pp. 19–28.

Walker, B. H. (2020): Resilience: What it Is and Is Not. In *Ecology and Society* 25 (2). DOI: 10.5751/ES-11647–250211.

Wolter, H.; Sievers-Glotzbach, S. (2019): Bridging traditional and new commons: The case of fruit breeding. In *International Journal of the Commons* 13 (1), p. 303. DOI: 10.18352/ijc.869.

Interim Conclusion

In Part III, it was discussed which fruit breeding approaches exist, what their differences are, and what effects they have on fruit cultivation and social-ecological resilience. Corporate-based, public, and commons-based apple breeding approaches have been analyzed in the empirical context of Germany (See Chapter 3) on the basis of the analytical framework that was developed in Section 6.3.

The German case studies reveal that commons-based apple breeding has the greatest potential for resilience – both regarding the resilience demands of the system-to-be-governed and the governance system. Consequently, commons-based organic apple breeding is defined as an approach that acknowledges the coupling of both breeding and cultivation. The case studies additionally show that although all evaluated breeding approaches propagate to breed varieties for sustainable cultivation, their specific understanding of sustainability largely differs. This emphasizes the conceptual vagueness of sustainability in practice, which can also be observed in other countries. Taking the examples of Plant & Food Research (New Zealand) and Agroscope (Switzerland) (See Section 3.3), both breeding organizations propagate sustainable fruit breeding. Plant & Food Research promotes a "smart green future" (Plant & Food Research 2022) through their breeding efforts, and Agroscope aims to breed "high-performing, site-adapted varieties [as] the basis for a successful, sustainable plant production" (Agroscope 2022). Additional research could further investigate different understandings of sustainability in apple breeding in a more global context. Moreover, the case studies showed very well to which extent breeding organizations depend on specific actors as key connectors (the chief breeder at ZIN, the apple breeder at ZO, Sattler as chairwoman, and farmer breeder at *apfel:gut*).

Because the findings in Part III are partly located on a conceptual level, they are also valuable for fruit breeding organizations outside the empirical case of Germany. This particularly applies to organic breeding initiatives that integrate participatory elements into their breeding approach, such as Koutis et al. (2020) indicate for several European initiatives. But the results of the case studies may also guide corporate-based or public fruit breeding organizations when they aim to achieve more resilience in their breeding efforts.

Overall, the evaluation shows that it is highly valuable to integrate commons-based solutions into the resilience discourse on a broader scale. So far, solutions

DOI: 10.4324/9781003355724-18

beyond traditional commons and Ostrom's design principles (see, for example, Schlüter et al. 2015) are underrepresented. Commons-based organic apple breeding could serve as an entry point to acknowledge the value and importance of new commons and consequently hybrid commons conceptions for finding resilient solutions in agricultural and food systems.

As concluded in Section 4.2, individual value chains in the sense of Tendall et al. (2015) serve as an entry point for building resilience. The next chapters in Part IV put the focus on current market structures in the German apple sector and discuss challenges for commons-based organic apple breeding.

References

Agroscope (2022): Plant Breeding. Agroscope. Available online at https://www.agroscope.admin.ch/agroscope/en/home/about-us/organization/competence-divisions-strategic-research-divisions/plant-breeding.html, checked on 8/16/2022.

Koutis, K.; Warlop, F.; Bolliger, N.; Steinemann, B.; Rodriguez Burruezo, A.; Mendes Moreira, P.; Messmer, M. (2020): Perspectives on European organic apple breeding and propagation under the frame of LIVESEED project. In FOEKO (Ed.): *Ecofruit. 19th International Conference on Organic Fruit-Growing: Proceedings.* Ecofruit. Hohenheim, 17.-19.02.2020. Fördergemeinschaft Ökologischer Obstbau e.V. (FOEKO), pp. 104–107. Available online at https://www.ecofruit.net/wp-content/uploads/2020/04/7_Koutis_104-107.pdf, checked on 8/16/2022.

Plant & Food Research (2022): Our Strategy. Available online at https://www.plantand-food.com/en-nz/our-strategy, checked on 8/16/2022.

Schlüter, M.; Biggs, R.; Schoon, M. L.; Robards, M. D.; Anderies, J. M. (2015): Reflections on building resilience - interactions among principles and implications for governance. In Reinette Biggs, Maja Schlüter, Michael L. Schoon (Eds.): *Principles for Building Resilience. Sustaining Ecosystem Services in Social-Ecological Systems.* Cambridge, U.K.: Cambridge University Press, pp. 251–282.

Tendall, D. M.; Joerin, J.; Kopainsky, B.; Edwards, P.; Shreck, A.; Le, Q. B. et al. (2015): Food system resilience. Defining the concept. In *Global Food Security* 6, pp. 17–23. DOI: 10.1016/j.gfs.2015.08.001.

Part IV

Implementing Commons-Based Organic Apple Breeding

Part IV

Introduction

This part adopts an action-oriented perspective to investigate market challenges for the implementation of commons-based organic apple breeding in Germany. This is a topic of particular relevance as the institutional logics of commons often collide with specific market logics (See Section 1.3). The investigation contributes to the fourth objective of this book: which factors inhibit the broad implementation of commons-based apple breeding and how can they be overcome to exploit its full potential? For a coherent overview of the organic apple market in Germany, the following aspects are evaluated: norms of market actors; their demands on apple varieties; promoting and inhibiting factors for the further expansion of organic apple farming and breeding; and market development and trends. The objective is to specifically elaborate on organic breeding in the market context and discuss the findings in the context of commons-based organic apple breeding.

Available literature does not provide many clues about the aspects mentioned above, especially about the organic sector in Germany. On the market level, few quantitative studies have been carried out: Garming et al. (2018) published a study on the geographical distribution of fruit yield and orchard areas, as well as an analysis of domestic trade structures. The most recent study with a focus on the organic apple sector was carried out by Zander (2011), who specifically analyzed production and sales structures, as well as the share of imports and domestic produce. Zander (2012, 2011) additionally investigated cooperation behavior and the quality of business relations. Another discourse focuses on the marketing management of apples. This includes the discussion of club concepts (Legun 2015; Schwartau 2010), marketing management and so-called managed varieties (Legun 2016; Weber 2008b), and the particular role of branding (Rickard et al. 2013; Weber 2008a). There is also a global discourse on consumer preferences for apple attributes and traits (Denver and Jensen 2014; Normann et al. 2019).

In addition to these market investigations, much literature on apple farming focuses on the economic performance and economic sustainability of fruit farms.[1] Following Mouron et al. (2012), the economic sustainability of orchards is defined by three attributes: profitability, production risk, and financial autonomy. The authors developed the tool Sustain OS to quantitatively assess the ecologic and economic sustainability of orchards. Success factors for the economic performance of apple production in the scope of these three attributes are widely

DOI: 10.4324/9781003355724-20

discussed in the literature (Gallardo and Garming 2017; Bravin et al. 2009). In particular, the choice and combination of cultivars from an economic perspective is a prominent topic (Gallardo and Garming 2017; Leumann and Bravin 2008; Waibel et al. 2001).

It can be concluded that the current literature does not provide any market study on the national level in the distinct context of organic apple breeding, and respective discussions only indirectly include this topic. However, an investigation of the potential of commons-based organic apple breeding requires an in-depth understanding of the specific national market and societal factors. A two-step Delphi study was carried out to get insights into these factors. The Delphi method was identified as a fitting method to collect detailed qualitative data by enabling the integration of different actors and their perspectives on the research topic (See Chapter 2). Two rounds of inquiry were carried out between November 2018 and April 2019.

In the first round (November 2018 to January 2019), 29 experts from the fields of organic fruit cultivation, breeding, marketing, research, and related fields from Germany, Austria, and Switzerland participated in an extensive online questionnaire. In the second round (March 2019 to April 2019), 22 experts from the first round participated in a follow-up online questionnaire. The general goal of this Delphi study is to get insights into (a) the general potential of organic apple breeding and cultivation, (b) accompanied market challenges, and (c) structures, norms, and developments along the value chain of organic apple production. A detailed overview of the sample, the procedure of the study, and the steps of data analysis are given in Appendix C.

All results only concern the national level of organic and general apple cultivation and breeding in Germany. The core results of this market study have been published in a conference paper (Wolter 2020). At the conference, results have been discussed with scientists and practitioners from organic fruit growing. In the following sections, a more comprehensive and detailed analysis of the study results takes place. In Chapters 11–13, the results of the market study are presented and discussed respectively. This involves insights into market structures, developments, and trends (See Chapter 11), promoting and inhibiting factors for organic apple cultivation and breeding (See Chapter 12), as well as business models for financing organic apple breeding (See Chapter 13). Based on these insights, challenges for commons-based organic apple breeding are discussed (See Chapter 14) to examine obstacles and possibilities for realizing the potential of this breeding approach.

Note

1 In general, many scientists from the Swiss organization *Agroscope* carry out research in this field.

References

Bravin, E.; Kilchenmann, A.; Leumann, M. (2009): Six Hypotheses for Profitable Apple Production Based on the Economic Work-package Within the ISAFRUIT Project.

In *The Journal of Horticultural Science and Biotechnology* 84 (6), pp. 164–167. DOI: 10.1080/14620316.2009.11512615.

Denver, S.; Jensen, J. D. (2014): Consumer Preferences for Organically and Locally Produced Apples. In *Food Quality and Preference* 31, pp. 129–134. DOI: 10.1016/j.foodqual.2013.08.014.

Gallardo, R. K.; Garming, H. (2017): The Economics of Apple Production. In Gayle M. Volk, Amit Dhingra, Sally A. Bound, Dugald C. Close, Peter M. Hirst, M. C. Goffinet et al. (Eds.): *Achieving Sustainable Cultivation of Apples*. 1st ed. Cambridge: Burleigh Dodds Science Publishing (Burleigh Dodds Series in Agricultural Science), pp. 485–510.

Garming, H.; Dirksmeyer, W.; Bork, L. (2018): Entwicklungen des Obstbaus in Deutschland von 2005 bis 2017: Obstarten, Anbauregionen, Betriebsstrukturen und Handel. Braunschweig, Germany (Thünen working paper).

Legun, K. A. (2015): Club Apples: A Biology of Markets Built on the Social Life of Variety. In *Economy and Society* 44 (2), pp. 293–315. DOI: 10.1080/03085147.2015.1013743.

Legun, K. (2016): Managed Apple Varieties Research Report. Available online at https://www.academia.edu/27324070/Managed_Apple_Varieties_Project_Report, checked on 4/4/2020.

Leumann, M.; Bravin, E. (2008): Obstbau: Entscheidungsgrundlage bei der Sortenwahl. In *AGRARForschung* 15 (5), pp. 214–219.

Mouron, P.; Heijne, B.; Naef, A.; Strassemeyer, J.; Hayer, F.; Avilla, J. et al. (2012): Sustainability Assessment of Crop Protection Systems: SustainOS Methodology and Its Application for Apple Orchards. In *Agricultural Systems* 113, pp. 1–15. DOI: 10.1016/j.agsy.2012.07.004.

Normann, A.; Röding, M.; Wendin, K. (2019): Sustainable Fruit Consumption: The Influence of Color, Shape and Damage on Consumer Sensory Perception and Liking of Different Apples. In *Sustainability* 11 (17), p. 4626. DOI: 10.3390/su11174626.

Rickard, B. J.; Schmit, T. M.; Gómez, M. I.; Lu, H. (2013): Developing Brands for Patented Fruit Varieties: Does the Name Matter? In *Agribusiness* 29 (3), pp. 259–272. DOI: 10.1002/agr.21330.

Schwartau, H. (2010): Liegt die Zukunft in den Club-Sorten? In *European Fruit Magazine* (5), pp. 21–22.

Waibel, H.; Garming, H.; Zander, K. (2001): Die Umstellung auf ökologischen Apfelanbau als risikobehaftete Investition. In *Agrarwirtschaft* 50 (7), pp. 439–450.

Weber, M. (2008a): Marken-Management - die neue Herausforderung für die Apfelbranche. In *Schweizer Zeitschrift für Obst- und Weinbau* (3), pp. 11–14.

Weber, M. (2008b): Sustainable Apple Breeding Needs Sustainable Marketing and Management. In FOEKO (Ed.): *Ecofruit. 13th International Conference on Organic-Fruit Growing: Proceedings*. Ecofruit. Hohenheim. Fördergemeinschaft Ökologischer Obstbau e.V. (FOEKO).

Wolter, H. (2020): Influencing factors for the further expansion of organic apple cultivation and breeding. In FOEKO (Ed.): *Ecofruit. 19th International Conference on Organic Fruit-Growing: Proceedings*. Ecofruit. Hohenheim, 17.-19.02.2020. Fördergemeinschaft Ökologischer Obstbau e.V. (FOEKO), pp. 19–26.

Zander, K. (2011): Ausländisches Angebot an ökologischen Äpfeln: Bedeutung für deutsche Öko-Apfelerzeuger. Universität Kassel, Fachgebiet Agrar- und Lebensmittelmarketing. Witzenhausen.

Zander, K. (2012): Zur Wettbewerbssituation bei Öko-Äpfeln in Deutschland. In *Jahrbuch der Österreichischen Gesellschaft für Agrarökonomie*, 21 (1). Wien: Facultas, pp. 13–22.

11 Market Structures, Developments and Trends

Current literature does not provide any comprehensive market study on the national level (Germany) in the distinct context of organic apple breeding, and respective discussions only indirectly include this topic. However, an investigation of the market potential of commons-based organic apple breeding requires an in-depth understanding of the specific national market and societal factors. This chapter presents all results of a market study that concerns market actor's norms, their general view on apple breeding in the market context, market demands of farmers and marketeers, as well as general market developments and trends in the apple sector. Subsequently, all empirical results are discussed in light of relevant scientific literature on market aspects and apple farming to uncover similarities, contradictions, and conflict areas. The market study has been carried out with experts from Germany, Austria, and Switzerland. Round 1 of the market study provided data on actor's norms, apple breeding in the market context, market demands of farmers and marketeers, the evaluation of club concepts, as well as market developments and trends. The raw qualitative data has been categorized and evaluated and is presented accordingly.

11.1 Results of the Market Study

11.1.1 Norms of Participants

For getting insights about the normative orientations of the study's participants in the field of (organic) fruit breeding and farming, they were asked to value their agreement on certain statements. Table 11.1 shows the results of these valuations.

High agreement among participants can be observed concerning the following norms: orientation of fruit cultivation on natural processes and cycles; high relevance of varietal diversity and genetic diversity; responsibility to protect this diversity; free access to knowledge on breeding and varieties; and participation of fruit farmers in fruit breeding. Even though the majority agrees with the inclusion of fruit farmers, the majority also tends to reject on-farm breeding. There is also an indifferent format of opinion on the reduction of plant protection products in organic fruit cultivation. While the majority rejects the use of genetically modified organisms in fruit cultivation, some participants disagree with that – although

DOI: 10.4324/9781003355724-21

Table 11.1 Ratings of statements on (organic) fruit cultivation and breeding (n = 29)

Statement	Absolutely disagree	Disagree	Neither	Agree	Absolutely agree	Don't know
	n	n	n	n	n	n
Organic fruit cultivation should be oriented on natural processes and cycles (e.g., closed loops)	0	0	0	24	5	0
Organic fruit cultivation should reduce the use of plant protection products as far as possible	0	5	3	17	4	0
Varietal diversity and genetic diversity inside varieties are important in organic fruit cultivation	0	0	0	13	16	0
Fruit cultivation should not use genetically modified organisms	1	3	0	6	18	1
Fruit varieties should categorically be no private property of businesses or individuals	2	11	2	5	7	2
Fruit farmers and breeders have a responsibility to protect and develop varietal diversity and genetic diversity inside varieties.	0	1	0	13	15	0
Fruit farmers should be included in fruit breeding processes	0	1	2	14	11	1
Fruit breeding including cultivation and selection of seedlings should take place on-farm	2	10	7	5	3	2
Knowledge on fruit breeding and fruit varieties should be openly accessible to anyone interested	1	1	1	1	12	1

Note: Grey coloring illustrates the tendencies of the ratings (own table).

organic cultivation generally rejects the usage of genetically modified organisms (IFOAM 2014). Concerning the role of private property rights, the valuations of proponents and opponents are balanced.

11.1.2 Relationship Between Breeding and Cultivation

Participants were asked an open question about how they perceive the relationship between fruit farmers and fruit breeders, especially of those cultivating or breeding apples. Results show that this perception and the actual arrangements between fruit farmers and breeders are very different. They depend on (a) regional characteristics, (b) individual preferences of farmers and breeders, and (c) economic conditions. Many participants emphasize these differences (12).[1] However, every breeder usually is in contact with at least one fruit farmer as a "corrective" (Delphi study round 1, own translation 2019), and enough possibilities for interested fruit farmers exist to participate in breeding activities (1). Some participants describe the relationship as intimate (2). Other participants add that the relationship is even more intimate in organic fruit cultivation in comparison to conventional fruit production (2).

Participants mention the following forms of cooperation between breeders and farmers:

- testing of varieties on fruit farms (4) or at testing locations (2)
- financial participation of fruit farmers in breeding projects via lump sums or (indirectly) via variety licensee fees (3)
- networking activities via the FÖKO (works as a 'hinge') and its respective working groups (3)
- distribution of varieties with joint corporations (2)
- individual practical participation of fruit farmers in breeding projects (2)
- irregular informal meetings (1)
- cooperation with direct marketeers (1)
- breeders inform fruit farming associations on the status quo of their work and farmers get the opportunity to give feedback (1)

Participants even mention specific examples: the cooperation of fruit farmers in the region of Lake Constance with the KOB (3) (See Section 8.1); the association *apfel:gut e.V.* (1) (See Section 9.2); the JKI that works together with practitioners (1) (See Section 8.2); and the breeding association *poma culta e.V.* from Switzerland that cooperates with bio-dynamic fruit farmers (1). However, the participation of fruit farmers does not guarantee a more organic form of breeding (1). Many fruit farmers are "caught in their narrow corset of spraying sequences[2]" (Delphi study round 1, own translation) and would make demands of breeders that do not promote more sustainable and organic fruit cultivation – for example, the demand for fruit size as an important selection criterion (1). Nevertheless, strengthening the relationships between farmers and breeders is beneficial for sensitizing farmers to the challenges and problems of breeding (1).

Many participants emphasize that close relationships between breeders and farmers are an exception (10). In general, fruit breeding is carried out in institutes or privately without contact with farmers (1). Breeding is rather a hobby task for farmers besides their usual farm work (1). Only an "elite of interested fruit producers and opinion leaders" (Delphi study round 1, own translation) is in contact with fruit breeders (1). It is criticized that not all varieties are freely available (e.g., the variety *Bonita*) and the market introduction of new varieties is only possible with the help of large corporations or large investments, which has negative effects on the general relationship between breeding and farming (1).

11.1.3 Market Demands of Farmers and Marketeers

According to the participants, organic apple farmers have a range of demands on apple varieties. In general, this includes sales guarantee and marketability (3) and as many positive traits for organic farming as possible (2). Some participants stated that the demands are similar to those of conventional farmers (3). However, the following traits were specifically stated as important:

- taste (14)
- good storability (11)
- robustness (11)
- visual appeal (10), for example, appealing color or color structure
- high yields and pack-out[3] (10)
- resistance against specific pests and diseases (5), for example, apple scab, fire blight, blood lice, powdery mildew
- good shelf life (4)
- vitality (3)
- healthy tree with growth characteristics that need little effort of cutting (3)
- appealing fruit size and format, not too small and uniform (2)
- low inputs of plant protection products are necessary (2)
- frost tolerance (2)
- little alternation (2)
- other traits: firmness; crispness; no hard shell; low-pressure sensitivity; site adaptation; high level of awareness (all 1)

One participant is particularly skeptical if farmers are actually interested in vital varieties as their usage of plant protection products is perceived as a natural component of apple farming (1). Moreover, professionals in organic apple farming would only be interested in such varieties if currently permitted plant protection products will be banned. One participant perceives these professionals as conservative in their choice of varieties, in comparison to direct marketeers that are more innovative in this respect (1). Another participant sees a general dichotomy between theoretically valuable traits (vitality, robustness, high inner quality of the product, minor role of visual aspects) and practical traits (shelf life, visual appeal, storability) that are more demanded by farmers, marketeers, and consumers, and perceived as the current market reality (1).

Participants were also asked about the demands of marketeers[4] on organic apple varieties. In general, marketeers have very high demands (3) and the same (3) or similar (2) on conventional apples or rather on apples out of integrated production. Organic varieties should not be more difficult to store than conventional varieties (1). Demands are very different across conventional food retailing, organic food retailing, or direct marketing (1). For example, the variety *Allurel* is suitable for direct marketing channels but would be difficult to distribute via conventional retailing channels. Participants see high potential in the direct marketing of organic varieties. Here, the introduction of new varieties and the general "presentation of diversity" (Delphi study round 1, own translation) is perceived as much easier as in other distribution channels (2). It is possible to offer a diverse range of varieties "throughout all ripening, taste, and storage groups" (Delphi study round 1, own translation) outside mainstream offers (1). The special taste and visual traits or the vitality of the tree could be an incentive for a direct marketeer to plant a niche organic variety (1). However, the following traits were specifically stated as important for marketeers in general:

- good storability (13)
- good taste (12); more explicit sweet (1), mainstream taste (crispy, juicy, sweet-sour) (1), or different from popular varieties (1)
- visual appeal (12), which should not differ from conventional apples (1)
- shelf life (8)
- low-pressure sensitivity (4)
- awareness of the variety among consumers (4)
- uniform outer and inner quality of the product (3)
- hard flesh (3)
- crispness (3)
- unique features or special characteristics (2)
- juiciness (2)
- regular yields (2) and regular ripening (1)
- red color (1) or at least 50% red coloring (1)
- low allergenic potential (1)
- healthy apple (1)
- low potential for fouling (1)
- free of pests (1)
- robust variety (1)
- free of faults (1)
- general suitability for commercial cultivation (1)
- cheap production (1)
- normal size, not too small or too large (1)
- more tolerance of smaller fruits in organic farming (1)
- suitable for regional cultivation, no "variety for the world" (1)
- suitable for transport (1)
- organic label (1), specifically of organic associations (1)
- exclusive variety solely for organic production to differentiate the variety from conventional varieties (1)

Participants were openly asked if marketeers influence apple breeding. On the one hand, the majority (59%) agree with that. By demanding specific characteristics, they influence the direction of breeding (5) and directly decide if a variety will be offered or not (2). Variety licenses to get the permission to trade or plant the variety are only bought if the variety corresponds with the demands of marketeers (1). Marketeers have very high demands that partly exceed the legal requirements, for example regarding residues of plant protection products (1). Newly developed varieties are only economically feasible if their characteristics generate enough turnover and guarantee a sufficient level of pack-out (2). In general, marketeers are perceived as an important link between producers and consumers (1). They are responsible for communicating the benefits (1), or specifically the organic benefits (1), of apple varieties. They often participate in fruit commissions (1) or breeding programs financially (1) or as consulters (2).

On the other hand, 41% of participants state that marketeers have no influence on breeding. They are not interested in the process of breeding (3) but only in the final product/variety (2). This is due to missing knowledge (1), time resources (2), and insufficient short- and mid-term benefits because the consumer demand for specific varieties is hard to predict (2). Fruit farmers would decide independently if they plant a new variety or not (1). Marketeers are only interested in low risks by marketing widely known and popular varieties and rather not in the development of new varieties (1). Especially organic breeding is not an issue for marketeers (2).

11.1.4 Evaluation of Club Concepts

Participants were specifically asked how they evaluate the pros and cons of club concepts. The following advantages have been mentioned:

- target-oriented and coordinated management of cultivation, marketing, and quality guidelines (4)
- price advantages (2) and thus cheaper prices for consumers (1)
- possibly high sales prices for farmers (1)
- breeders are economically independent of state funding and can conduct market-oriented breeding with low risks (1)
- actors along the whole value chain work together to introduce a new variety into the market (1)
- active engagement of club members because of their high investment costs (1)
- club concepts are the only possibility to successfully introduce a new variety in the highly saturated apple market (1)
- exclusion of free riders (1)
- better market position against the food retail sector (1)
- economic benefits for all club members if the club concept is consequently implemented (1)
- On the other hand, participants observe the following disadvantages and critical aspects of club concepts:
- limited access to club varieties (4)
- club members are highly dependent on the club (2)
- privatization of all parts of the value chain (1)
- financial aspects: license fees for farmers (1), high investment costs of farmers (1), low profits (1),[5] high marketing efforts (1)
- difficult cultivation of club varieties because of high quality standards (1)
- high loss of non-suitable yields (low pack-out), resulting in an inadequate ecological balance through necessary intensive cultivation practices (1)
- contract standards are challenging (1)
- profits and risks are not equally distributed: farmers have the highest risks (1)
- high risks for everyone participating in the club (1)

- concept is too static as it is directed on the horizontal level of the value chain: it would be better to use the concept of "managed varieties" (Delphi study round 1) where trademark owners have an overview of the value chain and give marketing impulses (1)
- skepticism about the success of the growing use of club concepts (1)
- natural product is "industrialized and converted to a standardized product" (Delphi study round 1, own translation) (1)
- incompatible with the norms of organic farming like solidarity, participation, and diversity (1) and also with the norms of organic consumers (1)
- club concept is perceived as unfair (1)

One participant sees club concepts as a role model for organizing and coordinating the breeding and final market introduction of a new variety in partnership with different actors along the whole value chain, which is a necessary strategic direction in the context of current market structures (1). Marketing, product placement, and packaging play key roles in this context (1). Club concepts are also a chance for organic apple varieties to further distinguish themselves from conventional varieties in a professional way (1). Moreover, one participant notes that the term club concept is old-fashioned. In consulting, the term trademark is established (1) (See Section 3.1).

Many participants conclude that the use of club concepts has increased worldwide (7), especially in the United States and New Zealand (1). Some see this as a contrary development to more diversity and regionality (2) as the only focus of club concepts lies on the outer quality of the product (1) and the access to club varieties is restricted (1). Not many new club varieties will prevail in the glutted market that is already dominated by successful club varieties (1). Powerful actors are needed to successfully introduce new club varieties, as it is the case for the variety *Bonita* (1). One participant observes an adaptation of club concepts on organic fruit farming when licenses for new varieties are only given to organic farmers (1).

11.1.5 Market Developments and Trends

Participants were asked about the developments and trends they observe along the organic and conventional apple value chain. The answers to this open question reveal insights into variety trends, the role of different actors along the value chain, the importance of marketing concepts, the relevance of genetics and patents, and specific developments in the organic sector.

11.1.5.1 Variety Trends

Some participants generally observe an increasing quantity and success of club varieties (4) whereas another participant sees several new and innovative varieties besides these club concepts (1). The decisions on planting a specific variety are solely taken on the basis of marketing tactics (1) and, concerning visual traits and taste, this decision involves three uniform categories: green and sour, colorful and aromatic, red and sweet (1). Especially small red apples are perceived as a new trend because they are suitable for children and nibbling (2). In general, varietal and genetic diversity is desired and requested but no one wants to put much effort into realizing this diversity (1). Robust varieties with visual appeal and taste (1) as well as good storability (2) are wanted. In organic apple cultivation, especially scab-resistant varieties are demanded (1). However, new varieties often do not fulfill their promises of higher quality and stable yields (1). The risk in planting a new variety is high because forecasting of the cultivation characteristics and market acceptance is not possible (1). Some participants state that many new and promising varieties are not cultivated because marketing concepts and consumer awareness are missing (2). Others state that new varieties are simply not available in nurseries (2). Another participant observes that the "variety carousel" (Delphi study round 1, own translation) turns faster than the life span of apple trees (1).

11.1.5.2 Roles of Specific Actors along the Value Chain

In food retail, the product range is restricted to a small number of varieties – diversity in visual traits and taste is missing and many participants observe a growing standardization (6). Consumers are "trained on a specific image of apples" (Delphi study round 1, own translation) that has nothing to do with the natural product (1). However, other participants see a growing trend toward offering exclusive varieties and a diversity of apples (4), as well as a growing relevance of niche products such as apples with low allergenic potential (3). Food retail increasingly influences the choice of varieties in cultivation (1) and creates dependencies of farmers on them (1). In the future, food retail could cooperate more with producers by direct contracting (1) which is already implemented in the conventional sector.

In breeding, some participants see a growing privatization (2). Breeders take their decisions solely on marketing reasons (1) and "force" (Delphi study round 1, own translation) farmers into contracts[6] (1). Some participants state that legal variety testing (See Section 3.1) is often skipped (3) and the market introduction of new varieties is thus manipulated (1) – variety testers "get a muzzle" (Delphi study round 1, own translation) (1). One participant notes that breeding activities are more intensive than ever (1).

On the general apple market, one participant observes that marketing cooperatives, producer cooperatives, and food retail have the greatest market power. In the search for the "ultimate apple innovation" (Delphi study round 1, own translation), this power is utilized with different instruments such as high license fees, the acquisition of grants from EU funding programs, and the manipulation of variety testing. Thus, smaller marketing structures are hampered (1). Although many new and promising apple varieties already exist, it is difficult to place them on the market (2) because of high market saturation. However, one participant argues that the organic market is a demand market where good varieties like *Topaz*, *Natyra*, or *Santana* can be easily placed (1).

One participant observes that although a multitude of actors exist along the value chain, the degree of cross-linking is low (1). Professionalization and "conventionalization" (Delphi study round 1, own translation) increase (1) and every part of the value chain wants to drive up profits as standard apple varieties do not provide profits anymore (1). Actors increasingly invest in "closed systems" (Delphi study round 1, own translation) such as club concepts (1).

11.1.5.3 Importance of Marketing Concepts

The importance of marketing concepts for the introduction of a new variety continuously increases (11). They are particularly relevant because the saturated apple market is marked by predatory competition (1). Marketing concepts decide the success of all actors along the value chain that are involved in the market introduction of a new variety (1). They generate higher profits and offer exclusivity (1). In times when general apple consumption decreases, consumers rather pick products with professional advertising (1). One participant observes a trend to develop umbrella brand strategies (1). Another participant states that organic labels are more important than marketing concepts (1). However, marketing concepts are important to communicate values of regionality, nature, or health (1). In organic fruit farming, concepts that emphasize those values are especially important (1). One participant argues that marketing concepts are only realizable within club concepts because they are too expensive for single varieties (1). Moreover, marketing concepts are too similar at the moment. A step-by-step building of trademarks and identities does not take place (1). As more and more marketing concepts are introduced, the former beneficial effects decrease (1).

11.1.5.4 Relevance of Genetics and Patents

For many participants, genetic engineering plays no role in fruit breeding at the moment, and will not play any role in the future (5). However, outside Europe, the importance increases and it is unclear how long Europe will remain in this special status quo (1). One participant argues that genetic engineering is important (1). Patents will gain an important role in the future (3) because high market introduction costs and hard competition emphasizes the demand for secure profits (1). At the moment, molecular diagnostics and marker-assisted selection speed up the breeding and make it cheaper (2).

Some participants observe that genetically engineered fruit products will not find any acceptance among German consumers (2). However, others see genetic engineering as an instrument for consumer deception (2). For example, the *Arctic Apple* in the United States, a variety that was bred with methods of genetic engineering, pretends freshness due to the elimination of open apple flesh's oxidation (1). Participants essentially observe four aspects that could foster the legalization of genetic engineering in Europe. First, campaigns are financed by external market and political actors that build up pressure to conquer new market segments (2). Second, the call for further differentiation could pave the way for genetic engineering as it seemingly opens up more possibilities (1). Third, new varieties could be bred faster with genetic engineering (1). Fourth, as varieties are bred worldwide, one participant sees a "gradual infiltration" (Delphi study round 1, own translation) of conventional breeding with these ideas (1).

11.1.5.5 Specific Developments in the Organic Sector

Most of the participants do not observe any special developments and trends in the organic apple sector and state that developments are essentially the same as in the overall sector – except regarding genetic engineering (2). However, some aspects are still observed. Although many market actors ask for special varieties, most producers offer, and most consumers buy the standard varieties (3). Special organic varieties have difficulties establishing themselves in organic food retail (1) – despite the trend going to "own organic varieties" (Delphi study round 1, own translation) to further differentiate the organic from the conventional sector (2). The introduction of *Natyra* is mentioned as the most prominent example (1). Additionally, certification by organic associations increases the availability of products with organic labels on the market (1). In organic breeding, many new private initiatives are formed (1). Breeders experiment with crossings of traditional and heirloom cultivars to get robust varieties with high quality levels (1). However, some participants see a growing conventionalization of the organic sector, especially regarding marketing structures (1), the concentration on storage specialists (1), and the use of club concepts in organic breeding and farming (2).

11.2 Discussion of Results

11.2.1 Norms of Actors and the Relationship Between Breeding and Cultivation

The study results show a high level of agreement among participants for most normative statements, highlighting the binding character of organic farming's general norms – at least concerning ecological aspects. Differing opinions on the use of plant protection products probably stem from the notion that these products are perceived as necessary to meet the demands of marketeers and consumers by providing high yields and visual appeal of the product. This argument is part of a global debate about the dependence of the organic apple industry on organic-compliant pesticides, which Delate et al. (2008) predicted already more than a decade ago. This debate is perhaps most prominently reflected in the new model of *low-residue apple production* (Bravin et al. 2019), developed by the Swiss breeding organization Agroscope (See Section 3.3), that tries to combine elements of integrated and organic apple production to reduce the use and residues of pesticides.

Most differences can be observed in the opinions toward statements with a social or an economic character. With the majority rejecting on-farm breeding, it is emphasized that a certain separation of farming and breeding exists in reality (and is not an aim), resonating with the view that a close relationship between farmers and breeders is an exception. Except as a side element in the literature on *managed varieties*,[7] the literature does not discuss this missing link in the fruit and apple context. On the contrary, the focus is obviously on cooperation and interlinkages between farmers and production organizations or marketeers (Zander 2011). However, there is high agreement on the statement that fruit farmers should participate in the fruit breeding process. Subsequently, a range of cooperation arrangements are mentioned by the participants. This shows the potential to strengthen the relationship between breeders and farmers.

This potential seems to be particularly true for the organic sector. Some participants mention that relationships in the organic fruit breeding sector are more sensual, with the FÖKO being a networking platform for these actors. Reasons for that are probably that the group of relevant actors is much smaller than in conventional fruit cultivation and that breeding is more relevant for organic fruit farmers because current modern varieties miss important traits. With *apfel:gut e.V.* as a pioneering initiative that aims to combine breeding and farming more strongly (See Section 9.2), organic breeding actually has a role model function for close cooperation.

Different valuations of fruit varieties as private property show the general conflict between the economic necessity of using private property rights as financing instruments, and the desire to have "free varieties" (Delphi study round 1, own translation). General literature on marketing management of apples is more concerned with the optimal design of private property rights in form of club concepts, managed varieties, or brand development than with alternative constructions. Gallardo et al. (2016), for example, discuss those commercialization strategies for fruit varieties in detail for the US market. As Weber (2008b) puts it: "The fruit royalty flow is the only mean to guarantee a sustainable financial system in an increasing capitalized fruit market and privatized apple breeding programs." For many participants, this still seems to be the leading paradigm. A discussion of alternative business models takes place at a later point (See Chapter 13).

11.2.2 Market Demands of Farmers and Marketeers

There is a general consensus in the literature that the choice and combination of cultivars are crucial for the profitability of fruit farms across the world (Waibel et al. 2001; Legun 2015). It is a strategic decision as every planted tree or cultivar is a long-term investment[8] that influences the overall yield, pack-out rate, and production costs (Leumann and Bravin 2008). Bravin et al. (2009) showed that the number of planted cultivars significantly influences the economic success of an orchard: "[W]ith an advised choice of a suitable cultivar mix in an orchard, farmers can allocate work peaks, reduce yield losses, and invest in successful new cultivars" (p. 165). Market demands thus play a major role in the choice and combination of apple cultivars.

The Delphi study results show that the traits that are most often perceived as important for apple varieties are similar for farmers and marketeers. Rather classical demands like the taste, visual appeal, or storability are most important for organic fruit farmers and marketeers. At least for farmers, robustness is an additional important trait that is also mentioned as a breeding trend by Hanke and Flachowksy (2017). These most important traits are not very different from those of conventional apple farming. This is also particularly stated by some participants as a notable benchmark. Literature reflects this case in the recommendation to include conventional varieties like *Elstar*, *Jonagold*, or *Braeburn*, which guarantee relatively safe sales and prices, in the planted cultivar mix (Waibel et al. 2001; Leumann and Bravin 2008). As Legun (2016) puts it: "[T]here are mechanisms within the industry that reduce the ability to differentiate apples in the market, which means that more apples compete under the same basic quality rubric." She describes this global phenomenon as *commodity apples* that are available for low prices, have similar aesthetic parameters, and are suitable for mass production (Legun 2015).[9] A differentiation from conventional fruit production thus seems not relevant in many aspects. Only a few participants recommend a differentiation strategy by introducing exclusive organic varieties or niche varieties.

Comments by the participants show that production and distribution organizations (See also Sections 3.2 and 3.3) have a high level of power and (direct or indirect) influence on the direction of breeding. They demand a high level of professionalization of farmers and have specific demands on the inner and outer quality of apples. It is valuable to add the insights of Zander (2011) to this context. According to her study, production and distribution organizations play an important role as regional actors that bundle a great quantity of apples and activate economies of scale in storage, sorting, and packaging. High market transparency and a high level of interlinkage between fruit farmers and these organizations exist. This is due to the fact that many of these companies emerged out of fruit farms. These strong interlinkages on the lower levels of the value chain led to many forms of cooperation and explain the similar demands on apple varieties.

Some participants argue about the high potential of *direct marketing* that allows more varietal diversity and room for experiments. However, from a quantitative perspective, direct marketing plays only a minor role in comparison to sales via food retail or organic food trade (Zander 2011). Direct marketing is an established additional strategy to diversify distribution channels and generate higher prices for directly marketed products (Wirthgen and Maurer 2000). Nevertheless, the globally valid saying that "different types of crops generate different types of markets" (Legun 2016) has also some truth the other way around: results of the market study confirm that direct marketing creates different demands on varieties. For example, storability, which is identified as a major demand by farmers and marketeers, is not as relevant as for wholesale. This leaves room to cultivate and market varieties with other traits. In her book on design elements of fruits and vegetables, Kleinert (2020) describes it the following way:

> Producers that cultivate crops for direct marketing put more emphasis on a diverse range of products and high quality of taste. [...] The characteristics of the distribution channel and the respective relevant quality criteria thus work as design elements. They determine the actions of actors [...]
>
> (p. 175, own translation).

Participants did not mention the origin of products as important traits. According to Zander (2011), regional and German produce is highly preferred by food retail and consumers. This trend has not changed until today: according to a representative study of the SINUS Institute (SINUS 2017), the regionality of apples is important for 72% of the participants. For 74%, it is important that offered apples are cultivated in Germany. In comparison, only 54% mark the trait 'organically farmed' as important.

11.2.3 Evaluation of Club Concepts

Results of the market study show that club concepts are highly polarizing among the participants. A range of advantages and disadvantages are stated and many participants attest to a continuously increasing development of club concepts – not only in Germany but all across the world. Already a decade ago, Schwartau (2010) concluded that club apples will have a regular share of the European apple market, although they will not dominate the market due to their cap on supply and the limited ability of the market to bear new varieties (Legun 2015; Weber 2008a). This view is shared by some participants. Few participants see opportunities for organic apple varieties to adapt to these developments by introducing organic club varieties.

Assessments by the participants differ to some extent from the theory of Legun (2015) on club apples and the biology of markets.[10] She argues that club apples symbolize the resistance against the further commodification of the global apple market by fruit farmers using market mechanisms to gain (back) control: "Club apples can be seen as a response to trends in the food industry that involve increased standards paired with a concentrated and dominant retail sector" (ibid., p. 294). Thus, a re-distribution of power from food retail and marketing organizations to fruit growers shall take place. In her understanding, farmers organize the club in form of a co-operative by "creating social boundaries around a variety" (ibid., p. 309),[11] discussing variety characteristics and price strategies. However, participants of the market study seem to have a different understanding of club concepts and the actors involved. They perceive many risks for farmers, evaluate most club apples as commodity apples, and argue that powerful marketeers are necessary to successfully introduce club varieties into the market. These demands and challenges are confirmed by Weber (2008b).

11.2.4 Market Developments and Trends

Variety trends observed by the participants are in some cases confirmed by already discussed results (see above) and findings from the literature. For example, Inderbitzin et al. (2016) evaluated the potential of small-sized snack apples on the Swiss market and concluded that although they will not create a high level of demand, it would make sense for farmers to include such varieties in the cultivar mix to expand its diversity. The notion of one participant that the variety carousel turns ever faster is also a prominent argument in recent as well as older literature (Hanke and Flachowsky 2017; Weibel 1995). Overall, *variety trends* seem to incorporate essentially two dilemmas that have not really changed throughout the last decades: First, diversity and differentiation by exclusive or niche varieties are wanted but commodification demands and the glutted market often inhibit actions toward it. Second, new varieties could have the potential to enable a more sustainable production but investments are risky, professional marketing concepts are necessary, and placement in the highly saturated market is difficult.

With the majority of participants emphasizing the importance of marketing concepts, they are in line with arguments from the literature. According to Weber (2008a, 2008b), four parameters exist for successful marketing management on the general apple market: a) return on investment; b) use of intellectual property rights (variety protection and trademark); c) professional networking and communication; and d) a dedication to organic production, fruit quality, and ethics. *Variety managers* should coach the whole marketing management process along the value chain (Weber 2008b). However, Weber (2008a) agrees with some participants that a lot of financial, personal, and time resources are necessary to adopt professional marketing management. The use of intellectual property rights as the central element of marketing management concepts collides with the controversial debate on them (See above).

Branding was not prominently discussed by the participants, with some even mentioning that organic labels on apples are more important than trademarks. This is different from the limited scope of literature that discusses this topic. Weber (2008a) argues that branding would play a major role in the future to generate additional revenue and differentiate from competitors. In a consumer experiment in the US, Rickard et al. (2013) confirmed that branding has some potential to influence consumer responses and their willingness to pay for apples. In general, the discourse on trademarks and branding for apples is still fairly new. However, it has gained some momentum with the increasing development of club concepts that use trademarks, slowly developing a new paradigm by adopting these concepts and theories from modern business administration. While Weibel (1995) argued that sales categories should be guided by 'archetypes' that describe the specific taste and appearance categories, the new branding paradigm solely focuses on the communication of trademarks. It could be argued that today, image and marketing concepts thus play a more and more prominent role in comparison to the actual traits of the product.

Some developments discussed in the literature were not mentioned by participants. Technological developments on farms, particularly robotics, are an emerging topic in global literature. Here, the farm is changed from a passive asset to an active one, further optimizing the economic performance of orchards (Legun and Burch 2021). Moreover, Weber (2017) discusses that in the context of marketing management and branding, instead of firms, whole value chains would compete against each other. This development can be observed in the example of ZIN (See Chapter 7), where a whole value chain from breeding to marketing is involved in the process of developing new varieties.

Overall, *managerial and organizational professionalization* of the introduction of new varieties increases. Marketing concepts, club concepts, and managed varieties are highly relevant. Non-managed varieties would not prevail in the glutted market for apples. Solutions for successful organic varieties are on the one hand seen in an adaptation to these developments and to the conventional sector. On the other hand, the special characteristics of the organic sector and a further differentiation from the conventional market are perceived as a possibility. This hypothesis was already confirmed by Weber (2008b), who argues that especially in the organic sector, a differentiation strategy would lead to major returns in the mid- and long-term. However, the discussion on market development and trends also emphasizes the complex interplay of several developments from regional to global scales:

> No single trend will dominate consumer demand for apples, but multiple, interwoven trends will play out for different varieties, in different countries and regions. To prosper in this environment, the apple industry will need greatly enhanced ability to evaluate the wide array of variety choices it will face.
>
> (O'Rourke 2017, p. 10)

Notes

1 The numbers in brackets indicate how many participants mentioned the respective aspect described in the text.
2 Here, spraying sequences describes the frequency of using plant protection products.
3 The term pack-out describes the percentage of apples that are actually marketed after harvest.
4 In this context, marketing of apples includes the distribution via production and distribution organizations as well as direct- and self-marketing.
5 It should be noted that this statement contradicts the statement "possibly high sales prices for farmers" concerning the advantages of club apples.
6 It should be noted that it is not clear what is actually meant with forcing farmers into contracts. Possibly, this remark concerns club concepts.
7 Weber (2008b) describes a partnership model of managed varieties where the financial risk of breeding is split across the whole value chain in a collective royalty system on trees and fruits.
8 The average standing time of apple trees in commercial orchards is 12–15 years.
9 For example, on the Swiss apple market the market position of *Gala, Braeburn,* and *Golden Delicious* has not really changed in a decade (Leumann and Bravin 2008; Perren et al. 2017). However, Weber (2008a) argues that apples also have a product life cycle. He shows that the market shares of the Top 5 marketed cultivars in Germany have fallen from 81% in the 1970s to 15% in the middle of the 2000s.
10 Legun (2015) defines this theory as follows:

> Plants compel actors more continuously and aggressively to craft their produce as a market commodity, but plants may also behave with spontaneity and generate unintentional products or conditions of production. The compulsion and unprompted opportunities of animate matter are part of what I call the biology of markets. […] While agriculture presents a particularly vivid case to explore the role of plants in economics, all objects can be said to contain aspects of unpredictability, and in a sense the line between the biological and the inert is blurry.
>
> (p 295)

11 Legun's research is based on empirical data from the United States.

References

Bravin, E.; Perren, S.; Naef, A. (2019): Low Residue Apple Production: Higher Production Risk and Lower Profit. In *Acta Hortic.* (1242), pp. 217–222. DOI: 10.17660/ActaHortic.2019.1242.30.

Delate, K.; McKern, A.; Turnbull, R.; Walker, J. T.; Volz, R.; White, A. et al. (2008): Organic Apple Systems: Constraints and Opportunities for Producers in Local and Global Markets: Introduction to the Colloquium. In *HortScience* 43 (1), pp. 6–11. DOI: 10.21273/HORTSCI.43.1.6.

Delphi Study Round 1 (2019): Qualitative Data Obtained by Online-questionnaires as part of a market study in the context of the Research Project EGON. November 2018 – January 2019. Oldenburg, Germany.

Gallardo, R. K.; McCluskey, J. J.; Rickard, B. J.; Akhundjanov, S. B. (2016): Assessing Innovator and Grower Profit Potential under Different New Plant Variety Commercialization Strategies. Agricultural and Applied Economics Association (AAEA) Conference, 2016 Annual Meeting, July 31-August 2, Boston, Massachusetts. Conference Paper, pp. 1–33. DOI: 10.22004/ag.econ.235940.

Hanke, M.-V.; Flachowsky, H. (2017): *Obstzüchtung und wissenschaftliche Grundlagen.* Berlin, Heidelberg: Springer Berlin Heidelberg.

IFOAM (2014): The IFOAM Norms for Organic Production and Processing. Version 2014. Available online at http://www.ifoam.bio/sites/default/files/ifoam_norms_version_july_2014.pdf, checked on 3/23/2021.

Inderbitzin, J.; Schütz, S.; Perren, S.; Kellerhals, M. (2016): Snackapfel - Frucht mit Potenzial? In *Schweizer Zeitschrift für Obst- und Weinbau* (17), pp. 11–14.

Kleinert, J. (2020): Lebendige Produkte. Obst und Gemüse als gestaltete Dinge. Bielefeld: transcript (Design).

Legun, K. A. (2015): Club apples: A Biology of Markets Built on the Social Life of Variety. In *Economy and Society* 44 (2), pp. 293–315. DOI: 10.1080/03085147.2015.1013743.

Legun, K. (2016): Managed Apple Varieties Research Report. Available online at https://www.academia.edu/27324070/Managed_Apple_Varieties_Project_Report, checked on 4/4/2020.

Legun, K.; Burch, K. (2021): Robot-ready: How Apple Producers Are Assembling in Anticipation of New AI Robotics. In *Journal of Rural Studies* 82, pp. 380–390. DOI: 10.1016/j.jrurstud.2021.01.032.

Leumann, M.; Bravin, E. (2008): Obstbau: Entscheidungsgrundlage bei der Sortenwahl. In *AGRARForschung* 15 (5), pp. 214–219.

O'Rourke, D. (2017): Consumer Trends in Apple Sales. In Gayle M. Volk, Amit Dhingra, Sally A. Bound, Dugald C. Close, Peter M. Hirst, M. C. Goffinet et al. (Eds.): *Achieving Sustainable Cultivation of Apples.* 1st ed. Cambridge: Burleigh Dodds Science Publishing (Burleigh Dodds Series in Agricultural Science), pp. 511–522.

Perren, S.; Schönberg, A.; Inderbitzin, J.; Kellerhals, M.; Schmid, M. (2017): Neue Apfelsorten mit Mehrwert. In *Schweizer Zeitschrift für Obst- und Weinbau* (3), pp. 8–13.

Rickard, B. J.; Schmit, T. M.; Gómez, M. I.; Lu, H. (2013): Developing Brands for Patented Fruit Varieties: Does the Name Matter? In *Agribusiness* 29 (3), pp. 259–272. DOI: 10.1002/agr.21330.

Schwartau, H. (2010): Liegt die Zukunft in den Club-Sorten? In *European Fruit Magazine* (5), pp. 21–22.

SINUS (2017): Studie zum Apfel: Die Hälfte hat schon einmal Äpfel vom Nachbarsbaum gepflückt. SINUS-Institut. Available online at https://www.sinus-institut.de/

veroeffentlichungen/meldungen/detail/news/studie-zum-apfel-die-haelfte-hat-schon-einmal-aepfel-vom-nachbarsbaum-gepflueckt/news-a/show/news-c/NewsItem/, updated on 4/8/2021.

Waibel, H.; Garming, H.; Zander, K. (2001): Die Umstellung auf ökologischen Apfelanbau als risikobehaftete Investition. In *Agrarwirtschaft* 50 (7), pp. 439–450.

Weber, M. (2008a): Marken-Management - die neue Herausforderung für die Apfel-branche. In *Schweizer Zeitschrift für Obst- und Weinbau* (3), pp. 11–14.

Weber, M. (2008b): Sustainable Apple Breeding Needs Sustainable Marketing and Man-agement. In FOEKO (Ed.): *Ecofruit. 13th International Conference on Organic-Fruit Growing: Proceedings*. Ecofruit. Hohenheim. Fördergemeinschaft Ökologischer Obstbau e.V. (FOEKO).

Weber, M. (2017): Apple brands - factors of success. Vorlesung MBA International Mar-keting, ESB Reutlingen.

Weibel, F. (1995): Bioobstbau: Anpassung der Vermarktungskonzepte an zunehmende Sortenvielfalt. In: Erfahrungsaustausch über Forschungsergebnisse zum ökologischen Obstbau. Weinsberg, pp. 84–87.

Wirthgen, B.; Maurer, O. (2000): Direktvermarktung. Verarbeitung, Absatz, Rentabilität, Recht; 51 Tabellen. 2., neubearb. und erw. Aufl. Stuttgart (Hohenheim): Ulmer.

Zander, K. (2011): Ausländisches Angebot an ökologischen Äpfeln: Bedeutung für deutsche Öko-Apfelerzeuger. Universität Kassel, Fachgebiet Agrar- und Lebensmittel-marketing. Witzenhausen.

12 Promoting and Inhibiting Factors for Organic Apple Cultivation and Breeding

In round 1 of the market study, answers to the open questions revealed societal, economic, political, and legal factors that have current inhibiting and promoting effects (at the moment of data collection) or could be promotive in the future. Overall, 29 promoting and 36 inhibiting factors for the further expansion of organic apple cultivation have been identified. In round 2, all factors have been rated. This randomized assessment took place with a 7-point Likert scale where 1 means that the factor has low relevance and 7 means that the factor has very high relevance to promote or inhibit organic apple cultivation. Participants rated the factors accordingly.

12.1 Results of the Market Study for Organic Apple Cultivation

12.1.1 Definition and Understanding of Organic Apple Cultivation

In round 1 of the market study, the participants were confronted with the following definition of organic apple cultivation: *Organic apple cultivation preserves and promotes soil health as well as genetic and functional diversity of plants, animals, and other organisms in agroecosystems. Natural resources are treated with care and cultivation aims at a stable ecological equilibrium. Organic apple farmers have a special responsibility for protecting the environment and livelihoods for current and future generations.* For the verification of this definition, in round 2 participants were asked about their understanding (goals and procedures) of organic apple cultivation.

Many participants primarily understand organic apple cultivation as treating natural resources with care and farming in an environmentally friendly way (13). Biodiversity is promoted (4) and cultivation is carried out from a holistic perspective (4). The aim is to carry out fruit farming in line with nature (4). Moreover, no pesticides and chemical-synthetic fertilizers are used (7). The usage of organic plant protection products is minimized (3). Self-regulating ecological systems are created by promoting beneficial insects and plants (4). From a market perspective, a healthy (3) and tasty (2) product with high quality standards (4) shall be cultivated. In general, long-term marketable apples shall be produced to be economically sustainable (3).

DOI: 10.4324/9781003355724-22

Some participants state that the establishment of continuous learning processes and the improvement of cultivation standards are also important attributes (2). Organic apple cultivation takes place in line with legal standards by the respective EU directive (2) or rather with standards of cultivation associations such as *demeter, Bioland,* or *Naturland* (1). Organic cultivation should use varieties with high robustness (2), high vitality (1), and resistance against fungal diseases (1). Additionally, the following cultivation aspects are stated: high yields with low inputs; no monoculture farming; farming with low-crown cultures; usage of hail nets; regular tree cuttings; preservation of soil health; no genetic engineering (all 1). On a market level, the following additional aspects are stated: orientation on consumer expectations; long-term combination of environmental friendliness and economic efficiency; exchange of market information; strengthening of regional marketing structures; intensification of lobbying and Public Relations (all 1).

The general understanding of organic apple cultivation that participants stated in round 2 largely matches with the definition given in round 1. Treating natural resources with care and environmental friendliness is mentioned very often whereas other aspects such as the promotion of biodiversity are only mentioned by a few participants. Promoting soil health is only mentioned by one participant and responsibility for future generations by no one. Besides, a lot of specific aspects have been stated.

12.1.2 *Promoting Factors for Organic Apple Cultivation*

Overall, the results show that a higher general acceptance toward organic cultivation by consumers and food retailing is perceived as promoting. Further, a higher demand for organically produced apples and the accompanied cost-covering prices could have promotive effects on farmers' economic performance. From a legal perspective, legislation on conventional apple cultivation should be more ambitious regarding environmental protection. More farmers should change to organic cultivation practices and these conversions should be further supported by the public. In more detail, the following insights are summarized.

Societal factors that have promoting effects at the moment:

- conscious nutrition practices like regional and healthy food consumption (1) as well as consumer's pressure on agricultural actors to produce food products free of residues (1)
- global crises or scandals foster a trend toward more respect for nature and thus to a further transition toward ecological agriculture (1)
- low regard for conventional products (1)
- conventional farmers changing to organic practices (1)
- "Organic cities" (Delphi Study Round 1, own translation 2019), accompanied with support and actions (1)
- increasing demand for organic products (1)
- higher willingness to pay for food products among consumers (1)

Societal factors that could have promoting effects in the future:

- increase in the level of education of trade actors and customers to obtain higher acceptance of organic agriculture (5)
- higher acceptance of products with commercial category II (1)
- establishment of organic apples as mainstream (1)
- implementing organic issues into training programs for fruit-growing professionals (1)
- alliance of farmers to call for/demand common goals (1)

Economic factors that have promoting effects at the moment:

- generally better market position and prices of organic products (4) and an increasing general demand of organically certified products (3)
- use of organic niche as a marketing tactic (1)
- poor farmgate prices for conventional products (1)
- professionalization of organic farms (1)
- disadvantageous trade structures in the conventional sector (1)

Economic factors that could have promoting effects in the future:

- further rising demand (2) with cost-covering prices and better marketing structures (1)
- focus on product diversity in food retailing (1)
- better structures for marketing, consulting, and research (2) that would automatically promote organic cultivation (1)
- strengthening of organic marketing structures (1)

Legal factors that could have promoting effects in the future:

- binding and more strict soil-, river-, insects- and resource protection as well as more ambitious environmental regulation (3)
- more ambitious legislation and official requirements, particularly for plant protection in conventional cultivation (2)
- reform of EU Common Agricultural Policy (CAP) in 2020 (2)
- strict rules for organic certification (whole production process) and transparency for customers (1)
- strict rules for origin and production process that clearly communicate their benefits for society (1)
- inclusion of external effects (e.g., pollution) into product prices and explanation of the reasons for this inclusion to consumers (1)

Political factors that could have promoting effects in the future:

- more financial public support (2)
- goal of the federal government to increase the share of organic culture (1)

- more public support in the conversion process from conventional to organic farming (1)
- change from land-area-based to environmental-based subsidies to reward good environmental performances of farms (1)

All of these insights have been condensed into 29 promoting factors and rated by participants in round 2 of the market study regarding their relevance. The results of this rating are depicted in Figure 12.1.

Overall, factors that have promoting effects at the moment and concern specific market developments are rated highest; factors that concern cultivation or production are rated lowest. General developments such as the rising demand for organic products, changing consumption practices, or product diversity play major roles. These developments indirectly influence organic apple cultivation and could lead to its further propagation. The least important factors seem to be factors that concern conventional apple production or political aspects such as the reform of CAP in 2020, organic cities, and general political goals.

All listed factors are generally not rated of low relevance as no mean value lies below 3.5. This emphasizes the validity of the results from round 1 of the market study.

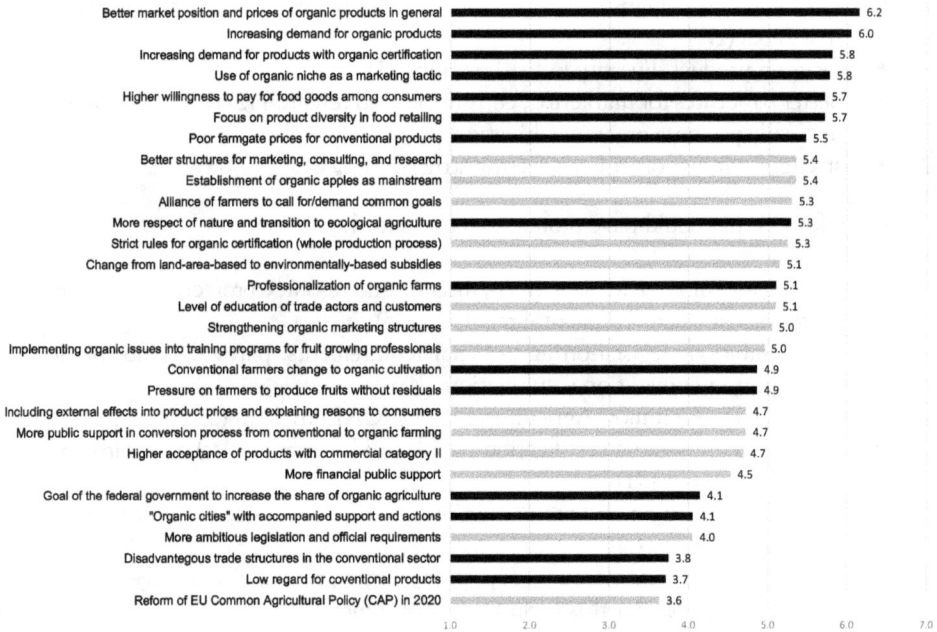

Figure 12.1 Promoting factors for organic apple cultivation. Mean value, rated on a 7-point Likert scale (n = 21). 1 = low relevance, 7 = very high relevance. Gray-colored boxes indicate factors that could be promoting. Black-colored boxes mark factors that are currently promoting (own figure).

12.1.3 Inhibiting Factors for Organic Apple Cultivation

Results on inhibiting factors show that the demands of food retailing and consumers for perfect quality and appearance are the biggest challenges for organic fruit cultivation. Simultaneously, public support for organic cultivation is perceived as too low. Missing or non-transparent solutions for the handling of plant diseases and plant protection as well as increasing demands of quality standards, certifications, and controlling mechanisms are perceived as inhibiting. Moreover, the competition with conventional farmers (particularly regarding product quality and price), high production risks, and unsafe guarantees of sale prove challenging. In more detail, the following insights are summarized.

Societal factors:

- consumers and food retailing expect excellent quality and visual traits of apples (7), but purchase prices have to be low, or rather cheaper products from conventional production are preferred (3)
- low appreciation of organic production or rather missing knowledge of its benefits among consumers (3)
- the same varieties as in conventional production are demanded (1)
- missing cost transparency for consumers (1)
- purchasing power of consumers is too low (1)
- consumption of apples as such is decreasing (1)

Political factors:

- state support for organic cultivation is too low, especially for research (3)
- no support for farmers when they try to change their product range and invest in alternative varieties (1)
- no support with advance payments, for example, to recruit organic fruit consultants (1)
- no clear position of political decision-makers or rather negative attitude against organic cultivation (2) combined with missing lobbyism for organic fruit cultivation (1)
- excessive support of organic apple cultivation leads to conversions of former conventional farmers that do not identify with organic norms (1)
- explicit statement that no political inhibiting factors exist (2)

Legal factors:

- high demands and excessive bureaucratical procedures for quality standards, certifications, and control mechanisms (6)
- lack of solutions and unclear rules for plant protection products (2), especially non-transparent and uncertain rules for the use of copper (3)
- legal requirements for conventional production are too low (1)
- missing explicit legal framework for 'correct' organic fruit production (1)

- general regulations are rather oriented on conventional production than on organic production (1)
- explicit statement that no legal inhibiting factors exist (2)

Economic factors:

- market saturation leads to higher production risks, fluctuating prices, and uncertain sale guarantees (5); production should not grow larger than demand (2)
- competition challenges with conventional production, especially regarding price (5) and quality (2)
- yield and productivity in organic apple cultivation are too low (3)
- missing professionalization and experience in the use of organic cultivation methods, which leads to high production costs (2)
- increasing pricing pressure with increasing cultivated land area (1)
- declining price differences between organic and conventional apples (1)
- overproduction of organic apples because more and more farmers converse to organic production (1)
- non-existent or rather slow transition from conventional varieties to new or alternative varieties, with the result that conventional varieties are not being pushed out of organic product ranges (1)
- difficulty of introducing "nameless" (Delphi Study Round 1, own translation) cultivars to the market (1)
- pressure to provide products with high aesthetic quality and thus low pack-out rate (1)
- commercial categories are the same for conventional and organic products (1)
- external effects ('true costs') are not included in the costs of production (1)

Other factors:

- changing the product range and thus changing farming structures takes at minimum three years (3)
- lack of willingness to change paradigms among farmers to implement organic cultivation although organic apples generate better prices on the market than conventional ones (1)
- conservative paradigms in associations, consulting, and professional schools (1)
- climate fluctuations (1) and late frost (1)

All of these insights have been condensed into 36 inhibiting factors and rated by participants in round 2 of the market study regarding their relevance. The results of this rating are depicted in Figure 12.2.

Overall, the factors that concern the high demands of different actors along the value chain on (outer) product quality are rated as most inhibiting. Moreover, cultivation challenges (yield and disease regulation, marketing of unknown or unpopular varieties) and the saturation of the organic apple market are of high

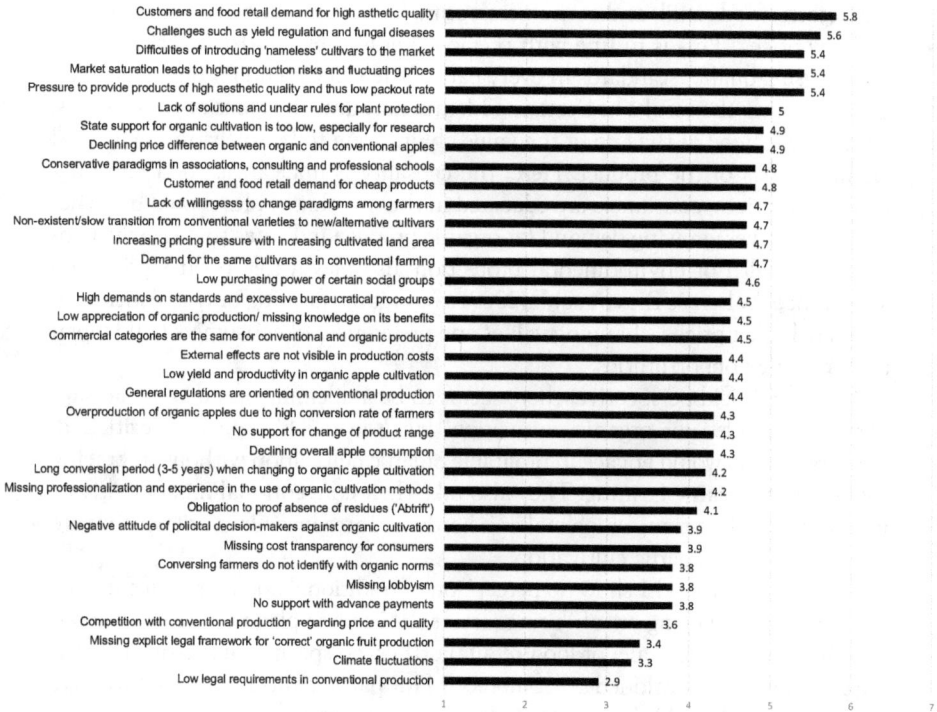

Figure 12.2 Inhibiting factors for organic apple cultivation. Mean value, rated on a 7-point Likert Scale (n = 21). 1 = low relevance, 7 = very high relevance (own figure).

relevance. The lowest relevance is assigned to low legal requirements in conventional relevance. The lowest relevance is assigned to low legal requirements in conventional production and the competition with those actors; climate fluctuations; as well as 'grey areas' in the legal framework. Thus, inhibiting factors concerning practical cultivation, marketing, and the general market context are perceived as most relevant.

On average, promoting factors are perceived as more relevant than inhibiting factors. Inhibiting factors for organic apple cultivation have a lower mean minimum (2.9) and maximum (5.8) than promoting factors (3.2 and 6.2).

12.2 Discussion of Results

In round 1 of the market study, the participants stated a range of promoting and inhibiting factors for the further propagation of organic apple cultivation. Congruent with the discussed results in Chapter 11, high trait-specific demands of farmers, marketeers, and consumers on apple varieties play a major role as an inhibiting factor. The economic conditions are challenging for organic cultivation, and are characterized by a lot of competitors in a saturated market. Especially marketeers, food retail, and consumers want a 'perfect product' with high quality and low prices. This is in line with previous findings by Zander (2011), who observed rising quality standards in every part of the value chain for organic apples. By also taking the results of Section 12.1 into account, the focus of these actors, which are located at the end of the value chain, seem to be solely on the general characteristics of the product. Here, the causality is unclear: as food retail demands low prices, consumers are educated in the way that apples should be rather cheap. However, growing demand for organic produce indicates that specific benefits like health or environmental protection also play a role besides price and appearance. Whereas Bravin et al. (2009) saw organic farming only as a means to generate higher prices, changing values and preferences of consumers build much more complex opportunities.

From a political perspective, there seems to be a dilemma: more financial support is demanded for organic cultivation, but 'too much' support is criticized. Large producers would engage in profitable organic cultivation without embracing the norms of organic farming. They also take their conventional mix of cultivars into organic cultivation, which makes it harder to introduce new varieties that are more suitable for organic cultivation.

The results of round 1 show no perception of traditional and heirloom cultivars as especially promoting or inhibiting factors. It seems to play no role in commercial production, wholesale, or food retail. Further, no specific alternative marketing structures are mentioned as promoters. Many participants rather want to leave their niches and establish organic apples as mainstream.

Participants do not mention elements of cooperation and networking as promoting factors at all. In her study a decade ago, Zander (2011) identified success factors for the organic apple market in Germany. The major factors were a high level of cooperation, trust, commitment, satisfaction, and knowledge sharing on a horizontal level (between producers and their trading partners):

> Interestingly, producers and their trading partners have a high level of commitment on both sides. For example, when experiencing supply bottlenecks, producers are buying produce from competitors to be able to continuously supply their trading partners. [...] As a result, trading partners have a high loyalty towards their suppliers

> (Zander, p. 19).

Probably because of the perennial character of apples, these actors prefer long-term, secure partnerships and follow common goals (Zander 2011). Additionally, vertical cooperation between traders and organic food trade or food retail is seen as a success factor, for example by designing product ranges together. It remains unclear if cooperation and networking structures have significantly changed or participants did not see it as important because it is a matter of fact.

By rating the relevance of promoting and inhibiting factors in round 2, promoting factors are rated marginally higher than inhibiting factors. In both categories, economic and market developments or demands are perceived as the most relevant. Thus, although overarching market trends promote organic apple cultivation, specific demands on aesthetic quality, yield, etc. are challenging for farmers. Although the market is highly saturated, fruit farmers see opportunities to change toward organic cultivation practices, which further challenges existing competitors. Political and legal factors are not perceived as highly relevant in both categories. In comparison to economic and market factors, they have no decisive influence. In conclusion, the ratings show that especially producers, production and distribution organizations, and marketeers are the most important market actors that decide about the future directions of organic apple cultivation.

12.3 Results of the Market Study for Organic Apple Breeding

In round 1 of the market study, answers to the open questions revealed societal, economic, political, and legal factors that have current inhibiting and promoting effects (at the moment of data collection) or could be promotive in the future. Overall, 17 promoting and 24 inhibiting factors for the further expansion of organic apple breeding have been identified. In round 2, all factors have been rated. This randomized assessment took place with a 7-point Likert scale where 1 means that the factor has low relevance and 7 means that the factor has very high relevance to promote or inhibit organic apple cultivation. Participants rated the factors accordingly.

12.3.1 Participation in Breeding Activities

In round 1 of the market study, participants were asked which specific actors or organizations along the value chain should generally participate (financially or advisorily) in apple breeding activities. They mention the following actors:

- fruit farmers (6) and production organizations (2)
- marketeers and distribution organizations (6)
- food retail and wholesale (5), specifically organic food trade (1)
- state actors (5), specifically public research institutes (2)
- others: fruit cooperatives (2), consumers (2), NGOs (2), private breeders (1), storage and plant protection specialists (1), organic associations (1)

A majority additionally state that the whole value chain should participate in breeding (8) to include the whole diversity of preferences and opinions in the breeding process. This should take place in an advisory and financial way. However, one participant acknowledges that the participation of different actors is generally difficult because consumer demands constantly change, and organic cultivation experiences high economic pressure (1).

12.3.2 Definition and Understanding of Organic Apple Breeding

In round 1 of the market study, the participants were confronted with the following definition of organic apple breeding: *Organic apple breeding is embedded in the general norms and guidelines of organic farming (promoting soil health and genetic diversity, treating natural resources with care, taking social responsibility). Breeding takes place on-farm und selections are carried out under organic conditions. Breeding goals match with the needs of organic fruit farming and the sustainable use of natural resources.* For the verification of this definition, participants were asked about their understanding (goals and procedures) of organic apple breeding in round 2.

Many participants state that breeding has to be in line with the methods (5) and norms (2) of organic farming. Genetic engineering is rejected (4). A wide pool of genetic resources is used for breeding (3), specifically wild and meadow orchards cultivars (1). Breeding should concentrate on varieties that specifically match with the needs of organic apple cultivation (3). Moreover, site adaptation is an important attribute (2). The following specific breeding goals are mentioned: individual taste (7); robustness (7); outer quality like visual appeal or texture (3); resistances (3); high yield security (2); healthy product (1); and good storability (1). Some participants mention further specific aspects that concern the breeding process: traditional cross-breeding; resource-preserving breeding; no use of chemical-synthetic products in the breeding process; acknowledging environmental conditions in the selection process; and on-farm breeding (all 1). On a market level, the following attributes are stated: no breeding of club varieties; free availability of bred varieties; organically bred varieties have to be marketable and suitable for extensive cultivation (all 1).

The term 'organic breeding' is criticized by a few participants: in general, there is no difference to the goals of conventional breeding except the non-use of chemical-synthetic products (1). Furthermore, the term is perceived as a bit exaggerated because conventional varieties can also be used for organic cultivation (1).

The given definition of organic apple breeding in round 1 partly matches with the mentioned attributes in round 2: organic breeding should take place under organic conditions and match with the needs of organic farmers. Robustness and individual taste are mentioned as important breeding goals. However, on-farm breeding plays no role for participants in round 2 (See also Section 12.1).

12.3.3 Promoting Factors for Organic Apple Breeding

The results show that not many factors exist that have promoting effects at the moment. Rather, most mentioned aspects refer to possible future promoting effects. A further communication to market actors and farmers on what organic apple breeding actually comprises (at least in the sense of the IFOAM guidelines) seems necessary. More financial and non-material state support of organic apple breeding is seen as promotive. In more detail, the following insights are summarized.

Societal factors that have promoting effects at the moment:

- trend toward higher demand for organic products (1)
- innovativeness among organic farmers and breeders (1)
- initiatives on open-source varieties (1)
- idealism among organic breeders as observed in the association *apfel:gut e.V.* (1)

Societal factors that could have promoting effects in the future:

- information campaigns on the importance of organic breeding (similar to those on the importance of traditional and heirloom varieties) to further strengthen its awareness among consumers (8)
- information campaigns on genetic engineering (3) and a clear dissociation from related techniques by enhancing unique features of organic breeding (1)
- establishing organically bred apples as mainstream (1)

Political factors that could have promoting effects in the future:

- more financial state support for organic breeding (7)
- political will to support organic breeding (1)
- state support for certain private organic breeders (1)
- consulting farmers on apple cultivation without any plant protection products (1)
- state-owned and independent trial areas (1)
- fully or partly state-run organic breeding (1)
- supporting education on fruit growing at universities (1)

Legal factors that could have promoting effects in the future:

- no patents on gene sequences (2)
- trends to limit pesticide use (2)

Economic factors that could have promoting effects:

- increasing the diversity of commercial categories instead of breeding new varieties (1)
- trade partners directly support breeding activities (1)
- working together in cooperatives to finance organic breeding activities (1)

All of these insights have been condensed into 18 promoting factors and rated by participants regarding their relevance in round 2 of the market study. The results of this rating are depicted in Figure 12.3.

Participants perceive the topic and goals of genetic engineering as well as further responsibilities by the state as the main promoting factors. A clear dissociation from genetic engineering is seen as a major opportunity to promote organic breeding. Additionally, more state support for organic apple breeding and further activities in this context are perceived as relevant. The lowest relevance has been assigned to educational support on fruit growing and an increase in the diversity of commercial categories.

All listed factors were generally not rated of low relevance as no mean value lies below 3.5. This emphasizes the validity of the results from round 1 of the market study.

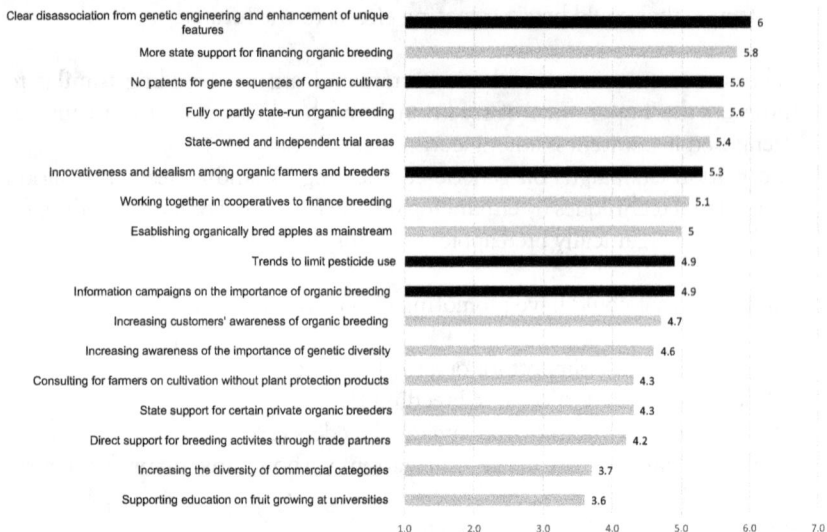

Figure 12.3 Promoting factors for organic apple breeding. Mean value, rated on a 7-point
Likert scale (n = 18). 1 = low relevance, 7 = very high relevance. Gray-colored
boxes indicate factors that could be promoting. Black-colored boxes mark factors that are currently promoting (own figure).

12.3.4 Inhibiting Factors for Organic Apple Breeding

Results show that especially economic factors and missing state support are perceived as inhibiting for the further propagation of organic apple breeding. In more
detail, the following insights are summarized.

Economic factors:

- high financial risks without secure profits (1)
- high long-term costs without short-term revenues, resulting in a dependency
  on private and public support (2) and sometimes short-term thinking (1)
- only pre-financed breeding is feasible (2)
- financial strength of conventional breeding (1)
- competition with club concepts and their professional marketing
  concepts (4)
- high demand from the whole supply chain for aesthetic quality (2)
- breeding goals like taste or robustness are only slowly gaining relevance (1)
- lack of time and resources for breeding (2)
- insufficient consumer orientation (1)
- commercialization of breeding and the focus on profit opportunities (2)
- missing specific organic orientation of fruit breeding (1)
- too small areas on farms available for breeding (2)

- professionalization of organic farmers to engage in breeding activities is insufficient (1)
- on-farm breeding could lead to pests building resistances or virus transmissions (1) and is not easy to implement in the "small-farmer reality" (Delphi Study Round 1, own translation) (1)

Political factors:

- lack of state support for organic breeding (5)
- too much state support for molecular genetics and too low support for the education of fruit-growing professionals (1)
- lobbying and power of agro-industrial companies (1)

Legal factors:

- European regulations support club concepts and impede a market-driven diversity of varieties (1)
- unclear approval of organic plant protection products (1)

Societal factors:

- knowledge on traditional and heirloom varieties, as well as organic breeding, is lost and too little fruit growing professionals exist (2)
- consumers do not strongly demand GMO-free food (1)
- too few organic breeders (1)
- "free breeders" (Delphi Study Round 1, own translation) are left out by private businesses (1)
- majority of market actors do not focus on organic breeding (1)

All of these insights have been condensed into 24 inhibiting factors and rated by participants regarding their relevance in round 2 of the market study. The results of this rating are depicted in Figure 12.4.

Overall, the rating shows that economic factors have the most relevant inhibiting effects on the further propagation of organic apple breeding. This includes the long-term costs of breeding, missing state support, and high financial risks. Moreover, many actors in the fruit sector do not focus on organic breeding. Insufficient consumer orientation, knowledge, and professionalization are rated lowest together with short-term orientation and the exclusion of so-called free breeders by private businesses.

All listed factors were generally not rated of low relevance as no mean value lies below 3.5. This emphasizes the validity of the results from round 1 of the market study. On average, promoting factors are perceived as slightly more relevant than inhibiting factors. Inhibiting factors for organic apple breeding have a slightly lower mean minimum (3.5) and maximum (5.9) than promoting factors (3.6 and 6.0).

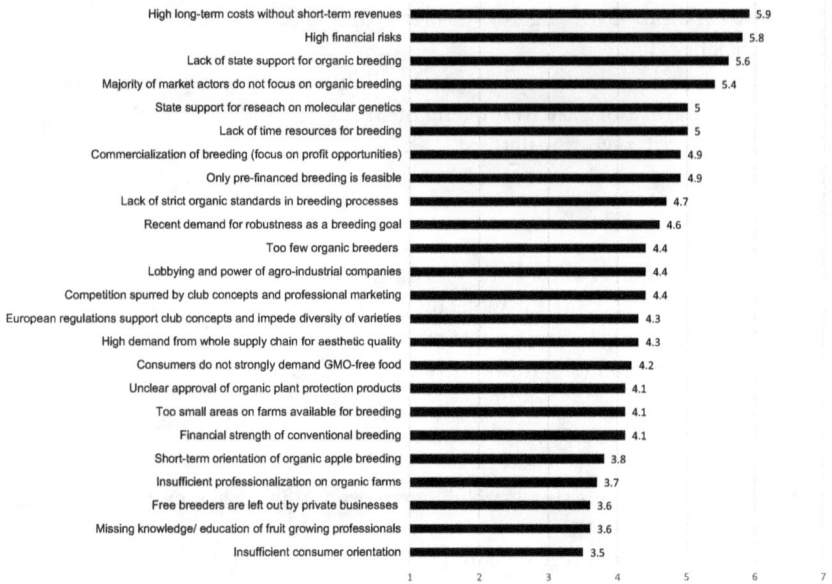

Figure 12.4 Inhibiting factors for organic apple breeding. Mean value, rated on a 7-point Likert Scale (n = 16). 1 = low relevance, 7 = very high relevance (own figure).

12.4 Discussion of Results

In general, there is some critique on the definition of organic apple breeding. In line with the rating of normative statements in Section 12.1, the majority does not define organic breeding as on-farm breeding. Results of this Section 12.3 should thus be interpreted carefully, as different understandings of organic breeding exist among participants. Although the participants identify a range of promoting and inhibiting factors for the further propagation of organic apple breeding in round 1, they identify not as many as for organic cultivation. This indicates that even in the organic fruit sector, specific knowledge on challenges for organic breeding is limited.

A diversity of general and specific factors, especially economic ones, inhibit organic apple breeding. Market developments and trends such as high competition, club concepts, and professionalized marketing concepts as already discussed in Section 12.2 are all perceived as inhibiting factors. An exception is a trend to limit pesticide use that is perceived as a promoting factor, also reflected in new cultivation models like low-residue apple production (Bravin et al. 2019). State support is lacking which is simultaneously perceived as a highly relevant factor that could have promoting effects in the future. At the moment, organic apple breeding still seems to be a niche with too few financial, time, and personal resources. However, innovativeness and idealism among organic farmers and breeders are perceived as an opportunity to slowly leave this niche and build opportunities.

Ratings of the promoting and inhibiting factors in round 2 primarily show that a lot of inhibiting factors stand against only a few promoting factors that have effects at the moment. Most of the identified promoting factors could have beneficial effects in the future. Based on the ratings, essentially three arguments can be derived that would further promote the further propagation of organic apple breeding. First, more political will on a national and EU level would promote and support organic breeding. The lack of state and public support was made very clear by the participants. This is particularly interesting as state support is not perceived as highly relevant in the promotion of organic apple cultivation. Second, more financial support from different sources is perceived as necessary because fruit breeding is a time-consuming and risky task that generates profits only in the long term. Organic breeders need financial incentives to carry out their tasks. This is not only a German and apple-specific phenomenon but applies to organic breeders in all major breeding centers, mainly across Europe (Kotschi et al. 2022). Third, further awareness of organic breeding as well as a clear dissociation from genetic engineering and conventional breeding could be major drivers.

Thus, besides the call for more state and financial support, new business models and alternative market mechanisms could be an opportunity to further promote organic apple breeding and establish an effective and long-term financing of breeding activities.

References

Bravin, E.; Kilchenmann, A.; Leumann, M. (2009): Six Hypotheses for Profitable Apple Production Based on the Economic Work-Package Within the ISAFRUIT Project. In *The Journal of Horticultural Science and Biotechnology* 84 (6), pp. 164–167. DOI: 10.1080/14620316.2009.11512615.

Bravin, E.; Perren, S.; Naef, A. (2019): Low Residue Apple Production: Higher Production Risk and Lower Profit. In *Acta Horticulturae* (1242), pp. 217–222. DOI: 10.17660/ActaHortic.2019.1242.30.

Delphi Study Round 1 (2019): Qualitative data obtained by online-questionnaires. Germany.

Kotschi, J.; Schrimpf, B.; Waters-Bayer, A.; Horneburg, B. (2022): Financing Organic Plant Breeding—New Economic Models for Seed as a Commons. In *Sustainability* 14 (16), p. 10023. DOI: 10.3390/su141610023.

Zander, K. (2011): Ausländisches Angebot an ökologischen Äpfeln: Bedeutung für deutsche Öko-Apfelerzeuger. Universität Kassel, Fachgebiet Agrar- und Lebensmittelmarketing. Witzenhausen.

13 Business Models for Financing Organic Apple Breeding

In round 2 of the market study, participants were asked about their opinion on two specific business models and perceived barriers to their possible market introduction. This qualitative data has been categorized and evaluated.

13.1 Results of the Market Study

13.1.1 Business Model A "Breeding Cent"

Participants were confronted with the following business model: *When buying organic fruits with a label from an association (Bioland, demeter, Naturland), 0.1 cents/ kg of fruits will be added to the sales price. This collected additional fee is invested in a fonds to promote organic fruit breeding. Initiatives and breeders that follow the guidelines of the associations and have a non-profit status can apply for funds out of this fonds to finance their breeding activities.*

The perceptions of this business model differ. Whereas half of the participants are critical toward this model and would not support it (8), the other half are interested and have a positive attitude toward it (7). Some of these participants state that this kind of model is already successfully established in other sectors: organic animal breeding (1); plant breeding in the United States (1); or grain breeding (1). Thus, they perceive it as a supportable and proven model. However, organic food retail (1) or rather the whole fruit production sector (1) should be included in the business model. One participant notes that probably too few associations and farmers would participate (1). Lastly, although the breeding cent could work, the fee would be too low to induce significant effects, thus it should be raised (1). However, participants also note criticisms, especially regarding the organizational and administrative efforts: administrative efforts to organize cash flows are high (4) and it is unclear who administers the fonds and who controls the administration (1). Moreover, the fee could be perceived as a "compulsory levy" (Delphi Study Round 2, own translation 2019) (1).[1] The system is also perceived as unfair because fees per kg are difficult to control in an exact way (1). Breeding would solely serve the interests of associations and their influence on breeding practices would be too large (1). It is also perceived as problematic that this model would further divide integrated

DOI: 10.4324/9781003355724-23

and organic production (1) because participation would only be possible for the organic sector.

These statements and valuations already include specific barriers for the possible market integration of the business model "Breeding Cent." When the participants were specifically asked about barriers, the following insights could be collected:

- Organizational and administrative barriers: the effort for administration is higher than the possible amount of funding; high effort for organizing the cash flow; high costs of control; it is unclear on which level (production, marketing) the breeding cent should be paid.
- Market-based barriers: difficult feasibility for direct marketeers; lack of motivation for marketeers because they do not directly participate in breeding activities; the influence of associations on the market is too low; higher production costs; compulsory levy could lead to complaints when no direct benefits are observed by the claimants (see footnote above).
- Cultural barriers: missing acceptance of market actors, specifically food retail; negative perception of "additional requirements" (Delphi Study Round 2, own translation); perception of breeding cent as a club concept.
- Other barriers: legal requirements (not further specified).

One participant further notes that a new organic breeding organization could be established that acquires all the funds (1).

13.1.2 Business Model B "Funding Alliance for Organic Fruit Cultivation"

Participants were confronted with the following business model: *Market actors from the organic fruit sector participate in a material (fixed allowances) and non-material way in a specialized organization that promotes organic fruit breeding. This organization funds non-profit breeders and breeding initiatives to develop varieties that meet the needs of organic fruit farming. The market introduction of developed varieties is professionally managed by the organization.*

Overall, the support for this model among the participants is higher than for model A (10). A minority would not support the model (4). Although the cooperation of all stakeholders would not be easy, it is perceived as the best solution (1). The alliance could fund breeding projects in an independent way (1) and the "organizational bundling" (Delphi Study Round 2, own translation) regarding the management of market introduction makes sense (1). The voluntary character of participation is key although it could be difficult to generate incentives for market actors that cannot see concrete benefits (1). However, voluntariness also creates opportunities for free riders (1). A further challenge is the realistic cost estimation of breeding projects to define the optimal level of fixed allowances (1). One participant argues that this business model could potentially be combined with model A to diversify funding sources (1). However, besides the observed challenges, participants that would not support this model further note some criticisms.

All beneficiaries of organic breeding should participate in the funding and not just a specific (voluntary) group of actors (1). The establishment of such an organization would be too time-consuming and requires too much coordination effort (1). Moreover, this kind of model would only work with absolute price transparency (1). Similar to model A, one participant notes that this model would further divide integrated and organic production (1).

These statements and valuations already include specific barriers for the possible market integration of the business model "Funding Alliance for Organic Fruit Cultivation." When the participants were specifically asked about barriers, the following insights could be collected:

– Organizational and administrative barriers: too many different actors, especially regarding their size; it is difficult to define authorized representatives; full transparency is necessary but difficult to establish; high efforts and establishment costs.
– Market-based barriers: no market transparency; the alliance could become a kind of cartel; it is difficult to acquire enough participants; the participation of all market actors including conventional food retailing is necessary to get acceptance for fixed allowances; uneven distribution of funds that could lead to competitive disadvantages for participating actors; professional market introduction of varieties could be disadvantageous for direct marketing.
– Cultural barriers: model could be perceived as unfair when free riders also benefit; fear of never-ending discussions about the level of fixed allowances; missing differentiation from club concepts.

Similar to model A, one participant notes that a new organic breeding organization could be established that acquires all the funds (1). Another participant argues that the FÖKO could become this kind of organization as many actors from the organic fruit sector are already represented in this organization (1).

13.1.3 *Further Ideas and Thoughts of Participants*

As the last question of round 2, participants were asked if they have further ideas or thoughts on future business models for funding organic fruit breeding independently from public funds. One participant demands a further alliance of organic associations on a European level and pre-financing instruments for organic breeding which are organized by production associations (1). Pilot projects with actors along the whole value chain should be introduced (1). Another idea is an "areal fee for all organic farmers" (Delphi Study Round 2, own translation) that is oriented at the land register on area payments (1). The fees would be invested in organic fruit breeding. A graduated fee for all farmers that are organized in organic associations is seen as another possibility to fund organic breeding (1). The norm of negative listing of varieties should change to a positive listing, which would further widen the range of varieties that are suitable for organic cultivation (1).

13.2 Discussion of Results

Both discussed business models encourage us to think about alternative ways to finance organic apple breeding activities. In the context of organic vegetable, grain, or animal breeding, different general funding sources have already been discussed in the literature that are also valid for other breeding organizations across the world (Kotschi et al. 2022; Schäfer and Messmer 2018; Kotschi and Wirz 2015; Wilbois 2011). According to Messmer (2019), five financing directions currently exist that are relevant for breeding in Germany:

1. Foundation funds: they are the most important funding source for most organic breeders. Examples for foundations are the *GLS Saatgutfonds*, the *Software AG Stiftung*, or the *Stiftung Mercator Schweiz*.
2. License fees: farmers or hobby gardeners pay license fees, re-seeding fees, or other fees that can be re-invested into breeding.
3. Public funding: here, only research on breeding can be funded, not the practical breeding. Administrative efforts are high. Possible funders are EU, BLE, or BMBF.
4. Participation of value chain actors: associations, processors, food retailers, or organic food traders participate via specific models in the funding of specific breeding projects. An example is the program FAIR-BREEDING in which several actors along the value chain invest small parts of their turnover to fund breeding projects (Fleck and Boie 2009).
5. Crowdfunding: adapted from other domains, the broader public gets the opportunity to fund specific breeding projects.

For the successful financing of long-term breeding activities, a combination of different funding sources is necessary (Schäfer and Messmer 2018). The proposed business models for fruit breeding (see above) need to be discussed in light of these findings from the literature.[2]

Overall, the proposed business model B is perceived as more attractive than business model A. Besides the general evaluations, participants identify several barriers to a possible market introduction of both models that are either model-specific or general. They assign high levels of necessary administrative and organizational efforts to both models. Additionally, in both cases, the question of how exclusive it should be designed arises (is the target group solely the organic or the organic and the integrated sector?) and if the model should be implemented into an existing organization or a new one should be founded. Exclusivity could also lead to the perception of the respective business model as a club concept, which is evaluated as a shortcoming.

Regarding model A it was not clear to the participants who actually pays the fee – trading partners, marketeers, or consumers. This resulted in some confusion as illustrated by the perception of the fee as a compulsory levy. The benefit of model A is that it already exists in other sectors, for example, at the German non-profit association *Kultursaat e.V.*, which has not been mentioned by the participants. *Kultursaat* breeds organic vegetable, herb, and flower varieties in a self-governed community of breeders. New varieties are introduced without claiming variety protection to make them freely available (Kultursaat e.V. 2021). Besides other funding sources, a certain percentage of retailers' revenues are invested in Kultursaat's breeding activities, described as *Sortenentwicklungsbeitrag* (Sievers-Glotzbach et al. 2020). As the association has much experience with this model, the organic fruit sector could learn from them.

Business model B would be a completely new model for plant breeding as it incorporates the whole value chain from breeding to food retail. Weber (2008) already proposed a different managed variety system along the value chain that relies on collective royalties on trees and fruits of newly developed varieties. Instead of a free participation in funding as proposed in model B, this model would acquire its financial resources from royalties:

> If the commercialization of the new apple varieties are distributed via accredited sales desks, a fruit royalty is taken. Some share of this royalty flow is getting re-invested into the ongoing breeding program without any options for future releases. The establishment of a B-to-B trademark on an international scale allows efficient marketing with relatively low cost involvement for the grower.
>
> (Weber 2008)

In this model, a club concept is applied that additionally involves the financing of breeding. This concept is actually more or less adopted by ZIN (See Chapter 7). In line with the paradigmatic orientation of corporate-based apple breeding, it aims to "become sustainable through profitability" (Weber 2008). However, there is a need for new business models beyond royalties that "require new relationships within the value chain including breeders as partners in the food chain or food networks" (Lammerts van Bueren et al. 2018, p. 16). These models are currently discussed in research and practice.

The association FiBL proposed a pool-funding concept similar to business model B for the organic breeding sector (Schäfer and Messmer 2018). In their model of a *cross-sector pool-funding*, an obligatory 0.1–0.2% of the whole turnover in the organic sector either on national levels or even on a European level would regularly be invested in a fonds for organic breeding. Organic breeders across all sectors (vegetables, fruits, animals, etc.) can apply for the funds and an independent steering committee decides on the distribution of the funds. This funding opportunity would add to already existing funding sources and build partnerships along value chains across sectors. First research results from a European stakeholder dialogue show that this form of value chain partnership could be a promising approach (Winter et al. 2021). At the moment, this model is being further discussed with actors from the organic sector in the context of the research project Engagement.Biobreeding Europe (FiBL 2020).

As an example, the Swiss association *Bioverita* is a partner in this research project. It aims to promote organic breeding and manage quality assurance.

The association wants to serve as a platform for building partnerships along the value chain by promoting networking, knowledge sharing, joint marketing actions, and awareness for organic breeding. As an important instrument, *Bioverita* certifies food producers with the *bioverita-Qualitätslabel* when they use organic varieties that are developed by organic breeders registered at *Bioverita*. Producers can print the label on their products (Bioverita 2021). This example shows opportunities for building partnerships. For both models A and B, this kind of association could work as an organization being responsible for implementing the business model.

Discussions show that there is a need for new partnerships and business models to secure funding for organic breeding activities in a sustainable way. Some new and alternative models are already implemented, albeit on a rather small scale. In line with the overall results of the market study, discussions on business models to finance organic breeding emphasize that the strategic choices of actors in organic apple cultivation and breeding are highly relevant for its further expansion:

> Among these choices are further collaboration with conventional farmers and breeders, a further demarcation from the conventional sector through new institutions or further development of existing institutional structures, or aiming for cooperation across the whole organic food sector. Results suggest that the latter two options in particular or even a combination of both might be successful strategic directions. The top promoting factors for organic cultivation and breeding encourage the further demarcation. Additionally, the positive overall development of the organic sector could be a window of opportunity for establishing reliable financial cooperation and provide the awareness for organic breeding that it needs for further expansion.
>
> (Wolter 2020, p. 25)

With these conclusions in mind, the market study provides the basis for the evaluation of the challenges for implementing commons-based organic apple breeding into the market.

Notes

1 It should be noted that some participants did obviously not fully understand the proposed model. The breeding cent would be paid by consumers and not by farmers, marketeers, or other actors.
2 In this scheme, the proposed business models A and B would fit to the financing directions 4 and 5.

References

Bioverita (2021): Vision und Mission. Available online at https://bioverita.ch/weiteres/vision-und-mission/, updated on 4/13/2021.
Delphi Study Round 2 (2019): Qualitative and quantitative data obtained by online-questionnaires. Germany.

FiBL (2020): Engagement der ökologischen Wertschöpfungskette zur Unterstützung der biologischen Züchtung in Europa. FiBL. Available online at https://www.fibl.org/de/themen/projektdatenbank/projektitem/project/1784.html, updated on 9/2/2020, checked on 4/13/2021.

Fleck, M.; Boie, K. (2009): FAIR-BREEDING. Wegweisende Partnerschaft zwischen Naturkostfachhandel und Gemüsezüchtern. In *Kritischer Agrarbericht*, pp. 116–120.

Kotschi, J.; Schrimpf, B.; Waters-Bayer, A.; Horneburg, B. (2022): Financing Organic Plant Breeding—New Economic Models for Seed as a Commons. In *Sustainability* 14 (16), p. 10023. DOI: 10.3390/su141610023.

Kotschi, J.; Wirz, J. (2015): Wer zahlt für das Saatgut? Gedanken zur Finanzierung ökologischer Pflanzenzüchtung. Edited by AGRECOL Verein für standortgerechte Landnutzung. Available online at https://opensourceseeds.org/sites/default/files/downloads/Wer_zahlt_fuer_das_Saatgut.pdf, checked on 4/17/2019.

Kultursaat e.V. (2021): Einblicke 2020- Jahresbericht Kultursaat e.V. Kultursaat e.V. Available online at https://www.kultursaat.org/aktuell/einzelansicht/news/einblicke-2020-jahresbericht-kultursaat-ev/?no_cache=1&cHash=40b21402ae467227f69d0bd8b58672e9, checked on 4/8/2021.

Lammerts van Bueren, E. T.; Struik, P. C.; van Eekeren, N.; Nuijten, E. (2018): Towards Resilience Through Systems-based Plant Breeding. A review. In *Agronomy for Sustainable Development* 38 (5), p. 42. DOI: 10.1007/s13593-018-0522-6.

Messmer, M. (2019): Möglichkeiten zur Finanzierung ökologischer Pflanzenzüchtung. EGON-Abschlussveranstaltung. Jork, Germany, 2019. Available online at https://uol.de/f/2/dept/wire/fachgebiete/oekogueter/EGON_Abschlussveranstaltung/EGON_Abschlussveranstaltung_Vortrag_Messmer.pdf, checked on 4/8/2021.

Schäfer, F.; Messmer, M. (2018): Eckpunktepapier für die Etablierung eines tragfähigen Finanzierungssystems der Biozüchtung. FiBL. Supportstelle ökologische Pflanzenzüchtungsforschung. Available online at https://orgprints.org/id/eprint/38440/1/schaefer-messmer-2018-FiBL_SOEPZ_Eckpunktepapier_VersionOktober_20181016_logos.pdf, checked on 4/8/2021.

Sievers-Glotzbach, S.; Tschersich, J.; Gmeiner, N.; Kliem, L.; Ficiciyan, A. (2020): Diverse Seeds – Shared Practices: Conceptualizing Seed Commons. In *International Journal of the Commons* 14 (1), pp. 418–438. DOI: 10.5334/ijc.1043.

Weber, M. (2008): Sustainable Apple Breeding Needs Sustainable Marketing and Management. In FOEKO (Ed.): *Ecofruit. 13th International Conference on Organic-Fruit Growing: Proceedings*. Ecofruit. Hohenheim. Fördergemeinschaft Ökologischer Obstbau e.V. (FOEKO).

Wilbois, K.-P. (2011): Ökologisch-partizipative Pflanzenzüchtung. Frankfurt am Main, Bochum: Forschungsanstalt für biologischen Landbau e.V. (FiBL), Zukunftsstiftung Landwirtschaft. Available online at http://orgprints.org/20574/1/1563-oekolog-partizipativ-pflanzenzuechtung.pdf, checked on 9/11/2019.

Winter, E.; Grovermann, C.; Aurbacher, J.; Orsini, S.; Schäfer, F.; Lazzaro, M. et al. (2021): Sow what you sell: strategies for integrating organic breeding and seed production into value chain partnerships. In *Agroecology and Sustainable Food Systems*, pp. 1–28. DOI: 10.1080/21683565.2021.1931628.

Wolter, H. (2020): Influencing factors for the further expansion of organic apple cultivation and breeding. In FOEKO (Ed.): *Ecofruit. 19th International Conference on Organic Fruit-Growing: Proceedings*. Ecofruit. Hohenheim, 17.-19.02.2020. Fördergemeinschaft Ökologischer Obstbau e.V. (FOEKO), pp. 19–26.

14 Challenges for Commons-Based Organic Apple Breeding

This chapter interprets barriers and possibilities for the integration of commons-based organic apple breeding into the current (German) market context. Many discussed results of a market study about organic apple breeding and cultivation that concern broader market structures and developments (such as promoting and inhibiting factors for organic apple cultivation and breeding) are also generally valid for commons-based organic apple breeding. They define the current regime any apple breeding approach is confronted with. The discussion in this chapter is structured along the main attributes of the commons-based organic breeding approach: (a) collective responsibility for agrobiodiversity; (b) collective management; (c) knowledge sharing; and (d) collective ownership. Based on the discussed results of the market study in Chapters 11–13, in this chapter barriers and possibilities for the implementation of commons-based organic apple breeding into the current market context are discussed. Although this discussion takes place in the empirical context of Germany (See Section 3.2) because of the market study's focus on this country, insights may also be applicable to other socio-economic contexts (See Section 3.3).

The following analysis is structured along the four attributes of the breeding approach that have been described in Section 9.1: (a) collective responsibility for agrobiodiversity; (b) collective, polycentric management; (c) knowledge sharing; and (d) collective ownership. Originally, these attributes have a descriptive and analytical function (Sievers-Glotzbach et al. 2020). However, to discuss the potential of commons-based organic apple breeding, it is beneficial to discuss the market study's results from the perspective of these attributes. In this way, the current norms, structures, and developments can be assigned to the relevant attributes and evaluated accordingly.

It must be noted that many discussed results that concern broader market structures and developments (like the promoting and inhibiting factors for organic apple cultivation and breeding) are also generally valid for commons-based organic apple breeding. They define the current regime and stability landscape any apple breeding approach is confronted with. To prevent content-related repetitions, the following analysis will concentrate on specific aspects that concern the four attributes.

DOI: 10.4324/9781003355724-24

14.1 Collective Responsibility for Agrobiodiversity

The discussion of the market study's results shows that several market-based barriers exist that hamper the promotion of agrobiodiversity in apple cultivation. Especially the current market developments (See Chapter 11) and the inhibiting factors for organic apple cultivation (See Chapter 12) show those barriers in detail. The dominant market reality calls for standardized products with high outer quality. Conventional and organic apple production still largely demand commodity apples with similar quality characteristics. Market developments such as growing standardization, conventionalization, and economic professionalization result in a low diversity of visual traits and taste across the range of cultivated varieties. This is especially true for varieties marketed via food retail. Even in organic food retail, special organic varieties have difficulties establishing themselves. Site adaptation was not stated as a major attribute by the participants when they were asked about the demands of farmers and marketeers on varieties (See Section 11.1) – even though this aspect has the potential to further promote diversity.

Another barrier is the ongoing dilemma that although diversity is desired by many actors, no common efforts are realized to actually supply it on a larger scale and thus solve this dilemma (See Section 11.2). This seems connected to conservative paradigms that dominate apple breeding and cultivation (See Section 14.4). In this respect, the whole sector generally seems to have low capacities to adapt and change. This is most illustrated by farmers changing from conventional to organic practices without changing their planted varieties.

Further, the current debate on genetic engineering (See Section 11.1) shows that access and diversity could be further hampered by a further propagation of those methods. However, as there are currently no specific developments in Europe concerning genetic engineering in the context of fruits, this aspect is difficult to discuss in detail.

Apart from the identified barriers, the market study also identifies the potential to foster the collective responsibility for agrobiodiversity. The analysis of participants' norms shows that there is agreement on the notion that diversity is important and farmers as well as breeders have a responsibility to protect and further develop it. Further, participants highly agree on the orientation of organic fruit cultivation on natural processes and cycles, thus site adaptation (See Section 11.1), and there is a growing trend to limit pesticide use (See Section 12.4). The main challenge seems the translation of these norms into large-scale actions.

Regionality is very important for consumers (See Section 11.2). Their changing values and preferences could build opportunities for more diversity (See Section 12.2). This could create a contradictory development against more standardization by fostering a more diverse product range in niche markets. The high potential of direct marketing as a catalyst of a diverse range of products was already discussed previously (See Section 11.2). Despite its current low relative importance in overall sales, this distribution channel can serve as an important space for experimentation and innovation.

The discussion of business models (See Section 13.2) indicates that actors are interested in mechanisms and models that foster collective responsibility. Especially the idea of a funding alliance (business model B) is well received. Such an alliance could build a voluntary institution that gives the opportunity to collectively take over responsibility for the further development of agrobiodiversity.

In sum, the general market structures and developments hinder a large-scale collective responsibility for agrobiodiversity. Possibilities to enact elements of this attribute could be niche distribution channels such as direct marketing or wholly new and innovative business models.

14.2 Collective Polycentric Management

A major element of collective polycentric management is that different actors (mainly breeders and farmers) work closely together to breed new varieties. Distinct breeding communities such as *apfel:gut* only formed because of this objective (See Chapter 9). However, the market study shows that a close relationship between breeders and farmers is an exception (See Section 11.1). Although the majority of the market study's participants agree that fruit farmers should be included in fruit breeding, the majority also tends to reject on-farm breeding (See Sections 11.1 and 12.4), which is of central importance for collective polycentric management.

The market study does not provide more specific insights into this phenomenon. In the discussion on inhibiting factors for organic apple breeding (See Section 12.4), participants state that there are generally too few organic breeders, and that fruit farmers do not have sufficient areas available for on-farm breeding. In the discussion on promoting factors, networking, cooperation, or collaboration between breeders and farmers are not even mentioned (See Chapter 12). All of this leads to the conclusion that collective management models like the one carried out by *apfel:gut* are not part of the mental models of many market actors. Even though *apfel:gut* and other organizations from vegetable or grain breeding have already been working with this approach for many years, many actors seem to lack knowledge about them.

Nevertheless, participants state a range of positive examples of cooperation modes between farmers and breeders that have certain potential (See Section 11.1). A few participants also state that relationships between farmers and breeders are more intimate in the organic sector. For a further propagation of commons-based organic apple breeding, this could be an especially important entry point that would connect with the preference of market actors for long-term relationships (See Section 12.2). Commons-based organic breeding could serve as a narrative to activate this potential. Similar to the collective responsibility discussed above, the results on business models (See Chapter 13) show a general interest of market actors to work together.

Overall, much skepticism exists among participants regarding on-farm breeding and the collective polycentric management of breeding activities. In addition, the overall number of organic fruit breeders is still very small, which leads to the fact that few human resources exist for such cooperation models. It seems crucial to raise the awareness of positive examples and further explore and explain on-farm breeding to market actors. This could solve the dilemma (similar to the promotion of agrobiodiversity) that there is a strong call for collaboration and, simultaneously, the idea of on-farm breeding is largely criticized. Embracing the social aspects of the norms of organic farming could further activate the potential of collective polycentric management.

14.3 Knowledge Sharing

A high consensus on the importance of knowledge on apple breeding seems to exist among the study's participants. Most participants agree that knowledge on fruit breeding and fruit varieties should be openly accessible to anyone interested (See Section 11.1). In practice, knowledge sharing between breeders and farmers takes place (to certain degrees) because usually every breeder has at least one farmer as a feedback partner (ibid.). The case studies in Part III show that knowledge sharing in apple breeding is carried out differently with different actors; depending on the norms, rules, and institutions of the respective breeding approach. However, institutions like the *Deutsche Genbank Obst* mark efforts to make knowledge on fruit varieties accessible as a public good. Working groups or round tables in federal institutions or associations also promote knowledge sharing on breeding activities, crossing combinations, or other aspects. Thus, many actors share their knowledge although breeding knowledge is important as a competitive asset.

Developments on the European level indicate that the goal to specifically increase organic breeding knowledge is pursued with joint efforts of organic breeders throughout many countries. Networking activities that emerged out of the research project LIVESEED recently led to discussions among European organic fruit breeders to establish a new institution for sharing experiences and plant material. A charter for a participatory network called European Robust Organic Fruits (EUROrganic FRUITS) was drafted that aims to link organic fruit breeders with organic growers, researchers, and fruit tree genetic resource holders to intensify their collaborations.[1]

Another key aspect seems to concern knowledge on how to cultivate traditional, heirloom, and new varieties besides the standard ones – especially in an organic way. Missing knowledge on organic fruit growing is specifically stated as an inhibiting factor for organic apple breeding (See Section 12.3). This fact could build opportunities to increase knowledge sharing between farmers and breeders and, hence, to promote on-farm breeding.

The only major barrier mentioned by the participants concerning the restriction of knowledge sharing is the increase of club concepts, which creates exclusive platforms and spaces. This trend toward exclusivity could also become more relevant in the organic sector (See Chapter 11). Further market developments will show how this trend counters the above-described general willingness of actors to share their knowledge.

In sum, there is a high level of interest among farmers and breeders in knowledge sharing regarding apple breeding and cultivation. What seems to be missing is the specific linkage of breeding and cultivation with knowledge on organic practices. Here, commons-based organic breeding could build opportunities for farmers and breeders to learn from each other in this particular context.

14.4 Collective Ownership

Results of the market study show that intellectual property rights are mostly perceived as an important and standard part of any apple breeding effort. This is illustrated by the standardized privatization of varieties, an increase in club concepts, and the relevance of trademarks (See Chapter 11). Variety protection and thus license fees are still the key instruments to generate income and finance breeding efforts. Private ownership of varieties marks the core business model of apple breeding.

Mental models of most actors do not seem to allow them to think outside the cosmos of intellectual property rights and the privatization of varieties. Participants of the market study specifically state the lack of willingness to change paradigms among farmers and conservative paradigms as inhibiting factors for the further propagation of organic apple cultivation (See Section 12.2). This insight could be transferred to the whole organic fruit sector. The case studies on corporate-based apple breeding (See Chapter 7) and public apple breeding (See Chapter 8) show that this mindset is omnipresent.

An approach toward collective ownership of varieties has to deal with this market situation and the accompanied challenges. NOVAFRUITS as a large participatory fruit breeding organization (See Section 10.1) also depends on license fees and public funding. This shows that although some aspects of other commons-based elements (collective responsibility, collective polycentric management, knowledge sharing) are implemented in this breeding approach, collective ownership of varieties is not realized.

The balanced valuation of the statement "Fruit varieties should categorically be no private property of businesses or individuals" (See Section 11.1) shows that not all market actors agree with the lack of alternatives that most actors assign to variety protection. This is illustrated by the polarized debate on club concepts (ibid.). New legal mechanisms like amateur varieties (See Section 3.1) or business models that build on collective values (See Section 13.2) build opportunities to realize the collective ownership of fruit varieties.

Members of *apfel:gut* as an example of commons-based organic apple breeding (See Chapter 9) currently debate about those challenges and the question of whether to use variety protection for future developed varieties or not.[2] Basically, using variety protection does not resonate with the norms of the breeding community. However, protecting future developed varieties would prevent their misuse as others could apply the variety or a variation of it for variety protection and thus privatize it. An option would be to sign the association *apfel:gut e.V.* as the holder of variety protection and not to enforce the license fees, which is possible within the legal framework (See Section 3.1). As the association has a non-profit character, the variety would remain in a form of collective ownership.

In sum, current market structures and developments prohibit the realization of forms of collective ownership and rather promote the use of private property rights. The results of the market study and the case of *apfel:gut* show that there is potential for alternatives. However, it remains open on which scale these alternatives can be established and if conventional paradigms can change. Possibly, the debate on collective ownership will only gain leverage until actors such as *apfel:gut* actually introduce a new and successful variety into the market with alternative forms of property rights.

Notes

1 The author is part of the mailing list where discussions among European organic fruit breeders are documented. It remains open how this community will further institutionalize.
2 The author had the opportunity to observe an internal online meeting of *apfel:gut* members in March 2021.

Reference

Sievers-Glotzbach, S.; Tschersich, J.; Gmeiner, N.; Kliem, L.; Ficiciyan, A. (2020): Diverse Seeds – Shared Practices: Conceptualizing Seed Commons. In *International Journal of the Commons* 14 (1), pp. 418–438. DOI: 10.5334/ijc.1043.

Interim Conclusion

In this part, a comprehensive look has been taken at market characteristics, developments, trends, norms, and prognoses regarding the fruit and apple sector, specifically the organic apple sector. With the in-depth discussion of the market study's results, a wide range of specific and general insights have been generated that concern not only developments in the German market but also globally relevant debates. Moreover, the results have been discussed in light of challenges for commons-based organic apple breeding.

The discussions of the market study's results show a lot of contradictions and dilemmas:

- Many actors call for more diversity and differentiation but no common efforts are realized to actually supply a diverse range of varieties on a large scale due to commodification demands and market saturation.
- There is an interest in more collaboration among farmers and breeders, but the idea of on-farm breeding is heavily criticized.
- New varieties with suitable characteristics could enable a more resilient apple production, but investments are risky and professional marketing concepts seem to be necessary.
- From a political perspective, more financial support is demanded for organic cultivation but, simultaneously, it is criticized that too many farmers change from conventional to organic practices.

In these conflict areas, market structures and developments build obstacles for the realization and implementation of elements for commons-based organic apple breeding. The main obstacles are the demand for standardized products, low capacities to adapt and change, a disconnect between farmers and breeders, and the dominance of intellectual property rights. However, some entry points could mark further possibilities to overcome these obstacles and realize the potential of commons-based organic apple breeding. These include the innovative potential of niche distribution channels, more collaboration between market actors, increased knowledge sharing, and new legal mechanisms and practices that could foster forms of collective ownership.

DOI: 10.4324/9781003355724-25

For club apples, Legun (2015) draws the following conclusion:

> While the biological materials themselves do not dictate social behavior, they create spaces for social change, and through their relationships with people they may gesture towards a direction for movement.
>
> (p. 296)

This could also be true for commons-based organic apple breeding: the market study shows several of those spaces for social change that have the potential to induce a regime change toward a more resilient apple breeding and cultivation system.

Reference

Legun, K. A. (2015): Club Apples: A Biology of Markets Built on the Social Life of Variety. In *Economy and Society* 44 (2), pp. 293–315. DOI: 10.1080/03085147.2015.1013743.

Conclusion

Central Results

A multitude of conceptual and empirical findings have been discussed within this book. The main question was: *How can commons-based organic apple breeding improve the social-ecological resilience of current apple breeding and the apple cultivation system, and how can the potential of this approach be applied to praxis?* This question was divided into four objectives that served as guidelines for the investigation of social-ecological resilience in fruit and apple breeding as well as cultivation systems.

Summary of Main Findings

Fruit breeding and cultivation take place in social-ecological systems, in which different actors use technological and knowledge resources to shape ecosystems in a specific institutional setting (see Chapter 4). The extent and character of these social-ecological interactions define the resilience of fruit breeding and cultivation. In the case of on-farm breeding, both breeding and cultivation overlap at the local level, as they are both situated in the same ecosystem. However, the conceptualization further reveals that fruit breeding and cultivation are also connected on regional, national, and even global levels. They depend on biodiversity as a central component for breeding and cultivating varieties, connected through direct, multi-level interlinkages, or telecoupled flows (e.g., economical or cultural). Furthermore, both breeding and cultivation are only segments of the fruit value chains, which are in turn embedded in larger food systems. Overall, the conceptualization demonstrates how fruit breeding and cultivation are located within a complex web of interactions, and how these interactions define their resilience.

Fruit breeding and cultivation provide and influence a broad set of ecosystem services (see Chapter 5). Both systems offer regulating services (water cycling and maintenance, pest and disease control), supporting services (soil nitrogen availability), and cultural services (aesthetic, ethical, and spiritual values, recreation, identity, learning opportunities). Fruit cultivation additionally provides the provisioning service of fruit production and further influences regulating

DOI: 10.4324/9781003355724-26

services (climate regulation, pollination). Building suitable buffer, adaptive, and transformative capabilities helps maintain and promote these vital ecosystem services. A total of 16 resilience attributes, derived from an extensive evaluation of relevant resilience literature, provide and foster these capabilities to promote and sustain the ecosystem services described above and in Chapter 6.

The resilience attributes describe ideal system and governance characteristics for resilient fruit breeding and cultivation systems. Using an analytical framework, the resilience attributes can be applied to analyze the resilience effects of fruit breeding and/or cultivation approaches in different national or rather socio-economic contexts. Comparing corporate-based, public, and commons-based apple breeding in Germany in this respective revealed a range of specific insights (see Part III). Overall, commons-based apple breeding best meets the resilience demands of the system-to-be-governed. While all approaches successfully integrate diverse social elements and exchange knowledge, corporate-based and public apple breeding lack ecocentric aspects (seeking site-specific solutions and intensifying ecological interactions) thus only resulting in a moderate integration of diverse ecological elements. Moreover, commons-based apple breeding delivers a wholly resilient governance system, and corporate-based apple breeding shows some promising elements. The public approach, however, does not achieve the characteristics of a resilient governance system. The key to resilience seems to be contingent on actors' normative orientation in the commons-based approach, corresponding with norms and practices of ecocentric breeding and farming approaches such as organic agriculture. These approaches show elements of CAS Thinking as they recognize the complexity of social-ecological systems and are potentially compatible with governance institutions like participatory structures or collective ownership. The empirical analysis further supports the conceptual value of commons-based organic apple breeding as an example that integrates breeding and cultivation in a single concept.

Market actors in the German apple sector are confronted with several dilemmas on strategic, economic, and political levels which are partly representative of the general (global) apple market (see Part IV). In these conflicts, commons logic collides to a certain extent with market logics, and market structures and developments present obstacles to the realization and implementation of elements for commons-based organic apple breeding. The main obstacles are the demand for standardized products, low capacities to adapt and change, a disconnect between farmers and breeders, and the dominance of intellectual property rights. The innovative potential of niche distribution channels, more collaboration between market actors, increased knowledge sharing, and new legal mechanisms and practices that could foster forms of collective ownership are all potential measures that would help overcome these obstacles.

Reflection on the Proposed Hypothesis

This book has investigated the following hypothesis: *Commons-based organic apple breeding, including the use of robust cultivars, testing in an organic setting, on-farm, participatory breeding methods, and treatment of resources as common goods, is preferable for the social-ecological resilience and long-term sustainability of apple breeding and cultivation as opposed to modern conventional breeding approaches.*

The hypothesis was evaluated using resilience concepts and insights as well as an in-depth empirical analysis of the German apple sector and existing breeding approaches. From a qualitative perspective, the comparison of the approaches shows significant benefits of a commons-based governance system over other approaches. However, commons-based organic apple breeding still needs to prove its ability to actually breed resilient cultivars and integrate them into the broader market. *Seed Commons* gives orientation on the challenges of market introduction as initiatives in vegetable and grain breeding have already introduced promising resilient cultivars (Sievers-Glotzbach et al. 2020).

Commons-based organic apple breeding is characterized by four elements: (a) collective responsibility for agrobiodiversity, (b) collective, polycentric management, (c) knowledge sharing, and (d) collective ownership. This book identified several challenges to the realization of these elements in the context of current market structures and developments. A successful integration of commons-based organic apple breeding into the apple and broader fruit sector faces challenges along several points of the value chain. Even if the commons-based organic approach successfully breeds and introduces resilient cultivars, their broad acceptance among market actors or if they are "forever niche" (Rohe et al. 2022) is unclear.

The results of this book emphasize the relative stability of the current regime that contributes to cement sustainability problems (see Introduction): Organic fruit farmers still cultivate varieties on a large scale that were not originally bred for their needs and demands. Commons-based organic breeding has the potential to initiate change toward another possible stability domain, in which breeding and cultivation sustainably provide and maintain ecosystem services. Not only market developments and the influences of different actors in the social-ecological systems, but also developments in the coupled broader food system determine the probability of a regime change toward more resilience and sustainability. However, a regime change does not necessarily need to be induced at the systemic level. Bennett et al. (2021) argue that diverse interacting regional pathways are necessary for sustainability transitions, and their aggregation could induce a systemic social-ecological transformation, slowly but gradually tipping the current regime over the threshold.[1] Commons-based organic breeding could play a vital role in the narrative of transformational change by fostering breeding communities with more diverse, regionally adapted fruit varieties for resilient regional cultivation combined with regional marketing channels.

Reflection of the Research Process and Applied Methods

Most forms of transdisciplinary research are confronted with typical challenges and conflicts and depend on specific criteria for success (Hirsch Hadorn et al. 2006; Adler et al. 2018). This book benefitted from its transdisciplinary embedding in that researchers from various disciplines and practitioners legit-imized the problem framing and research aims. The benefits of this iterative process were already described in the methodological framework (see chapter 2) and highlight the practical and scientific relevance of this research. However, the transdisciplinary research process also makes it difficult to transparently separate the argumentative structure of this book from the overall EGON pro-ject context – when were conclusions or the choice of scientific concepts and methods influenced by process dynamics, given the inherent normativity of sus-tainability science? This is particularly visible in the case study design, in that the data collection methods were not consistent throughout the cases. With established quality criteria as the basis for the reflection on the methodological framework (see Chapter 2.3) and a compact documentation of the process in the Appendix, the transdisciplinary setting has been extensively scrutinized and remains credible.

Conceptually, this book is mainly embedded in the discourse on social-ecological resilience, which has several theoretical shortcomings (see Section 1.1). Social science aspects such as human agency, power relations, and social thresholds are only briefly touched upon but not extensively discussed on a theoretical base. However, the disciplinary and conceptual boundaries of this book are discussed and weighed in the theoretical framework (see Chapter 1). Overall, the scientific concept of resilience proves pertinent for identifying the social-ecological problems and challenges of fruit breeding and cultivation, while also offering a platform for many fruitful insights and connecting factors for dis-cussing their sustainability.

Future Research Opportunities

This book serves as a gateway for further research by providing a multitude of different connecting points for the scientific discourses it is embedded in: social-ecological resilience, ecosystem services, and commons.

A solid foundation for this and further research are the general *Principles for Building Resilience* by Biggs et al. (2015). For further refinement of and the reflec-tions on these principles, case studies are a vital element for improving their "con-ceptual clarity" (Schlüter et al. 2015, p. 273). With apple breeding in Germany, this book not only offers such a case study but also provides future research op-portunities with the conceptualization of resilient fruit breeding and cultivation systems and the accompanied development of an analytical framework. Future case studies might investigate apple breeding approaches in other political-legal, economical, and cultural settings, specifically in the worldwide important breeding centers (see Section 3.3) or in the emerging European organic breeding community (Koutis et al. 2020). Further research could contribute more empir-ical insights into the resilience of apple breeding and continuously discuss and refine the proposed framework. Moreover, case studies from other fruit breeding contexts, such as pear or cherry breeding, would add further valuable insights for cross-case comparisons.

With the conceptualization of fruit breeding and cultivation as social-ecological systems as a starting point in this book, the aim was to acknowledge what Schlüter et al. (2015) coined as "an understanding that enables sensitivity to context but is not entirely context-dependent" (p. 275). Further research might add more detail and empirical observations of the interlinkages between fruit breeding and cultivation. Particularly a thorough investigation of telecoupling effects (a relatively young field of research) would be helpful to highlight and understand the influence of overarching dynamics and specific flows across the multilevel landscape. The conceptualization could also be transferred to other objects of investigation such as vegetables or potatoes, and discussed respectively. Additionally, this conceptualization may trigger more research on the interlinkages between breeding and marketing, breeding and consumption, etc. – the better these interlinkages are understood on a conceptual scale, the more comprehensively empirical phenomena can be observed and explained.

Literature on ecosystem services in the context of fruit largely neglects cultural services and is missing the influence of breeding. In this book, a set of ecosystem services is proposed that aims to close these gaps. Further research is needed on the influence of breeding on the maintenance and promotion of ecosystem services in different contexts. In particular, cultural ecosystem services need further investigation to unravel the full social-ecological potential of breeding for shaping resilient agroecosystems.

It has been shown that fruit breeding is a valuable object of investigation for the commons discourse as it shows elements of a "hybrid commons" (Wolter and Sievers-Glotzbach 2019, p. 332) that might benefit further theoretical conceptualizations. However, the case study on *apfel:gut e.V.* (see Chapter 9) illustrates how commons-based breeding captures characteristics of commoning as a concept from the social-anthropological perspective on commons. Similar to *Seed Commons* as a social practice (Sievers-Glotzbach et al. 2020), commons-based apple breeding highlights the active creation of a community and the importance of social relations and processes: commoning as such forms the commons-based structures and processes. These conceptual overlaps create opportunities for further research on how to combine the institutional perspective with the social-anthropological perspective on commons.

One remarkable feature of commons-based organic apple breeding is its coupling of two nominally different social-ecological systems – cultivation and breeding. Commons elements thus appear across system boundaries and scales; an idea that would be valuable to integrate in further commons studies. Similar to global natural resource systems where defined boundaries (one of Ostrom's design principles) play no role, in this case, they melt away on local and regional scales. These blurred lines call for re-thinking analytical frameworks and design principles for commons when they lose their specific categorization and instead move as a hybrid phenomenon across the commons continuum. Further case studies would be beneficial for investigations of hybrid commons.

Overall, the discussion of commons-based organic apple breeding provides a case typical of sustainability science: an alternative resource management model displays promising, beneficial characteristics but is not yet widely implemented enough to significantly contribute to a social-ecological transformation. Whether this typical dilemma of upscaling (Augenstein et al. 2020) prevails, depends particularly on the actions of both practitioners and decision-makers.

Implications for Policy and Praxis

Beyond possibilities for further research, the results of this book can be translated into policy and praxis recommendations that could support the further propagation of commons-based organic apple breeding not only in Germany but worldwide. Overall, seven key messages are derived from the conducted research.

1. Establishing legal instruments suitable for collective ownership: The global legal framework solely focuses on intellectual property rights as a standard instrument for financing breeding. In combination with patents that could probably enter the legal framework for fruits, these instruments potentially threaten the breeder's exemption and thus the freedom of breeders. Amateur varieties as a new instrument for fruits were recently established in Europe as an alternative to promote more diversity in registered marketable fruit varieties. Existing alternatives should be supported and further instruments that allow for collective ownership of varieties without discriminating breeders who renounce variety protection as a financing option should be developed. Global discussions on open-source seed systems (Kloppenburg 2014; Kotschi and Horneburg 2018) and creative commons licenses (Deibel 2013) could provide potential connecting points.
2. More state support for organic breeders: Organic breeding is still a niche worldwide, whereas public breeding, which does not necessarily adopt an ecocentric and commons-based approach, and other breeding approaches get funding sufficient for establishing a long-term financial base. Fruit breeding is a long-term process that needs partially stable funding to lower risks and provide incentives for (new) breeders to get involved. Public funding of commons-based organic breeding initiatives could ensure this stable base and build opportunities for organic breeders. More state support could foster the necessary security for commons characteristics to evolve.
3. Involving the whole value chain: Many market actors along the value chain have a high interest in further collaboration. Literature and praxis initiatives provide many opportunities for innovative business models where a diversity of market actors participate in the financing of organic breeding activities, like the value chain partnership model by FiBL (see Section 13.2). Such cooperative business models might build opportunities for establishing and maintaining commons-based breeding initiatives in addition to or besides public funding. Moreover, cooperative business models could foster collective responsibility, management, and ownership along the whole value chain.
4. Fostering the exchange of knowledge: Knowledge sharing is a central element of commons-based breeding approaches. Especially in the field of apples,

much knowledge is lost or not included in current discussions as, for example, Bannier (2011) showed in his paper by integrating literature from the 1930s. Organic breeders should institutionalize platforms for knowledge exchange and development across multilevel scales. The establishment of EUROrganic Fruits or the breeding approach of NOVAFRUITS (see Chapter 14) is building the first nodes to foster enhanced connectivity in this respective.

5. Changing paradigms of fruit farmers: Fruit farmers play an important role in the value chain directly and indirectly influence breeding. To overcome the common decoupling with breeding, skeptical farmers should be more open to on-farm breeding and new varieties with traits beyond the standard demand. This would create opportunities for adopting alternative farming practices as a necessary prerequisite for sustainable fruit farming. After all, "the existence of a variety creates cultivators and non-cultivators, as well as a politics of economic practices surrounding the variety" (Legun 2015, p. 297). Making the decision to plant a specific cultivar is not only an economic and strategic one but also a normative one. Farmers need to be more open to participation and ecocentric approaches requiring a departure from the current (mainstream) mindset.

6. Investments in diversity: Marketing and food retail influence the earlier segments of the value chain, as breeders and farmers choose to breed and cultivate varieties they can actually sell, consistent with demand traits. Food retail and marketers should invest in offering a more diverse range of varieties, which in turn would open up opportunities for new varieties to enter the market and potentially reach new market segments or customer groups. The apple market seems saturated primarily because major actors do not intend to change, although the cultivation of standard varieties leads to negative effects on sustainability.

7. Educating consumers: Consumers exercise steering power and collectively influence the demand for and subsequent supply of food products. The first information campaigns on the role and importance of organic breeding for sustainable food consumption have recently emerged in the German organic food sector. Public and private actors as well as education institutions should promote further understanding of these aspects so that (future) consumers might think beyond standardized products and embrace the value of diversity in the food supply. Apples offer an exemplary base for experimentation because a range of diversity (though narrow) in varieties is already a standard in food retail, which often is not the case for many vegetables.

Commons-based organic apple breeding presents a potential approach for a more resilient and sustainable breeding and cultivation system. Its integration into current political, economic, and societal settings on national and international levels is challenging but a valuable undertaking to promote and maintain important ecosystem services: "[w]hen economies are built around a living thing, keeping the market alive requires keeping that thing alive through compulsive and attentive care" (Legun 2015, p. 298). Commons-based organic apple breeding has the potential to sustain the social-ecological system by fostering measures that promote resilience and sustainability, thus building capacities to "ride the waves of unexpected change" (Darnhofer 2021).

Note

1 Transitions describe deliberate changes in specific systems or sectors with clearly defined goals. They are limited on a time and geographical scales and are often perceived as navigable. Transformations refer to fundamental systemic changes of socio-economic, political, or cultural practices, that take place over long time periods (Brand et al. 2013).

References

Adler, C.; Hirsch Hadorn, G.; Breu, T.; Wiesmann, U.; Pohl, C. (2018): Conceptualizing the Transfer of Knowledge across Cases in Transdisciplinary Research. In *Sustainability Science* 13 (1), pp. 179–190. DOI: 10.1007/s11625-017-0444-2.

Augenstein, K.; Bachmann, B.; Egermann, M.; Hermelingmeier, V.; Hilger, A.; Jaeger-Erben, M. et al. (2020): From Niche to Mainstream: The Dilemmas of Scaling Up Sustainable Alternatives. In *GAIA – Ecological Perspectives for Science and Society* 29 (3), pp. 143–147. DOI: 10.14512/gaia.29.3.3.

Bannier, H.-J. (2011): Moderne Apfelzüchtung. Genetische Verarmung und Tendenzen zur Inzucht. In *Erwerbs-Obstbau* 52 (3–4), pp. 85–110. DOI: 10.1007/s10341-010-0113-4.

Bennett, E. M.; Biggs, R.; Peterson, G. D.; Gordon, L. J. (2021): Patchwork Earth: Navigating Pathways to Just, Thriving, and Sustainable Futures. In *One Earth* 4 (2), pp. 172–176. DOI: 10.1016/j.oneear.2021.01.004.

Biggs, R.; Schlüter, M.; Schoon, M. L. (Eds.) (2015): Principles for Building Resilience. Sustaining Ecosystem Services in Social-Ecological Systems. Cambridge, UK: Cambridge University Press.

Brand, U.; Brunnengräber, A.; Andresen, S.; Driessen, P.; Haberl, H.; Hausknost, D. et al. (2013): Debating Transformation in Multiple Crises. In UNESCO, ISSC, OECD (Eds.): *World Social Science Report 2013. Changing Global Environments.* Paris: OECD Publishing, pp. 480–484.

Darnhofer, I. (2021): Resilience or How do we enable agricultural systems to ride the waves of unexpected change? In *Agricultural Systems* 187, p. 102997. DOI: 10.1016/j.agsy.2020.102997.

Deibel, E. (2013): Open Variety Rights: Rethinking the Commodification of Plants. In *Journal of Agrarian Change* 13 (2), pp. 282–309. DOI: 10.1111/joac.12004.

Hirsch Hadorn, G.; Bradley, D.; Pohl, C.; Rist, S.; Wiesmann, U. (2006): Implications of Transdisciplinarity for Sustainability Research. In *Ecological Economics* 60 (1), pp. 119–128. DOI: 10.1016/j.ecolecon.2005.12.002.

Kloppenburg, J. (2014): Re-Purposing the Master's Tools: The Open Source Seed Initiative and the Struggle for Seed Sovereignty. In *The Journal of Peasant Studies* 41 (6), pp. 1225–1246. DOI: 10.1080/03066150.2013.875897.

Kotschi, J.; Horneburg, B. (2018): The Open Source Seed Licence: A Novel Approach to Safeguarding Access to Plant Germplasm. In *PLoS Biology* 16 (10), e3000023. DOI: 10.1371/journal.pbio.3000023.

Koutis, K.; Warlop, F.; Bolliger, N.; Steinemann, B.; Rodriguez Burruezo, A.; Mendes Moreira, P.; Messmer, M. (2020): Perspectives on European Organic Apple Breeding and Propagation under the Frame of LIVESEED Project. In FOEKO (Ed.): *Ecofruit. 19th International Conference on Organic Fruit-Growing: Proceedings.* Ecofruit. Hohenheim, 17. 19.02.2020. Fördergemeinschaft Ökologischer Obstbau e.V. (FOEKO), pp. 104–107. Available online at https://www.ecofruit.net/wp-content/uploads/2020/04/7_Koutis_104-107.pdf, checked on 8/16/2022.

Legun, K. A. (2015): Club Apples: A Biology of Markets built on the Social Life of Variety. In *Economy and Society* 44 (2), pp. 293–315. DOI: 10.1080/03085147.2015.1013743.

Rohe, S.; Oltmer, M.; Wolter, H.; Gmeiner, N.; Tschersich, J. (2022): Forever Niche: Why Do Organically Bred Vegetable Varieties Not Diffuse? In *Environmental Innovation and Societal Transitions* 45, pp. 83–100. DOI: 10.1016/j.eist.2022.09.004.

Schlüter, M.; Biggs, R.; Schoon, M. L.; Robards, M. D.; Anderies, J. M. (2015): Reflections on Building Resilience – Interactions among Principles and Implications for Governance. In Reinette Biggs, Maja Schlüter, Michael L. Schoon (Eds.): *Principles for Building Resilience. Sustaining Ecosystem Services in Social-Ecological Systems.* Cambridge, UK: Cambridge University Press, pp. 251–282.

Sievers-Glotzbach, S.; Tschersich, J.; Gmeiner, N.; Kliem, L.; Ficiciyan, A. (2020): Diverse Seeds – Shared Practices: Conceptualizing Seed Commons. In *International Journal of the Commons* 14 (1), pp. 418–438. DOI: 10.5334/ijc.1043.

Wolter, H.; Sievers-Glotzbach, S. (2019): Bridging Traditional and New Commons: The Case of Fruit Breeding. In *International Journal of the Commons* 13 (1), p. 303. DOI: 10.18352/ijc.869.

Appendix A
Glossary of Central Terms

Agrobiodiversity	Agrobiodiversity is understood as diversity across ecological and spatial scales (Kremen and Miles 2012). Ecological scales include genetic diversity, varietal diversity, multiple intercropped species, and non-crop plantings that enable communities of plants and animals. Spatial scales include the implementation of agrobiodiversity within, across and around field level and at the landscape-to-regional level.
Commons	Commons describe collective action situations, in which a distinct community of actors collectively own, share, manage, and/or develop a (potentially) diverse range of goods and resources in a particular institutional setting. They encompass material (goods and resources), social (user community), and regulative (rules and norms) dimensions (Helfrich and Stein 2011).
Commons-based organic apple breeding	Commons-based organic apple breeding aims to consequently implement the IFOAM principles by adopting collective responsibility for agrobiodiversity, sharing knowledge between a diversity of actors, breeding resilient cultivars with collective and polycentric management practices, and enabling collective ownership of important resources and varieties. The breeding community breeds cultivars for fruit farmers that adopt the IFOAM principles in their farming practices.
Ecosystem Services	Ecosystem services are the direct and indirect contributions of ecosystems to human well-being (TEEB 2012). They are categorized into provisioning, regulating, supporting, and cultural services (MEA 2005; TEEB 2012).
Food Systems	Food systems encompass all activities along the value chain (from food production to processing and packaging, distribution and retail, and consumption) on different scales (social, economic, political, institutional, and environmental) (Tendall et al. 2015).
Institutions	Institutions are defined as sets of both formal and/or informal norms, rules, and shared strategies (Crawford and Ostrom 1995).
Participatory Plant Breeding	In participatory plant breeding, farmers actively participate in plant breeding by conducting on-farm breeding. They are involved in every step of the breeding process – from the formulation of breeding goals throughout the selection steps to the choice of cultivars for legal assessment (Ceccarelli 2012).

Resilience — This book adopts the understanding of resilience as social-ecological resilience. It is defined as "the capacity of a social-ecological system to sustain human well-being and a desired set of ecosystem services in the face of disturbance and change, both by buffering shocks and by adapting or transforming in response to change" (Biggs et al. 2015, p. 9). Table 1.1 in Section 1.1 depicts a glossary of specific resilience terms.

Social-ecological system — Social-ecological systems are nested multilevel systems that provide essential ecosystem services for human well-being (Berkes and Folke 1998).
Social and ecological systems are linked and influence each other through feedback mechanisms that culminate in complex, dynamic, and reciprocal social-ecological interactions, which create and influence ecosystem services.

Telecoupling — Telecoupling refers to social or environmental processes between spatially, socially, or institutionally distant social-ecological systems (Hull and Liu 2018).

Transdisciplinary research — Transdisciplinary research aims at generating and integrating knowledge through (a) the collaboration of different scientific disciplines, and (b) science-practice interactions where participating actors collaborate on equal levels (Brandt et al. 2013; Lang et al. 2012). It is a solution- and action-oriented research approach that aims to strengthen the science-society interface, specifically targeted at solving complex real-world problems (ibid.).

(Fruit) Variety — A variety is a cultivar that is either officially registered in the national list, commonly known (traditional and heirloom cultivars), or holds variety protection.

References

Berkes, Fikret; Folke, Carl (Eds.) (1998): *Linking Social and Ecological Systems. Management Practices and Social Mechanisms for Building Resilience.* Cambridge, UK: Cambridge University Press.

Biggs, Reinette; Schlüter, Maja; Schoon, Michael L. (Eds.) (2015): *Principles for Building Resilience. Sustaining Ecosystem Services in Social-Ecological Systems.* Cambridge, UK: Cambridge University Press.

Brandt, P.; Ernst, A.; Gralla, F.; Luederitz, C.; Lang, D. J.; Newig, J. et al. (2013): A Review of Transdisciplinary Research in Sustainability Science. In *Ecological Economics* 92, pp. 1–15. DOI: 10.1016/j.ecolecon.2013.04.008.

Ceccarelli, S. (2012): Plant breeding with farmers - a technical manual. Aleppo: International Center for Agricultural Research in the Dry Areas (ICARDA).

Crawford, S. E. S.; Ostrom, E. (1995): A Grammar of Institutions. In *American Political Science Review* 89 (03), pp. 582–600. DOI: 10.2307/2082975.

Helfrich, S.; Stein, F. (2011): Was sind Gemeingüter. In Bundeszentrale für politische Bildung (Ed.): *Aus Politik und Zeitgeschichte.* Gemeingüter, vol. 61 (61), pp. 9–15.

Hull, V.; Liu, J. (2018): Telecoupling. A New Frontier for Global Sustainability. In *Economy and Society* 23 (4). DOI: 10.5751/ES-10494-230441.

Kremen, C.; Miles, A. (2012): Ecosystem Services in Biologically Diversified versus Conventional Farming Systems. Benefits, Externalities, and Trade-Offs. In *Ecology and Society* 17 (4). DOI: 10.5751/ES-05035-170440.

Lang, D. J.; Wiek, A.; Bergmann, M.; Stauffacher, M.; Martens, P.; Moll, P. et al. (2012): Transdisciplinary Research in Sustainability Science: Practice, Principles, and Challenges. In *Sustainability Science* 7 (S1), pp. 25–43. DOI: 10.1007/s11625-011-0149-x.

MEA (2005): *Ecosystems and Human Well-being. Synthesis; A Report of the Millennium Ecosystem Assessment.* Washington, DC: Island Press.

TEEB (Ed.) (2012): *The Economics of Ecosystems and Biodiversity. Ecological and Economic Foundations. The Economics of Ecosystems and Biodiversity (TEEB).* London: Routledge.

Tendall, D. M.; Joerin, J.; Kopainsky, B.; Edwards, P.; Shreck, A.; Le, Q. B. et al. (2015): Food System Resilience. Defining the Concept. In *Global Food Security* 6, pp. 17–23. DOI: 10.1016/j.gfs.2015.08.001.

Appendix B

Case Studies: Data Collection and Analysis

Table Appendix B.1 Data overview for case studies on apple breeding (own table).

Case studies: data overview	Remarks
Züchtungsinitiative Niederelbe GmbH & Co. KG (See Chapter 7)	
Scientific and gray literature: • Dierend, W. (2017): Beurteilung von Nachkommen alter Apfelsorten im Rahmen der Züchtungsarbeit der Züchtungsinitative Niederelbe. Symposium Obstsortenvielfalt. Berlin, 11/14/2017. • Dierend, W.; Schacht, H. (2009): Züchtungsinitiative Niederelbe. In *Erwerbs-Obstbau* 51 (2), pp. 67–71. DOI: 10.1007/s10341-009-0083–6. • ZIN (2020): Website of ZIN - Züchtingsinitiative Niederelbe. Available online at http://www.zin-info.de/, checked on 11/3/2020 Empirical data: interview with a breeder from ZIN (in-text-references are marked as 'Interview ZIN').	The interview took place on September 02, 2019, in person.
Julius Kühn-Institute – Institute for Breeding Research on Fruit Crops (See Chapter 8)	
Scientific and gray literature: • JKI (2019): Institut für Züchtungforschung an Obst. Julius Kühn-Institut. Available online at https://www.julius-kuehn.de/media/Veroeffentlichungen/InstitutsbroschFlyer/JKI_Institutsbroschuere_ZO.pdf, checked on 7/16/2019. • JKI (2020): Institut für Züchtungsforschung an Obst. Edited by Julius Kühn-Institut. Available online at https://www.julius-kuehn.de/zo/, checked on 11/17/2020. • Peil, A. (2017): Nutzung alter Sorten in der Pillnitzer Obstzüchtung. Symposium Obstsortenvielfalt. Berlin, 2017. Empirical data: an interview with a breeder from ZO (in-text-references are marked as 'Interview ZO').	The interview took place on October 08, 2019, via telephone.
apfel:gut e.V. (See Chapter 9)	
Scientific and gray literature*: • apfel:gut (2020): Was wir wollen. Unsere Ziele. Available online at http://www.apfel-gut.org/was_wir_wollen.cfm, updated on 2020, checked on 12/31/2020.	The interview took place on May 15, 2017, via telephone.

- Wolter, H.; Sievers-Glotzbach, S. (2019): Bridging traditional and new commons: The case of fruit breeding. In *International Journal of the Commons* 13 (1), p. 303. DOI: 10.18352/ijc.869
- Ristel, M.; Bornemann, M.; Sattler, I. (2018): Apfel:gut – preliminary results. In FOEKO (Ed.): Ecofruit. Proceedings of the 18th international congress on organic fruit growing. Ecofruit. Hohenheim, Germany. Fördergemeinschaft Ökologischer Obstbau e.V. (FOEKO), pp. 96–99.
- Ristel, M.–Sattler, I. (2014): Apfel: gut–Participatory organic fruit breeding. In FOEKO (Ed.): Ecofruit. 16th International Conference on Organic Fruit-Growing: Proceedings. Ecofruit. Hohenheim. Fördergemeinschaft Ökologischer Obstbau e.V. (FOEKO), pp. 158–161.
- Ristel, M.; Sattler, I.; Bannier, H.-J. (2016): Apfel: gut– More vitality, genetic diversity and less susceptibility as an organic fruit breeding strategy. In FOEKO (Ed.): Ecofruit. 17th International Conference on Organic Fruit-Growing: Proceedings. 2016. Ecofruit. Hohenheim. Fördergemeinschaft Ökologischer Obstbau e.V. (FOEKO), pp. 136–139.

Empirical data: an interview with chief breeders and project leaders Inde Sattler and Matthias Ristel; focus group with apfel:gut members (in-text-references are marked as 'Focus group 2017')

* A document analysis additionally served as a source of data. It is further explained in Wolter and Sievers-Glotzbach (2019).

The focus group took place on June 19, 2017, in Kassel and took 105min. Six members of *apfel:gut e.V.* participated.

Table Appendix B.1 provides an overview of relevant data sources for the case studies on apple breeding in Germany (Part III). The structure and content of the used interview guidelines for data collection differed across the individual interviews. For the interview with apfel:gut's project leaders, a different guideline was used than for the interviews with the representatives of ZIN and JKI.

This difference is due to two reasons. First, as already explained in the methodological design of this book (See Chapter 2), the investigation of commons-based breeding approaches needed a different methodological approach to capture its social dynamics and specifics. The content and structure of the interview guideline thus had to be shaped in line with those specifics, making several questions and structural elements irrelevant to the investigation of the other cases. Second, the project dynamics of EGON (See Preface) produced a time lag between the conduction of the interviews. One of the first goals of the project had been to collect empirical data of a commons-based fruit breeding initiative to conceptualize the approach in an iterative process. This process included a review of the literature on commons combined with an empirical investigation. Both tasks were carried out in close cooperation with the project RightSeeds that aimed to conceptualize *Seed Commons*. Thus, the empirical investigation of *apfel:gut* took place with the start of the project EGON in January 2017. Comparing the commons-based approach with other breeding approaches was a later goal in the project's transdisciplinary process. As a result, the empirical investigation of the other cases took place in late 2019 with different guidelines adjusted to the cases and based on new insights, which had been collected in the time between.

In the empirical analyses of the cases, different data material was of relevance and different analytical steps have been taken. All case study analyses were generally oriented on the general process model of a qualitative content analysis by Mayring (2015).

For the case study of *apfel:gut*, a twofold step-wise approach was chosen. First, relevant scientific and gray literature was evaluated and the interview with the project leaders was carried out to get basic insights about the organizational structure, norms, and rules of the initiative. This enabled to prepare the content and methodological structure of the focus group. The interview data was analyzed along the categories of the interview guideline and coded by the author. Second, the first focus group was carried out to dive deeper into the same thematic aspects as in the interview and collect additional insights on social dynamics. The focus group was also analyzed along the categories of the interview guideline and coded by the author. Afterward, both analyses were combined and again analyzed with a thematic category-based scheme as depicted in Figure Appendix B.1. Categories of both interview guidelines served as main categories and the sub-categories have been developed inductively on the base of the individual analyses.

The conduction of the other case studies followed a different analytical procedure due to the above-described reasons. First, relevant scientific and gray literature was evaluated to identify knowledge gaps in the cases. Based on that and the guideline of the *apfel:gut* interview, an interview guideline was prepared. The interview data was analyzed with a thematic category-based scheme as depicted

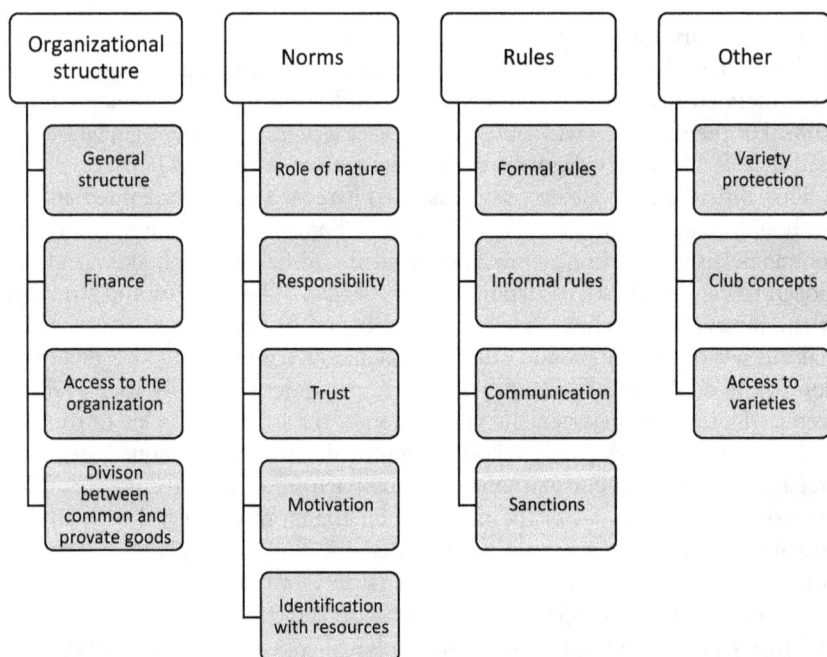

Figure Appendix B.1 Coding scheme for the case study apfel:gut e.V. (own figure).

in Figure Appendix B.2 on the base of transcriptions that have been manually coded by the author. Categories of the interview guideline served as main and sub-categories. This scheme differed from the coding scheme for *apfel:gut* because some sub-categories played no role in those cases (e.g., division between common and private goods; sanctions) and a particular emphasis was put on the breeding process as there have been large knowledge gaps.

After analyzing both cases, insights from the *apfel:gut* case study were also included in the new coding scheme to allow for an overview and comparison by carrying out a summarizing content analysis.

Critical Reflection of Used Methods and Analysis

In general, the data collection and analysis provide semantic, sample, and constructive validity. All category-based schemes have been validated by another researcher, the data material sufficiently reflects the cases, and the methods proved

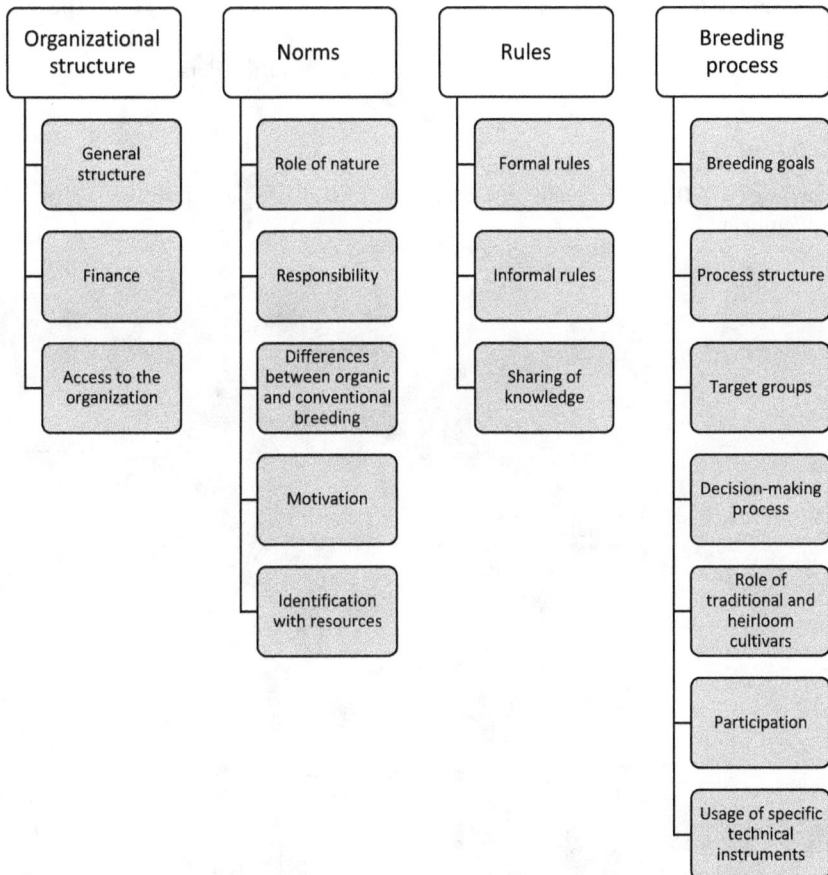

Figure Appendix B.2 Coding scheme for the case studies ZIN and ZO (own figure).

fitting for answering the research questions. Sample validity can be criticized because the sample of conducted interviews is small and limited. An imbalance exists because the database for the case study on *apfel:gut* is much larger than for the other two cases. However, as both interviews in the cases of ZIN and ZO have been carried out with major decision-makers on breeding issues, this limitation can be partly neutralized. Moreover, the time lag in the data collection (See above) has to be recognized. This time lag influences constructive validity because the social, economic, and ecological context probably changed in the time between, hampering comparability of the data material. However, as the apple sector is generally slow to change, this time lag should not play a significant role for the validity of the data material. Lastly, correlative validity could be criticized because the telephone interview for the case study on JKI is significantly shorter than the other interviews. It is possible that a personal interview would have generated more detailed insights.

Reliability understood as stability of the data analysis was achieved with a double-check of the manual or digital codings by the author. Reproducibility and thus intercoder reliability cannot be calculated as only the author coded the data material. However, as explained above, the category systems have been discussed with another researcher, strengthening the accuracy of the data analyses.

Reference

Mayring, P. (2015): *Qualitative Inhaltsanalyse. Grundlagen und Techniken.* 12th ed. Weinheim: Beltz

Appendix C
Delphi Study
Data Collection and Analysis

The Delphi study included two waves of data collection and analysis whereby each wave had a specific research purpose.

The first wave of data collection took place from November 19 to January 20, 2019. Before the start of the first wave, the online questionnaire was pre-tested with several colleagues and adjusted accordingly. The targeted sample size was 25–35 participants. Forty-seven potential experts were identified and invited to participate in the survey. Selection criteria for the identification of experts were either the authorship in the relevant literature and/or a recommendation by the research partners and consortium practitioners of EGON. The invitation was primarily given to German experts in organic fruit cultivation, breeding, or marketing. Because of their exceptional expertise, one expert from Austria, one from Switzerland, and two from Northern Italy were additionally invited. First, every potential participant got an invitation via email with information on the goals, purpose, and benefits of the Delphi study, politely asking for their participation. Second, all experts who did not respond to the mailing were phoned repeatedly or, when this was not successful, again invited via email. In the end, 37 experts gave their commitment to participate in the study, four experts refused, and six experts were not reached by repeated mailings or phone calls. After the survey period, 29 experts actually filled out the questionnaire completely. The list of participants is depicted in Table Appendix C.1. Six participants wanted to stay anonymous.

After the survey period, the qualitative data was summarized and clustered by the author according to the categories of the questionnaire. In the whole analysis process, all parts were frequently discussed with researchers from the project EGON. Qualitative data on promoting and inhibiting factors for organic apple cultivation and breeding served as the database for parts of the online questionnaire in the second wave.

The second wave of data collection took place from March 19 to April 30, 2019. Again, the questionnaire for this wave was pre-tested with several colleagues and adjusted accordingly. All 29 experts that participated in the first wave were invited to participate in the second wave. Because the average response rate in the second wave of Delphi studies is typically 70% (Häder 2014), the targeted sample size was 20 participants. After the survey period, 22 experts filled out the

Table Appendix C.1 Participants of the Delphi study

No.	Last name	First name	Expert category (self-assigned)
1	Adrion	Georg	Fruit cultivation; Associations
2	Altherr	Klaus	Fruit cultivation; Research on organic fruit cultivation; Consulting; Associations
3	Arp	Kirsten	Associations
4	Bannier	Hans-Joachim	Research on organic fruit breeding
5	Bentele	Johannes	Fruit cultivation; Marketing of fruits
6	Danzeisen	Bernhard	Other
7	Frölich	Guido	Other
8	Guerra	Walter	Research on organic fruit breeding
9	Gutberlett-Geisinger	Birgit	Marketing of fruits; Consulting
10	Heyne	Peter	Research on organic fruit cultivation; Consulting; Associations
11	Höfflin	Christoph	Fruit Cultivation
12	Johann	Markus	Associations
13	Kelderer	Markus	Fruit cultivation; Research on organic fruit cultivation
14	Krämer	Hubert	Fruit cultivation; Fruit breeding; Marketing of fruits
15	Mager	Andreas	Fruit cultivation; Marketing of fruits; Associations
16	Mayr	Ulrich	Fruit cultivation; Fruit breeding; Research on organic fruit breeding; Research on organic fruit cultivation
17	Rolker	Peter	Fruit cultivation; Marketing of fruits; Associations
18	Rueß	Franz	Fruit cultivation; Fruit breeding; Research on organic fruit breeding; Research on organic fruit cultivation
19	Schockemöhle	Ursula	Other
20	Schraff	Markus	Marketing of fruits; Consulting
21	Schüller	Elisabeth	Research on organic fruit cultivation; Other
22	Weber	Michael	Other
23	Wichmann	Thorsten	Fruit cultivation; Research on organic fruit cultivation; Research on marketing of fruits; Associations
24	Anonymous		Other
25	Anonymous		No categorization
26	Anonymous		Fruit cultivation; Research on organic fruit breeding
27	Anonymous		No categorization
28	Anonymous		Fruit cultivation
29	Anonymous		No categorization

Every expert who participated in both the first and the second waves is marked in gray (own table).

questionnaire. Six participants did not fill out the questionnaire completely. Their answers and ratings were, nevertheless, included in the evaluation of the results.

All qualitative data was summarized and clustered by the author according to the respective categories of the questionnaire. Quantitative assessments of promoting and inhibiting factors were descriptively analyzed by calculating the mean values of the ratings.

Critical Reflection of the Delphi Study

In general, the results of Delphi can be classified as valid because of two main reasons. First, sample validity is given as a diverse range of experts with different backgrounds and expertise have participated in the study. Second, all results have been extensively discussed with relevant comparable literature (See Part IV) to verify and reflect the results. The reliability of Delphi studies is generally difficult to assess as qualitative prognoses are usually volatile and uncertain (Häder 2014). All prognostic content was thus discussed sensitively. However, results that concern system knowledge on market structures can be categorized as reliable because they base on statements of a diverse range of experts.

Some aspects have to be reflected regarding the methodological design of the Delphi study. First, it probably would have been beneficial to conduct additional qualitative interviews with certain participants after the first and second wave for a better understanding of specific arguments and to increase the stability and accuracy of the analysis. Due to time and project constraints, it was not possible to conduct such interviews. Second, both questionnaires were not filled out completely by all participants. Especially in the quantitative assessment as part of the second wave, this created different sub-sample sizes for the assessed promoting and inhibiting factors. Thus, the quantitative assessments cannot be categorized as completely valid, which was considered in the interpretation of the results (See Chapter 12). Third, validity can be criticized as most parts of the online questionnaires were not designed to specifically examine the characteristics of commons-based organic apple breeding (See Chapter 14). Rather, the thematic framework was organic cultivation and breeding. It would have been conceptually difficult to directly include commons and their attributes in the study because it is still a rather unknown concept among practitioners and researchers in the fruit context. As a result, the challenges for commons-based organic apple breeding were indirectly interpreted from the collected data.

Reference

Häder, M. (2014): *Delphi-Befragungen. Ein Arbeitsbuch.* 3. Aufl. Wiesbaden: Springer VS (Lehrbuch).

Index

Note: **Bold** page numbers refer to tables; *italic* page numbers refer to figures and page numbers followed by "n" denote endnotes.

Achathaler, L. 5
adaptive capability **24**
Adler, C. 46, 255
aesthetic values 104
agriculture: conventional 95, **95**; diversified **95**, 96; industrial **95**; learning and 126–127; modeling ecosystem services to and from 92–95, *93*; organic **95**; supporting services in 93–94; trade-offs in 94–95
agrobiodiversity 178, 246–247
agroecology **95**, 96
Aligica, P. 130–131
Alspach, P. A. 3
Altieri, M. A. 2, 96
Anderies, J. M. 34
apfel:gut e.V. 49, 104, 156, 178–180
apple breeding: commons-based 6–8, *192*; comparing approaches in 191–197, *192*; conventional modern 3–4; corporate-based 157–166, *165*, *192*, *193*; cultivation and, relationship with 207–208; in Germany 53–65, *55*; in global context 62–65; organic 194–195, 203–220, 223–237; processes in 58–59; public 167–175, *174*, *192*; sustainability challenges in 2–4
apple consumption 62
apple cultivation: concepts and status quo in 60–61; in global context 62–65; organic 223–237; *see also* apple breeding
apple sector, value chain in *57*, 57–62
apple varieties: distribution of 59–60
Armitage, D. 126
Arnstein, S. R. 128
Augenstein, K. 258
Australia 64

Badenes, M. L. 9n3, 57–58
Baggio, J. A. 34
Bannier, H.-J. 3–4, 184, 259
Barde, M. 66n17, 195
Barnes, R. A. 37
Behringer, J. 49
Beichler, S. A. 25
Bennett, E. 9n1, 73, 92, 115–117, 121, 133, 137, 255
Berkes, F. 1–2, 23, 25, 36, 76–77, 78, 82
Bertschinger, L. 107n8
Bhamra, R. 18
Biasi, R. 104
Bieling, C. 6, 104
Biggs, R. 5, 7, 17–19, 24–25, 47, 73, 75, 82, 112–114, 118–119, 123, 256
Binder, C. R. 5, 76–77
biodiversity 93–94
Bohensky, E. L. 122–124
Böhm, C. 137
Bonny, S. 2
Braat, L. C. 27
Brand, F. 23–24, 38n3
Brandt, P. 46
Bravin, E. 204, 215–217, 236
breadth of inclusion 130
"breeding cent" 238–239
breeding goals 58–59
Brown, S. 2–4, 58–59, 62, 66n14, 103
Brummer, E. 101, 103
Bryman, A. 48–50
Buck, S. J. 39n15
buffer capability **24**
Burch, K. 219
Burr, V. 39n11
Bus, V. G. M. 3

business models 9, 140, 157, 204, 216,
    237–243, 249, 258
Byrne, D. H. 4, 9n3, 57–58

Cabell, J. F. 5, 73, 112, 115–116, 119, 122,
    132, 136
Cahenzli, F. 107n12
Campbell, A. J. 101
Carpenter, S. R. 23, 112
CAS *see* complex adaptive systems (CAS)
    thinking
Cassman, K. 9n1
Ceccarelli, S. 4, 84
Chable, V. 6
Chapin, F. 17, 119
Cheema, D. 58
China 62–64
Choi, Y. E. 115
Clark, J. R. 56, 64, 193
Cleveland, D. A. 2
climate regulation 103
Clothier, B. E. 97, 104
club concepts 211–212, 218
Colding, J. 37, 115
collective ownership 249–250
collective polycentric management
    247–248
collective responsibility 178, 246–247
Collins, K. 128
commons: defined 4; global 35;
    institutional perspective on 34–36;
    knowledge 35–36; resilience research
    and 36–37; in theoretical framework
    32–38
commons-based apple breeding 6–8,
    191–192, *192*, 199; case study 178–180;
    challenges for 245–250; defined
    177–178; diversity in 181; ecocentric
    breeding in 182; implementation of
    203–220; mental models in 183–184;
    organic 194–195, 203–220; redundancy
    in 182; social elements in 181; social
    learning and 185
complex adaptive systems (CAS) thinking
    122–124, 126, 137–138, 162–163, 165,
    191–192
connectivity 120–122, 136–137
consumption, of apples 62
controlling variables 20, 20–21, **24**
corporate-based apple breeding *192*,
    193; case study 157–159; defined
    157; diversity in 159–161; ecological
    elements in 159–160; polycentric
    governance and 164; redundancy in
    160–161; resilience of 159–164; social
    elements in 160

Costanza, R. 27
Cox, M. 33–34, 39n14
Cross, J. 100
crossbreeding 58
cultural ecosystem services 104–105
Cumming, G. 5, 77, 89n2
Cundill, G. 125–126

Dakos, V. 120–121
Dale, V. H. 4, 6, 73, 93–94
Damos, P. 60, 195
Danley, B. 6, 27
Dardonville 145n5
Darnhofer, I. 5, 20, 22, 106n1, 115, 117,
    122–124, 260
data analysis **48**, 48–49
data collection **48**, 48–49
Davidson, J. L. 24, 38n4
Davoudi, S. 22
de Groot, R. 27
Deibel, E. 258
Delate, K. 63, 215
Demestihas, C. 4, 6, 73, 97–98, *98*,
    100–101, 107n9, 135
Dessalegn, M. 37
Dhaliwal, M. S. 58
Dhingra, A. 64
Díaz, S. 27
Diedeurwardere, T. 4
Dierend, W. 159
Dirksmeyer, W. 57, 62
distribution: of apple varieties 59–60;
    marketing channels and 61–62
diversified farming systems 96
diversity 115–117, 134–135, 159–161,
    169–171, 181
Duit, A. 36
Dürrenberger, G. 49
Duru, M. 95, 97, 106n1

Eakin, H. 1, 5, 80, 82, 87, 115, 117, 119
ecocentric breeding 171, 182
ecological resilience 22–23
ecological-economic perspective *18*
economization 2–4, 62, 137
ecosystem services: approaches for study
    of 27–30, *28–29*; categorization of
    30; critical remarks on concept of
    30–32, *31*; cultural 104–105; farming
    approaches in, impacts of **95**, 95–97; in
    fruit breeding and cultivation 92–105,
    93, **95**, *98*, *106*; historical roots and
    conceptions of 26–27; modeling of,
    to and from agriculture 92–95, *93*;
    synergies between 101; trade-offs in 101
Edwards, V. M. 33

empirical context 53–65, *55*
engineering resilience 22
English, R. C. 36
Ericksen, P. J. 78, 132
Esopi, G. 37
Euler, J. 39n10
evolutionary resilience 23
experimentation 124–127, *125*, 138–139

Farina, A. 104
fast variables 118–119, 135
Faysse, N. 33, 38
feedbacks 117–119
Feeny, D. 34
Ficiciyan, A. 102–104, 107n13
financing 238–243
Fischer, J. 75
Fisher, B. 106n3
Flachowsky, H. 3, 9n3, 54, 56–59, 61, 87,
    167, 169, 175n6, 217, 219
Flick, U. 49
Foley, J. 96
Folke, C. 1–2, 5, 17–18, 22–23, 73, 76–77,
    78, 82, 112–113
food- and agroecosystems 1
food systems 78–80; learning and 126–127;
    multilevel 87–88, *88*
France 63
Francis, C. 2
Fraser, E. D. G. 121
Frewer, L. J. 128
Friedmann, H. 89n3
Frischmann, B. 36
fruit breeding and cultivation: broadening
    of participation in 139–140; complex
    adaptive systems thinking in 137–138;
    connectivity and 136–137; ecosystem
    services in 92–105, *93*, **95**, *98*, *106*;
    experimentation in 138–139; plant
    breeding *vs*. 58; polycentric governance
    systems in 140–141; resilience in
    112–144, *116*, *125*, *130*, *133*, *141*,
    143–144; as social-ecological system
    75, *75–89, *78–79*, *81*, *83*, *85–86*, *88*
fruit production 98–99
Fujita, N. 5
funding alliance 239–240

Galappaththi, E. K. 37
Galaz, V. 130
Gali, A. 6
Gallardo, R. K. 204
Garming, H. 57, 60–62, 66n19, 203–204
genetics 214–215
Germany: empirical context in 53–65, *55*;
    methodological framework in 53–65,
55, *57*; political-legal framework in
    53–57, *55*; regulatory framework in
    53–56, *55*
Gläser, J. 49
Glaser, M. 89n3
global commons 35
global context 62–65
Goddard, J. J. 115
Gómez-Baggethun, E. 26
Gonçalves, A. 115, 117, 119, 121, 124,
    127, 129, 132, 145n6
Gordon, L. J. 79, 89n5
governance, polycentric *130*, 130–132,
    140–141
Grafton, R. 39n15
Granatstein, D. 4, 60, 100
Green, J. 119, 129
Grove, G. G. 100
Gunderson, L. H. 20

Häder, M. 47, 49
Haines-Young, R. 27
Halewood, M. 2, 4
Hall, K. 37
Halliday, A. 89n3
Hanke, M.-V. 3, 9n3, 53, 56–59, 61, 87,
    167, 219
Hardin, G. 34
Haye, T. 100
Helfrich, S. 32
Hellin, J. 37
Herrero-Jáuregui, C. 75–76
Herrfahrdt-Pähle, E. 118
Hess, C. 6, 33, 36
heterozygosity 58
Hirsch Hadorn, G. 47, 255
Hochmann, L. 66n17
Hodbod, J. 1, 5, 115, 117, 119
Holling, C. 18–19, 22, 113, 120, 126
Horneburg, B. 258
Howard, N. P. 59, 185
Hull, V. 79

inbreeding 3, 62, 181, 184
Inderbitzin, J. 218
Institut für Züchtungsforschung an Obst
    167–169
institutional economics perspective
    37–38
institutions 162, 172, 184
Isik, F. 58–59
Ison, R. 128
Italy 63

Janick, J. 3
Jax, K. 23, 38n3

Jerneck, A. 17
Jondle, R. 56

Kates, R. W. 17
Kelderer, M. 63
Kellerhals, M. 103
Kienzle, J. 63
Kleinert, J. 217
Kliem, L. 177, 189n1
Kloppenburg, J. 2, 258
Kneen, B. 2
knowledge: commons 35–36; exchange 171–172; sharing 248–249; system 46; target 46
Koda, R. 5
Kotschi, J. 6, 241
Kotschy, K. 116, *116*, 117
Koutis, K. 194, 256
Krausmann, F. 80, 89n3
Kremen, C. 73, 92–96, 106n6
Kumar, S. 3

Lade, S. J. 25, 196
Lammerts van Buren, E. 4, 155–157, 167, 194–195
Lang, D. J. 46
Langthaler, E. 80, 89n4
Laudel, G. 49
Laurens, F. 66n14
learning: agriculture and 126–127; food systems and 126–127; loop 125, *125*; resilience and 124–127, *125*; social 125–126, 163, 172–173, 185
Legun, K. A. 9n4, 60, 203, 216–219, 220n10–220n11, 252, 259
Leitch, A. M. 128–129
Lesur-Dumoulin, C. 106n7
Leumann, M. 204, 216
Levin, S. 17
Liu, J. 79
loop learning 125, *125*
low-residue apple production 215

Maloney, K. E. 66n14
market demands 208–211, 216–218
market trends 212–215, 218–220
marketing channels 61–62
marketing concepts 214
Maurer, O. 217
McGinnis, M. D. 35
McMichael, P. 89n4
mental models 162, 172, 183–184
Messmer, M. 55, 241–242
methodological design 50–51
Meyers, J. A. 23
Mgbeoji, I. 2
Migliorini, P. 95–96

Miles, A. 73, 92, 94–95, 106n6
Milestead, R. 123
Misoch, S 49
Misselhorn, A. 121
monitoring programs, long-term 162–163
Montanaro, G. 97
Mosse, D. 33
Mouron, P. 203
Mustapha, A. B. 33, 38

Nerlich, K. 137
Ness, B. 17
Netherlands 63, 65
New Zealand 64
Nicholas, K. A. 104
Nicholls, C. I. 2
Niggli, U. 4
Noiton, D. A. 3
norms 206–207, **207**, 215–216
Nuijten, E. 56

O'Rourke, D. 220
Oberlack, C. 130
Oberthür, S. 53
Oelofse, M. 5, 73, 112, 115–116, 119, 122, 127, 129, 132, 136
Olsson, L. 17, 24–25, 114
organic apple breeding 194–195, 203–220, 223–237
organic apple production 61
organic farming **95**
organic sector, developments 215
Ostrom, E. 4–7, 32–36, 39n12, 77, 145n9

Pacilly, F. C. A. 5
Pahl-Wostl, C. 48, 125
Palomo-Campesino, S. 92, 97
Parisi 3
participation, broadening of 127–129, 139–140, 163–164, 185–186
participatory plant breeding (PPB) 4, 193
Pascual, U. 27
patents 214–215
Pautasso, M. 2
Peck, G. 4, 60, 100, 107n8
Peisley, R. K. 101
Penker, M. 37
Penny, G. 115
Penvern, S. 100
pest and disease regulation 103
Peterson, C. A. 1, 18
Petrescu, D. 37
Pfeiffer, B. 102
plant breeding: fruit breeding *vs.* 58; in global context 63–64; paradigmatic orientations of *155*; sustainability challenges in 2

Plieninger, T. 6, 104
Pohl, C. 47
Poland 63
Polasky, S. 4, 6, 73, 93–94
political-legal framework 53–57, 55
pollination 103
polycentric governance systems 130,
    130–132, 140–141, 164, 178, 247–248
Pörtner 38n6
Potschin, M. B. 27
Power, A. 4, 93–94
PPB *see* participatory plant breeding (PPB)
Preiser, R. 49
privatization 2–4, 62, 178, 211, 213, 249
property rights 81
provisioning service 98–99
public apple breeding *192*; case study
    167–169; defined 167; diversity in
    169–171; ecocentric breeding in 171;
    knowledge exchange with 171–172;
    mental models in 172; participation in
    173; redundancy in 170–171; resilience
    of 169–174; social elements in 170;
    social learning and 172–173; social self-
    organization in 173
public communication 128

qualitative research 46–51, 47, **48**
Quinlan, A. E. 5, 112, 114

Rapf, K. 6
Rasmussen, L. V. 2
redundancy 115–117, 134–135, 160–161,
    170–171, 182
Reed, M. 145n8
Reganold, J. P. 4, 95, 107n8
regime shift 21, **24**
research, qualitative 46–51, 47, **48**
resilience 18–26, *19–20*, **24**; analytical
    framework for 142–144, **143–144**;
    building, principles of 113–133, *116,
    125, 130, 133*; commons and 36–37;
    complex adaptive systems thinking and
    122–124; connectivity and 120–122,
    136–137; in corporate-based apple
    breeding 159–164; critical remarks on
    concept of 23–26, **24**; diversity and
    115–117, 134–135, 169–171, 181;
    ecological 22–23; engineering 22;
    evolutionary 23; experimentation and
    124–127, *125*; feedbacks in 117–119;
    in fruit breeding and cultivation system
    112–144, *116, 125, 130, 133, 141,*
    143–144; institutions and 162, 184;
    learning and 124–127, *125*; long-term
    monitoring programs and 162–163,
    172, 184–185; mental models and 162,

183–184; participation and 127–129,
    185–186; polycentric governance
    and *130*, 130–132, 164; polycentric
    governance systems and *130*, 130–132;
    in public apple breeding 169–174;
    redundancy and 115–117, 134–135,
    170–171, 182; slow variables in
    117–119; social learning and 125–126,
    163, 172–173, 185; thinking *19*,
    19–21, *20*; thresholds in *20*, 20–21;
    understandings of, in social-ecological
    systems 22–23; variables in *20*, 20–21,
    24, 117–119; *see also* social-ecological
    resilience
resilience principles 114–132
responsibility, collective 178, 246–247
Richardson, R. B. 102
Rickard, B. J. 203, 219
Ristel, M. 179–182, 184, 187
Robinson, T. L. 89n7
robustness 3–4, 87, 103, 134, 137, 158,
    161–162, 168, 171, 179–180, 184, 209,
    224, 232, 234, 236
Rodríguez-Gasol, N. 107n12
Rohe, S. 255
Ross, H. 25
Rotz, S. 121
Rowe, G. 128
Ruckelshaus, M. H. 38n6

Sakai, S. 5
Salt, D. 19–22, 25, 73, 113, 116, 122,
    126, 131
Samnegård, U. 101, 107n11
Sandhu, H. S. 4, 93
Sattler, I. 180–181
Schäfer, F. 241–242
Schauppenlehner-Kloyber, E. 37
Schlager, E. 34
Schlüter, M. 5, 36, 76, 118, 122, 132, *133*,
    193, 195, 196n3, 200, 256–257
Schoon, M. L. 130–132, 140
Schumpeter, J. 38n9
Schwartau, H. 218
Schweik, C. M. 36
SESF *see* Social Ecological Systems
    Framework (SESF)
Seufert, V. 95
Shackleton, S. 48
Shiva, V. 2
Sievers-Glotzbach, S. 3–4, 6, 177, 179,
    182–184, 186, 189n1, 189n4, 194,
    196n2, 242, 255, 257
Simon, S. 100, 107n8
Sinclair, K. 24–25
slow variables 114, 117–119, 127, 135–136,
    141, 165, *165*, 174, *174*, 188, *192*

Social Ecological Systems Framework (SESF) 34, 77–78
social learning 125–126, 163, 172–173, 185
social self-organization 164, 173, 186–187
social-ecological resilience 1, 17–26, *19–20*, **24**
social-ecological system 1; analytical framework for 78; breeding and cultivation as 82–84, *83*; fruit breeding and cultivation as *75*, 75–89, *78–79*, *81*, *83*, 85–86, 88; knowledge in 126; resilience thinking and *19*, 19–21, *20*; telecoupled 87–88, *88*; theoretical and empirical conceptualizations in 76–82, *78–79*, *81*; understanding of resilience in 22–23
soil nitrogen availability 103
Soleri, D. 2
Spengler, R. N. 2
spiritual values 105
Spornberger, A. 102
stability domain **24**
stability landscape 21, **24**
stability regime **24**
state variables *20*, 20–21, **24**
Steffen, W. 1
Stein, N. A. 32
Steins, N. A. 33
Stern, P. 35
Stokke, O. S. 53
Stowe, E. 64
Stringer, L. C. 127
supporting services 93–94
sustainability: challenges in apple breeding 2–4; challenges in plant breeding 2; ecological-economic perspective on *18*
Swinton, S. 6, 92–93
Switzerland 65
system knowledge 46

target knowledge 46
Tarko, V. 130–131
telecoupling 79, *79*, 87–88, *88*, 130, 137, 140–142, 160, 253, 257, **264**
Tendall, D. M. 5, 73, 78, 88, 115–117, 119, 121, 200
thresholds, in resilience *20*, 20–21, **24**
Tilman, D. 94–95
transdisciplinary research 32, 46, 50, 256, **264**
transformative capability **24**
trends, market 212–215
Tschersich, J. 53, 177, 189n1

Turner, B. L. 77
Turners Vulnerability Framework (TVUL) 77
TVUL *see* Turners Vulnerability Framework (TVUL)

Ulanowicz, R. E. 116
Umetsu, C. 5
Urruty, N. 78, 116, 121, 123–124, 126

value chain *57*, 57–62, 213–214
van de Wouw, M. 2
van Laerhoven, F. 34
variables: fast 118–119, 135; in resilience *20*, 20–21, 117–119; slow 114, 117–119, 127, 135–136, *141*, 165, *165*, 174, *174*, *188*, *192*
variety trends 212–213
vegetative propagation 58
vitality 3, 87, 179–180, 184, 209, 224

Wachter, J. M. 95, 97
Waibel, H. 204, 216–217
Walch, B. 102
Waldman, K. B. 102
Walker, B. 19–23, 25, 73, 76, 113, 116, 118, 122, 126, 131
Walsh-Dilley, M. 5
Wang, Y. 102
water cycle regulation and maintenance 100
Weber, M. 203, 216, 218–219, 220n7, 220n9, 242
Weibel, H. 219
Wezel, A. 95–96
Widmark, C. 6, 27
Wilbois, K. -P. 6, 241
Winkler, K. J. 104
Winter, E. 242
Winterfeld, U. V. 4
Wirthgen, B. 217
Wirz, J. 4, 6, 241
Wolter, H. 3, 177, 179, 182–184, 186, 189n4, 194, 204, 243
Wood, S. 9n1
Worstell, J. 119, 129

Yin, R. K. 47–48
Young, O. R. 34, 39n15, 75

Zander, K. 51, 57, 203, 216–218, 230
Zanotti, L. 38n4
Zhang, W. 2, 4, 6, 73, 92–93, *93*, 94
Zhu, Z. 63
Züchtungsinitiative Niederelbe (ZIN) 157–159